Basic Chemometrics for Analytical Chemists

Essential Textbooks in Chemistry   ISSN: 2059-7738

The *Essential Textbooks in Chemistry* explores the most important topics in Chemistry that all Physical Sciences students need to know to pass their undergraduate exams (years 1, 2 and 3 of the BSc).

Written by senior academics as well lecturers recognised for their teaching skills, they offer in around 200 to 250 pages a theoretical overview of fundamental concepts backed by problems and worked solutions at the end of each chapter.

Their lively style, focused scope and pedagogical material make them ideal learning tools at a very affordable price.

Published

Basic Chemometrics for Analytical Chemists
    edited by José Andrade-Garda and Riccardo Leardi

Orbitals: With Applications in Atomic Spectra
(Revised Edition)
    by Charles Stuart McCaw

The Essence of Crystallography
    by Mark Ladd

Problems of Instrumental Analytical Chemistry: A Hands-On Guide
    by JM Andrade-Garda, A Carlosena-Zubieta, MP Gómez-Carracedo,
    MA Maestro-Saavedra, MC Prieto-Blanco and RM Soto-Ferreiro

Astrochemistry: From the Big Bang to the Present Day
    by Claire Vallance

Atmospheric Chemistry: From the Surface to the Stratosphere
    by Grant Ritchie

Principles of Nuclear Chemistry
    by Peter AC McPherson

Orbitals: With Applications in Atomic Spectra
    by Charles Stuart McCaw

More information on this series can also be found at https://www.worldscientific.com/series/icpetc

Essential Textbooks in Chemistry

Basic Chemometrics for Analytical Chemists

Editors

José Andrade-Garda
University of A Coruña, Spain

Riccardo Leardi
University of Genoa, Italy

World Scientific

NEW JERSEY • LONDON • SINGAPORE • BEIJING • SHANGHAI • TAIPEI • CHENNAI

Published by

World Scientific Publishing Europe Ltd.

57 Shelton Street, Covent Garden, London WC2H 9HE

Head office: 5 Toh Tuck Link, Singapore 596224

USA office: 27 Warren Street, Suite 401-402, Hackensack, NJ 07601

Library of Congress Control Number: 2025003184

British Library Cataloguing-in-Publication Data
A catalogue record for this book is available from the British Library.

Essential Textbooks in Chemistry
BASIC CHEMOMETRICS FOR ANALYTICAL CHEMISTS

ISBN 978-1-80061-681-3 (hardcover)
ISBN 978-1-80061-682-0 (ebook for institutions)
ISBN 978-1-80061-683-7 (ebook for individuals)

For any available supplementary material, please visit
https://www.worldscientific.com/worldscibooks/10.1142/Q0495#t=suppl

Desk Editors: Kannan Krishnan/Gabriel Rawlinson/Shi Ying Koe

Typeset by Stallion Press
Email: enquiries@stallionpress.com

Preface

The seed of this book was sown in 2019, shortly before the onset of the pandemic caused by the SARS-CoV-2 virus. A group of teachers, responsible for educational duties at various European universities, felt that there was a gap in the resources available for students to handle studying chemometrics. It was true that there were many advanced books dealing with various aspects of this relatively new chemical discipline, even encyclopaedias, and of course, many reviews in scientific journals. But there were too few books aimed at presenting to students the very fundamentals of the most basic chemometric techniques without including overly advanced mathematical details. There was also a big gap in how students could practice the basic concepts they had studied. Commercial software was not always available to them or was difficult to handle. There were programming environments, but students had to spend too much time to get a working script, and of course, that was not the main objective of their subjects.

In 2013, after a meeting of some Italian chemometricians, Gian Marco Polotti got the idea of developing a user-friendly, menu-driven software based on R, a free environment used by scientists worldwide. This software, which was later named Chemometric Agile Tool (CAT), can be freely downloaded and used by anyone wishing to perform chemometrics. Another important point is that its scripts are open source; this means that anyone with some knowledge in R can read and eventually modify them. Furthermore, thanks to the help of Camillo Melzi, the software is quite regularly improved and updated, and new versions are made available every few months (this is the reason for the small differences that can occur between the outputs shown in the book and the current software that the reader can download from the website).

During 2020 and 2021, the COVID-19 pandemic situation was quite complicated worldwide. Education systems, including universities, struggled to do their best in maintaining a certain degree of continuity and assisting students and families. We all tried to adapt our teaching within a few days to the (electronic) possibilities available to teachers, universities, students, and families. It was not easy at all, and the severe lockdowns that were enforced in most cities will remain in our minds forever. Our deepest gratitude and sincere acknowledgements go to all those who fought so hard in one way or another during those difficult days.

We decided to keep the seed of the book in hibernation, although not forgotten. In those days, all teachers and, likely, every student also realised the crucial importance of books and, undoubtedly, the World Wide Web as well. They became two fundamental tools to help educators and students. Although young students are disappointingly not too accustomed to working closely with books, the latter are the most formidable (peaceful) weapons mankind has developed to advance our knowledge and to maintain and transfer it.

The seed began to germinate when the pandemic was almost over and classrooms became active again. Student-focused chemometric material was still needed. The CAT software proved valuable, so why not use it to develop a teaching tool focused on the students? A monograph that complemented the software (like an associated "hardware") seemed adequate to present the students the knowledge they (should) have to get some skills on the use of some fundamental multivariate chemometric tools.

Finally, the seed has now borne a nice fruit: a book aimed at teaching students through theoretical demonstrations, although not overly advanced (with a few exceptions, which are indicated), and practical examples. It can be used in classrooms or for self-training.

To the best of our knowledge, the student should (must?) possess most of the scientific background required to follow it because, in most cases, multivariate chemometrics is taught in advanced courses rather than introductory ones. If that is not the case, an initial chapter is dedicated to summarising the mathematical background required to understand the subsequent chapters.

The book is organised as follows:

A short initial introduction (Chapter 1) is devoted to guiding readers through downloading and installing the software, along with some basic operations.

Chapters 2 and 5 contain transversal, generic concepts that will be used in the other chapters. The former reviews and explains at an introductory level the mathematical background that may be required to follow certain explanations given throughout the text. Note that it can be read as is from the very beginning or consulted when certain concepts are mentioned in subsequent explanations. Chapter 5 deals with signal preprocessing – an issue that is neither trivial nor straightforward yet may determine the success of a given model development.

Chapter 3 should be most familiar to those who have studied basic analytical techniques or physico-chemical principles. Classical univariate modelling is briefly reviewed and extended to multivariate regression using an intuitive approach. However, such an approach presents many practical problems, most of which can be solved or addressed using the alternatives presented in subsequent chapters (mainly Chapters 6, 7 and 8).

Organising the experimental work to develop a method of measurement is really complex. In Chapter 4, devoted to the design of experiments (DoE), the most common designs are presented, together with some real examples. The goal is to underline the significant advantages of performing designed experiments compared to the usual approach in which one variable is changed at a time (still used by many researchers). DoE yields an effective and economical strategy for planning first and executing later, ensuring that all variables are studied simultaneously.

Chapter 6 deals with two tasks that scientists usually have to perform: to discriminate among different types of samples and/or to classify them. These two terms (discrimination and classification) are not always correctly differentiated. Chapter 7 is devoted to finding and describing patterns in multivariate data, as well as extracting relevant information from both samples and variables. This is done using descriptive methods in either two (typical data matrices) or three (a data cube) ways, using principal component analysis and parallel factor analysis. Finally, Chapter 8 studies advanced regression methods: principal components regression and partial least squares regression.

All chapters present worked examples (and propose additional ones for students) that demonstrate how to address difficulties (or highlight that they exist!) and how to apply and obtain information from the models. Important efforts were dedicated to validating the models for their use in real-world situations, a crucial task – not always emphasised strongly enough – the scientist must carry out carefully. The data files required to perform the examples can be downloaded with the CAT software.

Finally, as no human activity is perfect, we would greatly appreciate comments and suggestions regarding this volume and/or the software. A first edition is like a tree; it needs time to mature, improve, and become stronger. We hope you find this book useful. Thank you for using it.

About the Editors

José Andrade-Garda is a full professor at the University of A Coruña since 2011. He has been in charge of teaching different subjects in analytical chemistry since 1995. His main interests are in quality control and chemometrics, focusing on multivariate regression and pattern recognition methods (both unsupervised and supervised). In the field of atomic spectrometry, he has applied formal optimisation techniques to optimise analytical protocols and has developed multivariate regression models to cope with spectral and chemical interferences in ETAAS (traditional sample introduction and slurry-based). In the study of infrared spectrometry, he has developed analytical methods for the petrochemical field. Currently, he is working on environmental pollution, developing IR analytical methods for microplastic identification, and data mining. He has published around 130 papers in peer-reviewed journals, coauthored 12 book chapters, and edited three books.

Riccardo Leardi served as an associate professor in the Department of Pharmacy of the School of Medical and Pharmaceutical Sciences at the University of Genoa in Italy. Since 1985, he has been working in the Analytical Chemistry section of the Department of Pharmacy at the University of Genoa. His field of research is chemometrics, with interests mainly in problems of classification and regression (applied especially to food, environmental, and clinical data), experimental design, process optimisation, multivariate process monitoring, and multivariate quality control. He collaborated in developing the chemometric software CAT and BasiCAT. He has coauthored about 150 papers and more than 130 communications in national and international meetings, several as an invited speaker. He has also been invited to give talks and courses in several industries and research centres. He organised two courses on chemometrics (multivariate analysis and experimental design), each held twice a year at the University of Genoa. Since November 2002, he has been working as a chemometric consultant.

About the Contributors

Ricard Boqué was born in La Palma d'Ebre, Catalonia, Spain, in 1968. He graduated with a degree in chemical sciences from the University of Barcelona in 1991 and obtained his PhD in chemistry from the Universitat Rovira i Virgili (URV), Tarragona, in 1997. His thesis received the Extraordinary Doctorate Award. In 2002, he became an associate professor in the Department of Analytical Chemistry and Organic Chemistry of the URV, where he has since been teaching analytical chemistry and chemometrics for BSc and MSc students. He has published more than 130 scientific papers and supervised twelve doctoral theses. He has also taught more than 50 courses for companies and public administrations. His research interests include qualimetrics and multivariate calibration and classification methods. He is a member of the Spanish Society of Analytical Chemistry and the Catalan Society of Chemistry and is currently a review editor of the *Journal of Chemometrics*.

Richard Brereton was educated (BA, MA, PhD, and postdoc) at the University of Cambridge, UK, before moving to the University of Bristol, UK, where he has served as a Lecturer, Reader, and Professor and is now an Emeritus Professor. He is a fellow of the Royal Society of Chemistry, Royal Statistical Society, and Royal Society of Medicine. He has published over 400 articles, including nine books (six authored and three edited), 23 book chapters, and 228 articles listed in Clarivate. He is the Editor-in-chief of the journal *npj Heritage Science* (published in *Nature* Portfolio), columnist for the *Journal of Chemometrics*, and is involved in several other journals. He has given over 200 invited lectures in 31 countries and has been the primary or sole supervisor for 39 research students (33 PhDs). He has been a member of over 50 conference organising committees, and given evidence as an expert in 13 court. He is listed in Elsevier/Stanford in the top 2% of scientists worldwide of all disciplines, and is listed as sixth international (third living) scholar in chemometrics in ScholarGPS. He is cited more than 20,000 times in Google Scholar.

María de la Cruz Ortiz has a PhD in Chemistry from the University of Valladolid, Spain. She has served as a full professor of Analytical Chemistry since 2010 at the University of Burgos, Spain, where she is currently an emeritus professor. Her fields of expertise are the development of analytical methods, chemometrics and quality control, and the design of experiments. She has authored or coauthored around 170 papers (138 in the first quartile, JCR index) and nine chapters in international books, and has supervised 15 PhD theses. Her interests focus on developing analytical methods for food and environmental analysis and applying chemometric tools to implement them. Her research is also devoted to the application of three-way multivariate data analysis techniques, such as PARAFAC, to extract relevant information from analytical data of chromatography and fluorescence spectroscopy.

Emanuele Farinini received his PhD in Sciences and Technologies of Chemistry and Materials (University of Genoa) in 2024, with a specialisation in pharmaceuticals, cosmetics, and food sciences. He conducted his research in the Analytical Chemistry and Chemometrics research group (Department of Pharmacy, University of Genoa) in collaboration with manufacturing industries. Prior to that, he worked as an R&D scientist in Early Product Development (EPD) and the Chemical Manufacturing and Control (CMC) department at Chiesi Limited, located in Chippenham, Wiltshire, UK. Following this, he began a professional career as a chemometric consultant in collaboration with Professor Riccardo Leardi, a position he continues to hold. His work and consultancy focus on solving industrial problems using chemometric techniques. These techniques include experimental design, process optimisation, multivariate process monitoring, and multivariate quality control. He has coauthored 12 papers and 15 communications at national and international congresses. He holds courses in several industries and research centres. He also collaborated in developing the chemometric software Chemometric Agile Tool (CAT).

María Sagrario Sánchez has a PhD in Mathematics from the University of Valladolid, Spain. She currently works at the University of Burgos, Spain, where she is a full professor of Statistics and Operation Research. She has taught mathematics, statistics and chemometrics at undergraduate, master and doctorate levels. Her main research activities include modelling and analysis of n-way data, class modelling, design of experiments, and multi-response optimisation using classical methods as well as computationally intensive methods (such as neural networks and evolutionary algorithms). In the field of process control, she is working on the inversion of latent variable models with application to industrial processes and/or chemical problems.

Luis Antonio Sarabia has a PhD in Mathematics from the University of Valladolid, Spain. He has served as a full professor of Statistics and Operation Research since 2000 at the University of Burgos, Spain, and currently as emeritus professor. His areas of expertise are mathematics, chemometrics, quality control, and the design of experiments. He has authored and coauthored around 160 papers (126 in the first quartile, JCR index), 225 presentations in conferences, and nine chapters in international books. His interests are focused on multiway data analyses, optimisation of analytical methods, and design of experiments applied to chemical problems. He has also supervised 10 PhD theses.

Contents

Chapter 3. The Basics of Univariate and
Multivariate Calibration   31

José Andrade-Garda

Chapter 4. Statistical Experimental Design   77

Riccardo Leardi and Emanuele Farinini

Chapter 5. Data Pre-Processing: The Importance of Properly Transforming Your Data Before Further Insights

151

Ricard Boqué

Chapter 6. Classification   185

Richard Brereton

Chapter 7. Exploring Multivariate Data:
**   Models Based on Latent Variables**   235

María Sagrario Sánchez, María de la Cruz Ortiz,
and Luis Antonio Sarabia

*Luis Antonio Sarabia, María de la Cruz Ortiz,
and María Sagrario Sánchez*

Chapter 1

Guidelines to Install the Software

Riccardo Leardi and José Andrade-Garda

Objectives and Scope

This brief introduction describes the main steps to obtain the software package that is recommended for performing the examples and exercises presented throughout this textbook.

Despite there being many software devoted to chemometrics, not all of them are totally free and open source, which is why the authors decided to open up this possibility to our readers. From the basic underlying platform to the software itself, students do not have to pay for anything.

In addition, the well-known and ubiquitous Excel spreadsheet from Microsoft will also be used to develop some examples. Excel is so widely used that it is worth knowing some of its capabilities.

1.1. Some General Comments About CAT

CAT is the acronym for "Chemometric Agile Tool".

In 2013, during a meeting of Italian chemometricians, it was pointed out that one of the major obstacles to the diffusion of

1

chemometrics was the lack of free (or very inexpensive) software. All the commercial software available on the market are quite expensive; in the great majority of cases, it is possible to buy only yearly licences. While this is not a big issue at the academic level (as universities will be paying for it), it can become quite expensive for private individuals (those who have already finished their studies) or companies (who need to buy licences for multiple users).

So, Gianmarco Polotti got the idea of building a menu-driven interface based on the programming language R. This allows for keeping all the advantages of the well-known (and free) environment of R without having to learn (almost) anything about the language itself.

The software, originally aimed at beginners, has progressively expanded its target and now covers a very wide range of chemometric methods for both multivariate analysis and experimental design (including some univariate and multivariate tools).

A YouTube channel has also been created, in which several videos containing tutorials and demos are available (https://www.youtube.com/channel/UCVIUJAhMVR0a59m3BG_dR0A).

It is a completely open project, and everyone can contribute in various ways, such as suggesting new ideas or improvements, writing help files, producing videos, and finding bugs. Many people have provided their contributions, and their names are listed under the "Credits" section of the software.

CAT can be downloaded from the following link: http://gruppochemiometria.it/index.php/software. It is meant to be installed under a Windows (Microsoft) environment, but it can also be run under Linux, using the software Wine (64-bit), and on a Mac, with a proper Windows emulator.

Detailed instructions can be obtained on the website (in the file titled "Instructions.txt") and on the YouTube channel.

If you possess proprietary or administrative privileges, the only thing you need to do is download the executable file "setup_CAT.exe" to your computer. Then, click on it, and the installation will run automatically.

In many cases, the Windows operating system will pop up a window suggesting that you not run the application. This is because the software developers had not been introduced to Microsoft. So, do not panic; click on "more information" in that window, and then just click again on "execute anyway". Another window will open, asking you again whether you authorise the "unknown" program to modify your computer. (It is good that Windows takes care of undue installations, but this should not be the case.) Please accept the installation to proceed.

Then, you are in! CAT starts asking for acceptance of the licence agreement and afterwards the folder in which you wish to install it. The folder "CAT" is suggested by default, and we recommend accepting it. The process finishes when the shortcut appears on your desktop screen, and CAT asks you to complete the process.

In case you do not have administrative privileges, you need to run a step-by-step approach, as described in the "Instructions.txt" file (or on the YouTube channel).

Another very convenient possibility is to install CAT on a USB memory stick. This will allow you to work with the software on any computer. This option can be very useful for companies where users may not be allowed to install "external" software (of course, provided that one can connect an external USB memory stick).

At the end of the installation procedure, six subfolders will be created inside the folder "CAT".

CAT is updated quite frequently (on average, every couple of months). For installing the updates, it is not required to repeat the whole procedure. It is enough to download the updated folder (the date of publication of each folder is displayed on the website), delete the old one, and replace it with the new one (after having unzipped it).

Please note that, when started, CAT takes a while to retrieve all files to the computer's memory, so it will take a few seconds before the menu bar appears.

The first thing to know when using any software is how to load the data. CAT has four different options: from a .csv file, from a .txt file, from an Excel file, or by copying and pasting (see Figure 1.1). In the majority of cases (unless the size of the data matrix is huge), the last option is the most convenient.

For this, it is recommended to include labels (typically, short names) for the columns and rows. Copy the data you want to use (including the labels), and go to the "Data Handling" menu, then select the option "Copy and Paste" (see Figure 1.1). A name is requested for your data matrix, and then you enter the R built-in "spreadsheet". Right-click with the mouse on the cell that is visible and select paste. Your data will be pasted, but remember to tell CAT to use the row and column labels you have copied as well (indicated by the red circle in Figure 1.2).

Be careful with the decimal separator since CAT, as any scientific software, only accepts dots as decimal separators. If the decimal separator is a comma, as is common in many European countries, then one should go to the "Format" window and select "Germany".

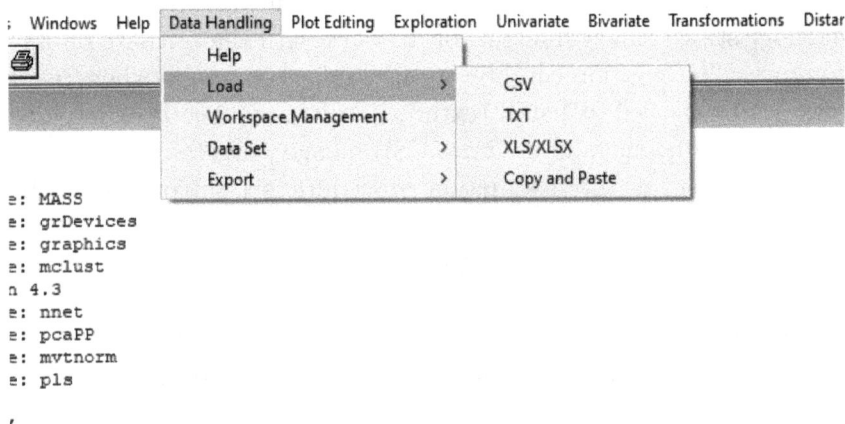

Figure 1.1. Data importing options available in CAT.

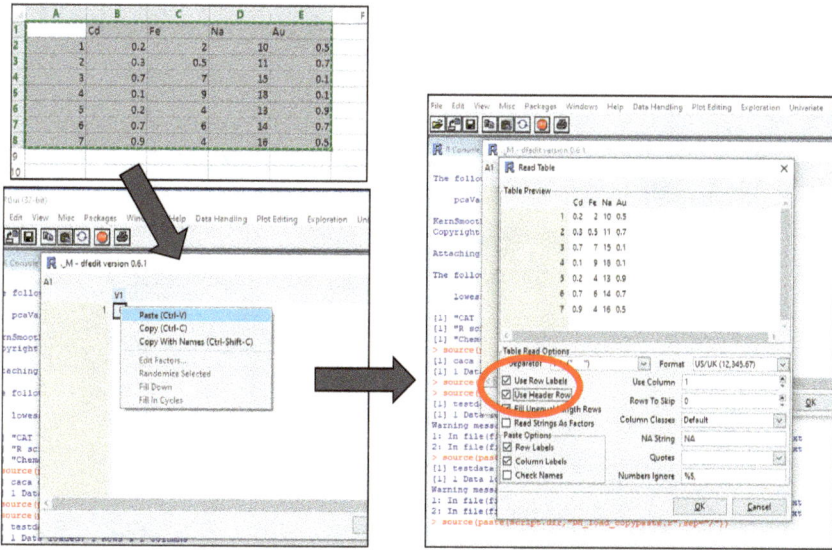

Figure 1.2. Copy and Paste option to retrieve data with CAT.

Now, you are almost ready to start working with CAT. To get the most out of it, you can continue and study the following chapters of this textbook.

Chapter 2

A Summary of Operations and Common Matrix Decompositions in Chemometrics

María Sagrario Sánchez, María de la Cruz Ortiz, and
Luis Antonio Sarabia

Objectives and Scope

Matrices are useful for representing multivariate numeric data, although they occur more naturally when working with linear equations or when expressing linear relationships among objects. Several algorithms for a variety of tasks involve matrix and vector arithmetic. Therefore, this chapter contains an introduction to the mathematical notation and a summary of matrix operations.

There are different precise ways of defining vectors and matrices, but for our purposes in this chapter, they will be considered merely as linear or rectangular arrangements of numbers that, unless otherwise stated, are all real numbers.

Nevertheless, this is not an impediment for adopting, as in some other chapters of this book, a geometrical perspective for vectors and/or considering matrices to define geometrical or linear transformations.

2.1. Matrices

2.1.1. *Matrices and vectors*

A *matrix* is a rectangular arrangement of numbers in rows and columns, which looks like a traditional table, enclosed between brackets. For example, Equation (2.1) shows a matrix named $\mathbf{X}$ with seven rows and three columns:

$$\mathbf{X} = \begin{pmatrix} 2.3 & 17 & 215.25 \\ 1.9 & 15 & 214.17 \\ 2.0 & 19 & 218.35 \\ 1.7 & 21 & 220.65 \\ 2.5 & 27 & 221.45 \\ 2.1 & 25 & 216.12 \\ 2.0 & 20 & 219.89 \end{pmatrix}. \qquad (2.1)$$

In general, with m rows and p columns, we have an $m \times p$ matrix. Usually, it is said that $m \times p$ is the *dimension* of such a matrix. However, to avoid confusion with the dimension of the spaces generated by the different arrays, we refer to $m \times p$ as the *size* of the matrix.

To specify the elements of a matrix, say $\mathbf{X}$, it is usual to write $\mathbf{X} = (x_{ij})$. Each element of such a matrix is identified by its position in the matrix: the first subscript refers to the row, while the second refers to the column. For example, $\mathbf{X}$ in Equation (2.1) is a 7×3 matrix, and x_{52} is the element in the fifth row and second column, that is, $x_{52} = 27$.

With this notation, a matrix with a single column ($p = 1$) is a vector, whereas a scalar is a 1×1 matrix (a single number). As usual, the matrices will be written in bold uppercase, vectors in bold lowercase, and scalars in italics. These characteristics are summarised in the following shaded box:

Matrix : $\mathbf{X} = (x_{ij})$, $i = 1, 2, \ldots m$, $j = 1, \ldots, p$ ($m \times p$ matrix).
Vector : $\mathbf{x} = (x_i)$, $i = 1, \ldots, m$ ($m \times 1$ matrix).
Scalar : x (1×1 matrix).

The *transpose* of a matrix, denoted as $\mathbf{X}^T$ or $\mathbf{X}$', is the matrix obtained by interchanging rows and columns. Therefore, $\mathbf{X}^T$ is a $p \times m$ matrix, and clearly, $(\mathbf{X}^T)^T = X$. The transpose of a column vector is a row vector, and vice versa.

For example, Equation (2.2) shows the transpose of $\mathbf{X}$ in Equation (2.1), which is a 3×7 matrix, and now 27 is the element x'_{25}. If we transpose again, we will get back $\mathbf{X}$.

$$\mathbf{X}^T = \begin{pmatrix} 2.3 & 1.9 & 2.0 & 1.7 & 2.5 & 2.1 & 2.0 \\ 17 & 15 & 19 & 21 & 27 & 25 & 20 \\ 215.25 & 214.17 & 218.35 & 220.65 & 221.45 & 216.12 & 219.89 \end{pmatrix}.$$

$$(2.2)$$

In the context of data analysis, the items, individuals, samples, etc., are arranged as rows and are called *objects*, while the columns contain the different measurements or property values determined for each object, which are generically known as *variables*. Hence, m objects for which p variables have been determined define an $m \times p$ *data matrix*. Each object is then represented by a p-dimensional vector (the transpose of the corresponding row); therefore, geometrically, the objects are seen as points in a p-dimensional space. Usually, $p > 1$, so the representation is multivariate.

For example, if $\mathbf{X}$ in Equation (2.1) were a data matrix, it displays the measured or observed values of three variables (the first one around 2, the second one around 20, and the third one around 220) obtained for the seven objects, which can be thought of as seven points in a three-dimensional space, as depicted in Figure 2.1, where the first object (first row in $\mathbf{X}$) is identified to show its coordinates in the graph. Variable $X1$ corresponds to the variable measured in the first column, $X2$ to the second one, and $X3$ to the third column of $\mathbf{X}$, which now act as the coordinate system.

2.1.2. *Matrix operations*

For matrix arithmetic or algebraic operations, the size of the matrices should always be taken into account. The simplest situation is

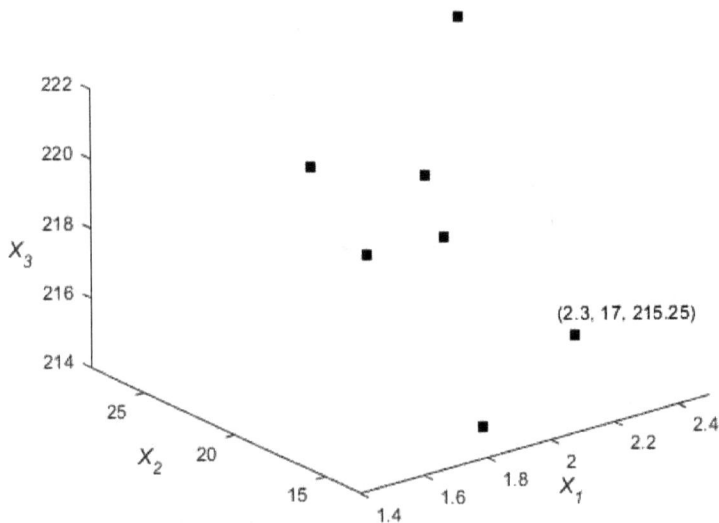

Figure 2.1. Geometrical representation of data matrix $\mathbf{X}$ in Equation (2.1).

multiplying a matrix by a scalar, which is defined for any matrix and consists of the product of the given scalar by every element of the matrix, so that $\lambda \mathbf{A} = (\lambda a_{ij})$ for any scalar λ and matrix $\mathbf{A} = (a_{ij})$.

Two matrices are *equal* if and only if they are of the same size and each element of one matrix is equal to the corresponding element of the other. Regarding arithmetic operations, the sum is also defined on a "per-element" basis: for matrices of the same size, $\mathbf{A} = (a_{ij})$ and $\mathbf{B} = (b_{ij})$, their *sum* is matrix $\mathbf{A} + \mathbf{B} = (a_{ij} + b_{ij})$ with the same size as $\mathbf{A}$ and $\mathbf{B}$. Therefore, for the particular case of $\lambda = -1$, $\mathbf{A} - \mathbf{B} = (a_{ij} - b_{ij})$.

The properties of the sum and product by scalar are those of the real numbers being operated in each element, so generalising, we have the following:

For $m \times p$ matrices $\mathbf{A} = (a_{ij})$ and $\mathbf{B} = (b_{ij})$, $i = 1, 2, \ldots, m$, $j = 1, \ldots, p$, and any scalars λ and μ,

$\lambda \mathbf{A} + \mu \mathbf{B} = (\lambda a_{ij} + \mu b_{ij})$, $i = 1, 2, \ldots, m$, $j = 1, \ldots, p$ ($m \times p$ matrix),

$$(\mathbf{A} + \mathbf{B})^T = \mathbf{A}^T + \mathbf{B}^T.$$

Although there is a so-called Hadamard product [1] (element-wise multiplication) that "works" similar to the addition in the previous paragraph, the usual matrix multiplication is not made on a per-element basis. Before defining matrix multiplication, some definitions related to vector multiplication are useful.

There are different products that can be defined for vectors. To obtain them, let $\mathbf{v} = (v_1, v_2, \ldots, v_m)^T$ and $\mathbf{w} = (w_1, w_2, \ldots, w_m)^T$ be two vectors of the same size. The *inner* product (or *dot* product or *scalar* product) is computed as the sum of the products between the corresponding elements of the vectors, as in Equation (2.3), where the last expression is the "condensed" form of the definition obtained by using the "sum" symbol (Σ) over all the coordinates. From the definition, $\mathbf{v}^T \cdot \mathbf{w} = \mathbf{w}^T \cdot \mathbf{v}$ holds.

$$\mathbf{v}^T \mathbf{w} = (v_1, \ldots, v_m) \begin{pmatrix} w_1 \\ \vdots \\ w_m \end{pmatrix} = v_1 w_1 + v_2 w_2 + \cdots + v_m w_m = \sum_{i=1}^{m} v_i w_i.$$

$$(2.3)$$

Note that the result of the dot product is a single number (a scalar, hence the name) irrespective of m, and thus for a single vector, $\mathbf{v}^T \cdot \mathbf{v}$ is the sum of the squares of its elements. The squared root of this sum is the geometric *length* of the vector or, more generally, the *norm* of the vector (the 2-norm, to be precise), as in Equation (2.4):

$$\|\mathbf{v}\| = \sqrt{\mathbf{v}^T \mathbf{v}} = \sqrt{\sum_{i=1}^{m} v_i^2}. \qquad (2.4)$$

Finally, $\mathbf{v}^T \cdot \mathbf{w} = \|\mathbf{v}\| \cdot \|\mathbf{w}\| \cos(\widehat{\mathbf{v}, \mathbf{w}})$, that is, the dot product is, geometrically, proportional to the cosine of the angle between the vectors.

For example, $\mathbf{v} = (1, 2)^T$ and $\mathbf{w} = (3, 1)^T$, then $\mathbf{v}^T \mathbf{w} = 1 \times 3 + 2 \times 1 = 5$, and the length of $\mathbf{v}$ is its 2-norm $\|\mathbf{v}\| = \sqrt{\mathbf{v}^T \mathbf{v}} = \sqrt{1 + 4} \approx 2.24$, which is geometrically the length of the arrow in Figure 2.2. Likewise, for $\mathbf{w}$, it is $\|\mathbf{w}\| = \sqrt{\mathbf{w}^T \mathbf{w}} = \sqrt{9 + 1} \approx 3.16$, a little longer. Therefore, the cosine of the angle between them, ϕ in Figure 2.2, is $\cos(\phi) = \cos(\widehat{\mathbf{v}, \mathbf{w}}) = \frac{\mathbf{v}^T \cdot \mathbf{w}}{\|\mathbf{v}\| \cdot \|\mathbf{w}\|} = \frac{5}{\sqrt{5}\sqrt{10}} = \frac{\sqrt{2}}{2} = 0.7071$, which is the known cosine of $45°$, i.e., the angle $\phi = \pi/4 = 0.785$ radians.

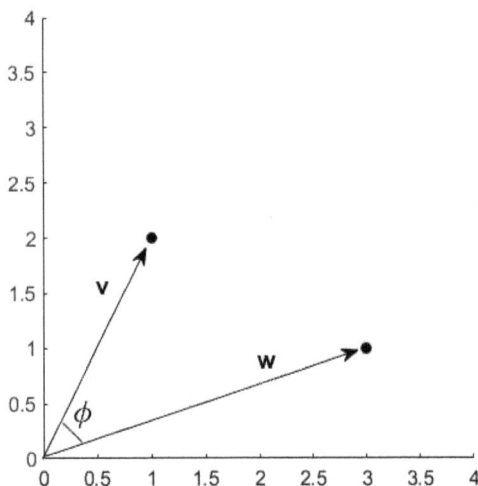

Figure 2.2.   Illustration of the geometrical interpretation of the scalar product between two two-dimensional vectors.

The scalar product has an immediate application in characterising *orthogonal or perpendicular* vectors: as $\cos(\frac{\pi}{2}) = 0$, two vectors form a right angle if and only if their scalar product is null.

For the dot product in Equation (2.3), transposing $\mathbf{v}$ (a row vector $\mathbf{v}^T$ multiplied by a column vector $\mathbf{w}$) is not strictly necessary. However, it is used to remain congruous with the matrix multiplication introduced in the following paragraph.

Let $\mathbf{A} = (a_{ij})$ be an $m \times p$ matrix and $\mathbf{B} = (b_{ij})$ a $p \times k$ matrix. The product matrix $\mathbf{C} = \mathbf{AB} = (c_{ij})$ is an $m \times k$ matrix, where c_{ij} is the dot product of the ith row in $\mathbf{A}$ and the jth column in $\mathbf{B}$. Note that the matrices do not need to be of the same size, but the number of columns in the preceding matrix $\mathbf{A}$ must coincide with the number of rows of the following matrix $\mathbf{B}$. The resulting matrix has the same number of rows as $\mathbf{A}$ and the same number of columns as $\mathbf{B}$.

Sometimes, to highlight the size of a matrix, it is annotated as sub-indexes of the own matrix. For example, we define $\mathbf{A}_{mp}$ instead of just $\mathbf{A}$. With this notation, with emphasis on the sizes of the matrices, the multiplication is $\mathbf{A}_{mp}\mathbf{B}_{pk} = \mathbf{C}_{mk}$.

For example, in Equation (2.5), the 4×2 matrix $\mathbf{A}$ can be multiplied with the 2×3 matrix $\mathbf{B}$ (the inner sizes coincide), resulting in a 4×3 matrix. The first two rows of the resulting matrix illustrate how the elements are computed. The reader can verify the rest of the numbers.

$$\mathbf{AB} = \begin{pmatrix} 2 & 0 \\ 5 & 4 \\ 1 & 3 \\ 7 & 6 \end{pmatrix} \begin{pmatrix} 3.1 & 2.2 & 4.5 \\ 5.3 & 5.0 & 3.7 \end{pmatrix}$$

$$= \begin{pmatrix} 2 \times 3.1 + 0 \times 5.3 & 2 \times 2.2 + 0 \times 5.0 & 2 \times 4.5 + 0 \times 3.7 \\ 5 \times 3.1 + 4 \times 5.3 & 5 \times 2.2 + 4 \times 5.0 & 5 \times 4.5 + 4 \times 3.7 \\ 19.0 & 17.2 & 15.6 \\ 53.5 & 45.4 & 53.7 \end{pmatrix}.$$

$$(2.5)$$

Contrary to the "usual" operations, the multiplication of matrices is not commutative, that is, $\mathbf{AB}$ is not the same as $\mathbf{BA}$. In the example in Equation (2.5), $\mathbf{BA}$ does not even exist (since a 2×3 matrix cannot be multiplied with a 4×2 matrix because the inner sizes are different).

Even if both $\mathbf{AB}$ and $\mathbf{BA}$ could be computed and were of the same size, in general, $\mathbf{AB} \neq \mathbf{BA}$, as shown in the following example with matrices 3×3 in Equation (2.6):

$$\mathbf{A} = \begin{pmatrix} 1 & 2 & 3 \\ 4 & 5 & 6 \\ 7 & 8 & 9 \end{pmatrix}, \quad \mathbf{B} = \begin{pmatrix} 2 & 8 & 1 \\ 4 & 7 & 9 \\ 3 & 6 & 5 \end{pmatrix},$$

$$\mathbf{AB} = \begin{pmatrix} 19 & 40 & 34 \\ 46 & 103 & 79 \\ 73 & 166 & 124 \end{pmatrix} \neq \mathbf{BA} = \begin{pmatrix} 41 & 52 & 63 \\ 95 & 115 & 135 \\ 62 & 76 & 90 \end{pmatrix}. \quad (2.6)$$

Not only is $\mathbf{AB} \neq \mathbf{BA}$, but also both are different from the matrix resulting from multiplying them on a per-element basis (Hadamard product), as shown in Equation (2.7) for the matrices in Equation (2.6), which is defined because the matrices have the same size. Furthermore, contrary to the property just mentioned, the Hadamard product is commutative, that is, $\boldsymbol{A} \odot \boldsymbol{B} = \boldsymbol{B} \odot \boldsymbol{A}$, where $\odot$ denotes the Hadamard product:

$$\mathbf{A} \odot \mathbf{B} = \begin{pmatrix} 1 \times 2 & 2 \times 8 & 3 \times 1 \\ 16 & 35 & 54 \\ 21 & 48 & 45 \end{pmatrix} = \mathbf{B} \odot \mathbf{A}. \tag{2.7}$$

Therefore, it is important to be aware of the restrictions on the size for matrix multiplication to be computed and the order of matrices. For example, when applied to vectors with m elements, $\mathbf{v}^T \cdot \mathbf{w}$ gives a number, whereas $\mathbf{v} \cdot \mathbf{w}^T$ results in an $m \times m$ matrix. This matrix contains copies of $\mathbf{w}$, by rows, multiplying each row with a scalar, the corresponding coordinate of $\mathbf{v}$, namely,

$$\mathbf{v} \cdot \mathbf{w}^T = \begin{pmatrix} v_1 \\ v_2 \\ \vdots \\ v_m \end{pmatrix} \begin{pmatrix} w_1 & w_2 & \cdots & w_m \end{pmatrix}$$

$$= \begin{pmatrix} v_1 w_1 & v_1 w_2 & & v_1 w_m \\ v_2 w_1 & v_2 w_2 & & v_2 w_m \\ & & \ddots & \\ v_m w_1 & v_m w_2 & & v_m w_m \end{pmatrix}.$$

Unless otherwise indicated, the product of two matrices is the one just defined, which is sometimes referred to as the Cayley product [2]. Some useful properties are summarised in the following shaded box.

Some useful properties of the product of matrices (Cayley product)

For matrices $\mathbf{A}, \mathbf{B}$, and $\mathbf{C}$, with sizes adequate for the corresponding multiplications, and any scalar λ:

$\mathbf{A}(\mathbf{BC}) = (\mathbf{AB})\mathbf{C}$.

$\mathbf{A}(\mathbf{B} + \mathbf{C}) = \mathbf{AB} + \mathbf{AC}$.

$(\mathbf{A} + \mathbf{B})\mathbf{C} = \mathbf{AC} + \mathbf{BC}$.

$$\lambda(\mathbf{AB}) = (\lambda\mathbf{A})\mathbf{B} = \mathbf{A}(\lambda\mathbf{B}).$$

$$(\mathbf{AB})^T = \mathbf{B}^T\mathbf{A}^T.$$

(In fact, this works for any finite number of matrices.)

Besides these two products (Cayley and Hadamard), there is yet another type of matrix multiplication: the Kronecker multiplication. The Kronecker product is defined for any two matrices, irrespective of their size. Thus, for $\mathbf{A}_{mp} = (a_{ij})$ and $\mathbf{B}_{qk} = (b_{ij})$, the resulting product matrix, denoted $\boldsymbol{A} \otimes \boldsymbol{B}$, is an $mq \times pk$ matrix, where each element of $\mathbf{A}$ is multiplied by all the elements of matrix $\mathbf{B}$, according to Equation (2.8):

$$\mathbf{A} \otimes \mathbf{B} = \begin{pmatrix} a_{11}\mathbf{B} & a_{12}\mathbf{B} & \cdots & a_{1p}\mathbf{B} \\ a_{21}\mathbf{B} & & & a_{2p}\mathbf{B} \\ \vdots & & \ddots & \\ a_{m1}\mathbf{B} & & & a_{mp}\mathbf{B} \end{pmatrix}. \quad (2.8)$$

For example, for $\mathbf{A} = \begin{pmatrix} 1 & 2 \\ 3 & 4 \end{pmatrix}$ and $\mathbf{B} = \begin{pmatrix} 2 & 3 & 0 \\ 1 & 8 & 1 \\ 5 & 0 & 9 \\ 7 & 2 & 4 \end{pmatrix}$, their Kronecker product is

$$\mathbf{A} \otimes \mathbf{B} = \begin{pmatrix} 1\begin{pmatrix} 2 & 3 & 0 \\ 1 & 8 & 1 \\ 5 & 0 & 9 \\ 7 & 2 & 4 \end{pmatrix} & 2\begin{pmatrix} 2 & 3 & 0 \\ 1 & 8 & 1 \\ 5 & 0 & 9 \\ 7 & 2 & 4 \end{pmatrix} \\ 3\begin{pmatrix} 2 & 3 & 0 \\ 1 & 8 & 1 \\ 5 & 0 & 9 \\ 7 & 2 & 4 \end{pmatrix} & 4\begin{pmatrix} 2 & 3 & 0 \\ 1 & 8 & 1 \\ 5 & 0 & 9 \\ 7 & 2 & 4 \end{pmatrix} \end{pmatrix}$$

$$= \begin{pmatrix} 2 & 3 & 0 & 4 & 6 & 0 \\ 1 & 8 & 1 & 2 & 16 & 2 \\ 5 & 0 & 9 & 10 & 0 & 18 \\ 7 & 2 & 4 & 14 & 4 & 8 \\ 6 & 9 & 0 & 8 & 12 & 0 \\ 3 & 24 & 3 & 4 & 32 & 4 \\ 15 & 0 & 27 & 20 & 0 & 36 \\ 21 & 6 & 12 & 28 & 8 & 16 \end{pmatrix}.$$

$\mathbf{A} \otimes \mathbf{B}$ is different from $\mathbf{B} \otimes \mathbf{A}$, so the Kronecker product is also not commutative. However, contrary to the usual matrix multiplication (last line in the previous shaded box), there is not change of order of matrices when transposing, that is, $(\mathbf{A} \otimes \mathbf{B})^T = \mathbf{A}^T \otimes \mathbf{B}^T$.

The operations covered so far are easy to compute by hand but tedious; the greater the size of the matrices, the more tedious the computation. This issue will become worse in the following sections; therefore, we will focus on conceptual ideas rather than computations, assuming that a suitable software is available for the actual computations. However, the reader should be aware that, as stated in Ref. [2], "the form of a mathematical expression and the way the expression should be evaluated in actual practice may be quite different".

2.1.3. *Matrix rank*

In brief, terms such as linear independence for vectors also apply to rows or columns of a matrix (for a deeper discussion and proofs of the properties stated in this section, see, for example, Ref. [2]). The maximum number of linearly independent vectors (either the rows or the columns of the matrix) is called the *rank* of the matrix.

Geometrically, we consider the vector space generated by the columns of a $m \times p$ matrix $\mathbf{A}$, that is, m-dimensional vectors spanning a subspace of dimension less than or equal to p, which is called the column space of $\mathbf{A}$, whose dimension is the rank of $\mathbf{A}$.

It is outside the scope of the current chapter to discuss matrices as linear applications between vector spaces, but the rank considering the column space is the same as the rank considering the row space, that is, the rank is a unique characteristic of the matrix.

Therefore, $0 \leq \text{rank}(\mathbf{A}) \leq \min \{m, p\}$. Only the zero matrix (a matrix filled with zeroes) has rank equal to 0. On the other hand, when $\text{rank}(\mathbf{A}) = \min \{m, p\}$, it is said that $\mathbf{A}$ is of *full rank*; otherwise, it is *rank deficient*.

Some properties of the rank of matrices for different operations

$\text{rank}(\mathbf{A}) = \text{rank}(\mathbf{A}^T)$	for any matrix $\mathbf{A}$.
$\text{rank}(\mathbf{A}) = \text{rank}(\lambda\mathbf{A})$	for any $\mathbf{A}$ and scalar $\lambda \neq 0$.
$\text{rank}(\mathbf{A} + \mathbf{B}) \leq \text{rank}(\mathbf{A}) + \text{rank}(\mathbf{B})$	for any $\mathbf{A}$ and $\mathbf{B}$ of the same size.
$\text{rank}(\mathbf{AB}) \leq \min\{\text{rank}(\mathbf{A}), \text{rank}(\mathbf{B})\}$	for any $\mathbf{A}$ and $\mathbf{B}$ that are conformable for the multiplication (the number of columns of $\mathbf{A}$ equal to the number of rows in $\mathbf{B}$).

2.2. Inversion of Matrices

Particular importance is given to *square matrices*, which are those with the same number of rows and columns. The a_{ii} elements of a matrix $\mathbf{A} = (a_{ij})$ are called *diagonal elements*, whereas a_{ij}, $i \neq j$, are the *off-diagonal elements*. The vector consisting of all of the diagonal elements is called the principal or main diagonal or, simply, the *diagonal*.

Although the definition of principal diagonal applies whether or not the matrix is square, it makes more sense for square matrices.

A square matrix whose off-diagonal elements are all zeros is called a *diagonal matrix*.

If $\mathbf{S}$ is a square matrix, $\mathbf{S}^T$ is also square. The matrix is said to be *symmetric* when $\mathbf{S} = \mathbf{S}^T$ and *skew symmetric* when $\mathbf{S}^T = -\mathbf{S}$ i.e., ($a_{ij} = -a_{ji}$, for all i, j, and hence, the diagonal elements are all zero). In particular, a diagonal matrix is always symmetric. For example, $\mathbf{S}_1$ in Equation (2.9) is symmetric, whereas $\mathbf{S}_2$ is skew-symmetric:

$$\mathbf{S}_1 = \begin{pmatrix} 2 & -1 & 5 \\ -1 & 0 & 7 \\ 5 & 7 & 3 \end{pmatrix}, \quad \mathbf{S}_2 = \begin{pmatrix} 0 & -1 & 5 \\ 1 & 0 & 7 \\ -5 & -7 & 0 \end{pmatrix}. \tag{2.9}$$

The diagonal of $\mathbf{S}_1$ in Equation (2.9) is $(2, 0, 3)^T$, the null vector for $\mathbf{S}_2$. Both matrices have different off-diagonal elements, not all of them zero, so none of the matrices is a diagonal matrix.

The *identity* matrix is a diagonal matrix with all its diagonal elements equal to one. The $m \times m$ identity matrix is denoted as $\mathbf{I}_m$. For example, $\mathbf{I}_3$ is

$$\begin{pmatrix} 1 & 0 & 0 \\ 0 & 1 & 0 \\ 0 & 0 & 1 \end{pmatrix}.$$

A nice property of the identity matrix is that it can be multiplied with any matrix without modifying it, i.e., for any $m \times p$ matrix $\mathbf{A}$, $\mathbf{A}\mathbf{I}_p = \mathbf{A}$ and $\mathbf{I}_m \mathbf{A} = \mathbf{A}$.

2.2.1.   *Square matrices and inverses*

In relation to general square matrices, the *trace* of a square matrix $\mathbf{S} = (s_{ij})$ is the sum of its diagonal elements (Equation (2.10)):

$$\mathrm{Tr}(\mathbf{S}) = \sum_i s_{ii}. \tag{2.10}$$

For example, the trace of $\mathbf{S}_1$ in Equation (2.9) is five, the trace of $\mathbf{S}_2$ is zero, and $\mathrm{Tr}(\mathbf{I}_m) = m$. Clearly, $\mathrm{Tr}(\mathbf{S}) = \mathrm{Tr}(\mathbf{S}^T)$, and taking the definition of matrix multiplication into account, for any two square matrices $\mathbf{A}$ and $\mathbf{B}$ of the same size, $\mathrm{Tr}(\mathbf{AB}) = \mathrm{Tr}(\mathbf{BA})$. In fact, this is true even if $\mathbf{A}$ is an $m \times p$ matrix and $\mathbf{B}$ a $p \times m$ matrix, and thus, $\mathbf{AB}$ is $m \times m$ and $\mathbf{BA}$ is $p \times p$ [1, p. 12].

Therefore, for any $m \times p$ matrix $\mathbf{A}$, $\mathbf{AA}^T$ and $\mathbf{A}^T\mathbf{A}$ are both square matrices (of size $m \times m$ and $p \times p$, respectively) and with the same trace. This property is used to define the *squared Frobenius norm* of any matrix $\mathbf{A} = (a_{ij})$, as in Equation (2.11), which is the generalisation of the norm of a vector in Equation (2.4):

$$\|\mathbf{A}\|^2 = \mathrm{Tr}(\mathbf{AA}^T) = \mathrm{Tr}(\mathbf{A}^T\mathbf{A}) = \sum_{i,j} a_{ij}^2. \tag{2.11}$$

An $m \times m$ square matrix $\mathbf{S}$ is *invertible* or *nonsingular* if there is another $m \times m$ matrix $\mathbf{T}$ such that $\mathbf{ST} = \mathbf{TS} = \mathbf{I}_m$. Be aware that the conditions to be met are not trivial: both matrix products must exist, be the same, and be equal to the identity matrix.

The definition implies that nonsquare matrices do not have inverses and also that not all square matrices have inverses. When such a matrix exists, it is unique [3] and is called the *inverse* of $\mathbf{S}$, denoted as $\mathbf{S}^{-1}$.

For example, $\begin{pmatrix} -2 & 1 \\ 1.5 & -0.5 \end{pmatrix}$ is the inverse of $\begin{pmatrix} 1 & 2 \\ 3 & 4 \end{pmatrix}$, and vice versa, because

$$\begin{pmatrix} 1 & 2 \\ 3 & 4 \end{pmatrix} \begin{pmatrix} -2 & 1 \\ 1.5 & -0.5 \end{pmatrix} = \begin{pmatrix} 1(-2) + 2(1.5) & 1 + 2(-0.5) \\ 3(-2) + 4(1.5) & 3 + 4(-0.5) \end{pmatrix} = \begin{pmatrix} 1 & 0 \\ 0 & 1 \end{pmatrix}$$

$$= \begin{pmatrix} -2 & 1 \\ 1.5 & -0.5 \end{pmatrix} \begin{pmatrix} 1 & 2 \\ 3 & 4 \end{pmatrix}.$$

A (square invertible) matrix $\mathbf{S}$ such that $\mathbf{S}^{-1} = \mathbf{S}^T$ is called an *orthogonal* matrix. That means the columns are unit length vectors and mutually orthogonal (perpendicular). Clearly, the matrix in the previous example is not orthogonal. Sometimes, to stress the unit length of column vectors, they are called *orthonormal*.

From the definition of inverse and transpose, $(\mathbf{S}^{-1})^{-1} = \mathbf{S}$, and $(\mathbf{S}^{-1})^T = (\mathbf{S}^T)^{-1}$. Regarding matrix multiplication, if $\mathbf{S}$ and $\mathbf{T}$ are invertible matrices of the same size, $(\mathbf{ST})^{-1} = \mathbf{T}^{-1}\mathbf{S}^{-1}$, and, irrespective of the sizes, $(\mathbf{S} \otimes \mathbf{T})^{-1} = \mathbf{S}^{-1} \otimes \mathbf{T}^{-1}$. There is no equivalent property for the Hadamard product.

Examples of invertible matrices include diagonal matrices with all the diagonal elements non-null. In this case, its inverse is also easily computed: the inverse of diagonal $\mathbf{S} = (s_{ij})$ is simply the diagonal matrix with elements $1/s_{ii}$.

In general, for any square matrix, the characterisation of invertible matrices is in terms of the rank. As previously stated, the concept of rank is related to the structure of the subspace spanned by a matrix.

> Invertible matrices are full-rank square matrices. In other words, they are square matrices $(m \times m)$ for which the subspace generated by the m-dimensional columns has the maximum dimension, m. Indeed, the rank is exactly m.

Particular methods of computing the rank of a matrix include the use of determinants. The definition and computation of determinants are outside the scope of this chapter.

2.2.2. *Symmetric matrices and distances*

For the purposes of the current book, real symmetric matrices play an important role in several procedures that will be introduced in the following chapters. In this section, we only focus on its use in defining distances.

A symmetric $m \times m$ matrix $\mathbf{S}$ is said to be *positive definite* when the result of $\mathbf{v}^T \mathbf{S} \mathbf{v}$ is a positive scalar for any non-null m-dimensional vector $\mathbf{v}$. If the scalar is only non-negative, it is called *positive semi-definite or non-negative definite*. For example, a correlation matrix is always a symmetric non-negative definite matrix whose diagonal elements are all one.

A simple example is the Euclidean distance, denoted as d_2 when it is convenient. In that sense, Equation 2.12 shows different ways of writting the squared Euclidean distance between two m-dimensional vectors $\mathbf{w} = (w_{ij})$ and $\mathbf{z} = (z_{ij})$.

$$d^2(\mathbf{w}, \mathbf{z}) = d_2^2(\mathbf{w}, \mathbf{z}) = \|\mathbf{w} - \mathbf{z}\|_2^2 = \sum_{i=1}^{m} (w_i - z_i)^2$$

$$= (\mathbf{w} - \mathbf{z})^T (\mathbf{w} - \mathbf{z}) = (\mathbf{w} - \mathbf{z})^T \mathbf{I}_m (\mathbf{w} - \mathbf{z}). \quad (2.12)$$

Equation (2.12) shows that the distance between two vectors can be written as the particular matrix product $\mathbf{v}^T \mathbf{S} \mathbf{v}$ for the vector of the differences, $\mathbf{v} = \mathbf{w} - \mathbf{z}$, and with the identity matrix $\mathbf{I}_m$ as a positive definite matrix $\mathbf{S}$.

The fact is that $\mathbf{I}_m$ in Equation (2.12) can be replaced by any other positive definite matrix and we would still be calculating a distance between vectors, just not the length of the segment connecting them. Particularly useful is *the distance of Mahalanobis,* which is a probabilistic distance, computed when using a full-rank variance–covariance matrix $\mathbf{C}$ that is always positive definite.

It is a known property that any positive definite matrix is nonsingular, and its inverse is also positive definite [1]. Then, the inverse of a full-rank covariance matrix $\mathbf{C}$ is positive definite, and the precise definition of the Mahalanobis distance is given in Equation (2.13):

$$d_{\text{Mah}}^2(\mathbf{w}, \mathbf{z}) = (\mathbf{w} - \mathbf{z})^T \mathbf{C}^{-1}(\mathbf{w} - \mathbf{z}). \quad (2.13)$$

For example, if $\mathbf{X}$ is an $m \times p$ data matrix with centred columns (i.e., all the columns have a null mean), its variance–covariance matrix, or simply covariance matrix, is $\mathbf{C} = 1/(m-1)\mathbf{X}^T\mathbf{X}$, and the Mahalanobis distance defined with $\mathbf{C}^{-1}$ will weigh the length in each direction by the correlation in this variable to take into account the mutual behaviour among the variables.

2.2.3. *General matrices and pseudoinverses*

For nonsquare matrices of full rank, for which the concept of invertible matrix does not apply, there is still some kind of "pseudo" inversion that can be defined. With this goal, suppose $\mathbf{A}$ is an $m \times p$ matrix of full rank.

If $m \leq p$, $\mathbf{A}$ being of full rank means that $\text{rank}(\mathbf{A}) = m$ (the number of rows). In these conditions, the *right inverse* of $\mathbf{A}$ is a $p \times m$ matrix $\mathbf{B}$ of full rank m (its number of columns) such that $\mathbf{AB} = \mathbf{I}_m$.

On the other hand, for $m \geq p$, $\text{rank}(\mathbf{A}) = p$ (now, there are fewer columns than rows if the matrix is nonsquare), and then a matrix $\mathbf{B}$ is the *left inverse* of $\mathbf{A}$ if $\mathbf{BA} = \mathbf{I}_p$. Matrix $\mathbf{B}$ is still $p \times m$, but now $\text{rank}(\mathbf{B}) = p$.

For square matrices ($m = p$), the left and right inverses coincide, and the corresponding matrix $\mathbf{B}$ is indeed $\mathbf{A}^{-1}$.

Typical uses of these inverses include the least-squares solution of an overdetermined linear system of equations, which is required, for example, to build linear regression models.

In that case, the usual system will look like $\mathbf{Xb} = \mathbf{y}$, where $\mathbf{X}$ is an $m \times p$ data matrix (i.e., m objects for which p variables have been measured), $\mathbf{y}$ is the (m-dimensional) response vector, and $\mathbf{b}$ is the p-dimensional vector of coefficients. It is assumed that $\mathbf{X}$ is of full rank and, to facilitate the notation, that there is no offset in the model.

If $m = p$, which is rather infrequent, it is clear that the unique solution is $\mathbf{b} = \mathbf{X}^{-1}\mathbf{y}$. Otherwise, if a solution exists, it is not unique, and hence we use additional criteria to select one of the solutions. When $m > p$ (more objects than variables), the least-squares criterion of fitting is, in fact, solving $\mathbf{X}^T\mathbf{X}\mathbf{b} = \mathbf{X}^T\mathbf{y}$. Now, $\mathbf{X}^T\mathbf{X}$ is a $p \times p$ square (and symmetric) matrix of full rank (p), and as such, it is invertible; therefore, the unique least-squares solution is $\mathbf{b} = (\mathbf{X}^T\mathbf{X})^{-1}\mathbf{X}^T\mathbf{y}$. The matrix $(\mathbf{X}^T\mathbf{X})^{-1}\mathbf{X}^T$ is precisely the left inverse of $\mathbf{X}$.

Although it is not related to the least-squares solution, note that the other product possible would be $\mathbf{X}\mathbf{X}^T$, which is also a square symmetric matrix, $m \times m$, but its rank is still p, so it is a rank deficient matrix and hence has no inverse, neither left nor right.

A more general pseudoinverse can be defined for rank deficient matrices, including singular square matrices. Although there are several definitions under the same name, most commonly, the term "pseudoinverse" refers to the so-called *Moore–Penrose pseudoinverse*.

For real-valued matrices, the Moore–Penrose inverse (or pseudoinverse) of $\mathbf{A}(m \times p)$, usually denoted as $\mathbf{A}^+$, is a $p \times m$ matrix such that $\mathbf{A}\mathbf{A}^+\mathbf{A} = \mathbf{A}$, $\mathbf{A}^+\mathbf{A}\mathbf{A}^+ = \mathbf{A}^+$, and both $\mathbf{A}\mathbf{A}^+$ and $\mathbf{A}^+\mathbf{A}$ are symmetric. With these four conditions, the Moore–Penrose pseudoinverse exists, and it is unique.

If $\mathbf{A}$ is of full rank, then either $\mathbf{A}\mathbf{A}^+ = \mathbf{I}_m$ if rank($\mathbf{A}$) = m or $\mathbf{A}^+\mathbf{A} = \mathbf{I}_p$ if rank($\mathbf{A}$) = p. The definition is also consistent with inversion in the sense that $\mathbf{A}^+ = \mathbf{A}^{-1}$ for invertible $\mathbf{A}$. Some more properties are provided in the following shaded box.

Some properties of the (Moore-Penrose) pseudoinverse

$(\mathbf{A}^+)^+ = \mathbf{A}$.

$(\mathbf{A}^+)^T = (\mathbf{A}^T)^+$.

rank($\mathbf{A}$) = rank($\mathbf{A}^+$) = rank($\mathbf{A}\mathbf{A}^+$) = rank($\mathbf{A}^+\mathbf{A}$) $\leq$ min$\{m, p\}$.

$\mathbf{A}^+ = \mathbf{A}^T \Leftrightarrow \mathbf{A}^T\mathbf{A} = \mathbf{I}_p$.

For example, $\mathbf{A} = \begin{pmatrix} 1 & 2 \\ 3 & 6 \end{pmatrix}$ is a noninvertible (singular) square matrix because the second column is twice the first, so they are linearly dependent, and rank($\mathbf{A}$) = 1 < 2. Its Moore–Penrose pseudoinverse is $\mathbf{A}^+ = \begin{pmatrix} 0.02 & 0.06 \\ 0.04 & 0.12 \end{pmatrix}$ because

$$\begin{pmatrix} 0.02 & 0.06 \\ 0.04 & 0.12 \end{pmatrix}\begin{pmatrix} 1 & 2 \\ 3 & 6 \end{pmatrix} = \begin{pmatrix} 0.20 & 0.40 \\ 0.40 & 0.80 \end{pmatrix} \quad \text{SYMMETRIC,}$$

$$\begin{pmatrix} 1 & 2 \\ 3 & 6 \end{pmatrix}\begin{pmatrix} 0.02 & 0.06 \\ 0.04 & 0.12 \end{pmatrix} = \begin{pmatrix} 0.10 & 0.30 \\ 0.30 & 0.90 \end{pmatrix} \quad \text{SYMMETRIC,}$$

$$\begin{pmatrix} 1 & 2 \\ 3 & 6 \end{pmatrix}\begin{pmatrix} 0.02 & 0.06 \\ 0.04 & 0.12 \end{pmatrix}\begin{pmatrix} 1 & 2 \\ 3 & 6 \end{pmatrix}$$

$$= \begin{pmatrix} 0.10 & 0.30 \\ 0.30 & 0.90 \end{pmatrix} \begin{pmatrix} 1 & 2 \\ 3 & 6 \end{pmatrix} = \begin{pmatrix} 1 & 2 \\ 3 & 6 \end{pmatrix},$$

$$\begin{pmatrix} 0.02 & 0.06 \\ 0.04 & 0.12 \end{pmatrix} \begin{pmatrix} 1 & 2 \\ 3 & 6 \end{pmatrix} \begin{pmatrix} 0.02 & 0.06 \\ 0.04 & 0.12 \end{pmatrix}$$

$$= \begin{pmatrix} 0.20 & 0.40 \\ 0.40 & 0.80 \end{pmatrix} \begin{pmatrix} 0.02 & 0.06 \\ 0.04 & 0.12 \end{pmatrix} = \begin{pmatrix} 0.02 & 0.06 \\ 0.04 & 0.12 \end{pmatrix}.$$

In the case of solving the underdetermined linear system $\mathbf{Xb} = \mathbf{y} (m > p)$ using least squares, irrespective of the rank of $\mathbf{X}^T\mathbf{X}$, it has a solution: $\mathbf{b} = \mathbf{X}^+\mathbf{y}$.

2.3.   Matrix Decomposition and Diagonalisation

There are several types of factorisations that can be computed for matrices, such as LU, QR, and Jordan decomposition. In this section, we focus on diagonalisation of square matrices and singular value decomposition (SVD) of any matrix because they are used for the development of common chemometrics methods, such as principal component analysis discussed in Chapters 6 and 7.

2.3.1.   *Eigenanalysis (eigenvector-eigenvalue problem)*

For a given square matrix $\mathbf{S}$ $(m \times m)$, a nonzero m-dimensional vector $\mathbf{v}$ is said to be an *eigenvector* if there is a scalar λ such that $\mathbf{Sv} = \lambda\mathbf{v}$. Such a scalar is called the *eigenvalue* corresponding to eigenvector $\mathbf{v}$. As in the rest of the chapter, only real eigenvalues are considered.

For example, $\mathbf{v} = (1, 0, 0)^T$ is an eigenvector of $\mathbf{S} = \begin{pmatrix} -2 & 4 & 2 \\ 0 & 1 & 3 \\ 0 & -1 & -2 \end{pmatrix}$ because

$$\mathbf{Sv} = \begin{pmatrix} -2 & 4 & 2 \\ 0 & 1 & 3 \\ 0 & -1 & -2 \end{pmatrix} \begin{pmatrix} 1 \\ 0 \\ 0 \end{pmatrix} = \begin{pmatrix} -2 \\ 0 \\ 0 \end{pmatrix} = -2 \begin{pmatrix} 1 \\ 0 \\ 0 \end{pmatrix} = -2\mathbf{v},$$

and thus, $\lambda = -2$ is the eigenvalue that makes $\mathbf{v}$ the corresponding eigenvector. In practice, first the eigenvalues are found, and then we compute the corresponding eigenvectors. For example, all the elements in a diagonal matrix $\mathbf{D}$ are eigenvalues of $\mathbf{D}$. Then, for each of them, λ_i, the eigenvector(s) $\mathbf{v}$ are found as solutions to the linear system $(\mathbf{D} - \lambda_i \mathbf{I})\mathbf{v} = \mathbf{0}$, where $\mathbf{I}$ is the identity matrix of appropriate size and $\mathbf{0}$ is a vector filled with zeros, also in the appropriate amount.

Let $\mathbf{D} = \begin{pmatrix} -2 & 0 & 0 \\ 0 & 1 & 0 \\ 0 & 0 & -2 \end{pmatrix}$. It has two eigenvalues, namely $\lambda_1 = 1$ and $\lambda_2 = -2$, which appears twice (and is said to be an eigenvalue with an algebraic *multiplicity* of 2). An eigenvector corresponding to $\lambda_1 = 1$ is a solution of the homogeneous underdetermined system in Equation (2.14):

$$(\mathbf{D} - \mathbf{I}_3)\mathbf{v} = \begin{pmatrix} -2 - 1 & 0 & 0 \\ 0 & 1 - 1 & 0 \\ 0 & 0 & -2 - 1 \end{pmatrix} \begin{pmatrix} v_1 \\ v_2 \\ v_3 \end{pmatrix} = \begin{pmatrix} 0 \\ 0 \\ 0 \end{pmatrix}. \quad (2.14)$$

The infinitely many solutions of Equation (2.14) are of the form $(0, a, 0)^T$ for any scalar a. As such, $(0, 1, 0)^T$ is an eigenvector corresponding to the eigenvalue 1.

However, a more interesting question is whether and how a square matrix $\mathbf{S}$ can be factorised, or decomposed, including in the decomposition a diagonal matrix with the eigenvalues of $\mathbf{S}$.

Precisely, an $m \times m$ matrix $\mathbf{S}$ is *diagonalisable* if there exists an invertible matrix $\mathbf{P}$ such that $\mathbf{P}^{-1}\mathbf{S}\mathbf{P}$ is a diagonal matrix, $\mathbf{D}$. Equivalently, $\mathbf{S}$ is decomposed as in Equation (2.15) for invertible $\mathbf{P}$ and diagonal $\mathbf{D}$:

$$\mathbf{S} = \mathbf{P}\mathbf{D}\mathbf{P}^{-1}. \quad (2.15)$$

The necessary and sufficient condition for $\mathbf{S}$ to be diagonalisable (see, for example, Theorem 7.5 in Ref. [3, p. 437]) is that it should have m linearly independent eigenvectors. Furthermore, in that case, $\mathbf{P}$ in Equation (2.15) is a matrix whose columns are these m eigenvectors, and $\mathbf{D}$ is the diagonal matrix with the eigenvalues (repeated as many times as indicated by their multiplicity) in the same order used to form $\mathbf{P}$.

In particular, any matrix with m different eigenvalues is diagonalisable and so is a symmetric matrix; remember that $\mathbf{A}^T\mathbf{A}$ is always symmetric for any $\mathbf{A}$. In fact, every non-null eigenvalue of $\mathbf{A}^T\mathbf{A}$ is also an eigenvalue of $\mathbf{A}\mathbf{A}^T$, although they might not share the eigenvectors. (For nonsquare $\mathbf{A}$, they might even be of different sizes.) This property is used for the SVD, which is introduced in the following section.

Among the remarkable properties of diagonalisable matrices and the related eigenpairs (eigenvector-eigenvalue), two are of interest here, namely, $\mathrm{Tr}(\mathbf{S}) = \mathrm{Tr}(\mathbf{P}\mathbf{D}\mathbf{P}^{-1}) = \mathrm{Tr}(\mathbf{D})$ and $\mathrm{rank}(\mathbf{D}) = \mathrm{rank}\,(\mathbf{S})$. The second property implies that the number of nonzero eigenvalues of a diagonalisable matrix $\mathbf{S}$ is equal to its rank, or more precisely, the sum of the multiplicities of the unique nonzero eigenvalues of a diagonalisable matrix is equal to the rank of the matrix.

Besides, in the case of a full-rank symmetric matrix $\mathbf{S}$, any eigenvectors corresponding to distinct eigenvalues are orthogonal. Therefore, $\mathbf{P}$ in Equation (2.15) can be chosen to be orthogonal, and the decomposition is then written as

$$\mathbf{S} = \mathbf{P}\mathbf{D}\mathbf{P}^T.$$

It is said that any square symmetric matrix is *orthogonally diagonalisable*. In fact, orthogonally diagonalisable matrices are only the symmetric matrices [2, p. 154].

2.3.2. *Singular value decomposition*

Singular value decomposition is one of the most useful decompositions that can be applied to any $m \times p$ matrix $\mathbf{A}$. The decomposition or factorisation is written as in Equation (2.16):

$$\mathbf{A} = \mathbf{U}\mathbf{D}\mathbf{V}^T, \qquad (2.16)$$

where $\mathbf{U}$ and $\mathbf{V}$ are orthogonal matrices of size $m \times m$ and $p \times p$, respectively, and $\mathbf{D}$ is an $m \times p$ diagonal matrix with non-negative elements in the diagonal d_{ii}. A nonsquare diagonal matrix refers to an $m \times p$ matrix with $\min\{m, p\}$ elements in the diagonal and zeros in every other entry.

The SVD is unique for $\mathbf{A}$, and thus the elements on the diagonal of $\mathbf{D}$, which are also unique, are called the *singular values* of $\mathbf{A}$. There are $\min\{m, p\}$ singular values, and it is common to order them along the diagonal of $\mathbf{D}$ from largest to smallest. If they are not, it suffices to rearrange them and the columns of $\mathbf{U}$ accordingly.

As $\mathbf{U}$ and $\mathbf{V}$ are full-rank matrices, $\text{rank}(\mathbf{A}) = \text{rank}(\mathbf{D})$, which is the number of positive singular values of $\mathbf{A}$, which in turn coincides with the number of non-null eigenvalues of $\mathbf{A}^T\mathbf{A}$ (or of $\mathbf{A}\mathbf{A}^T$ because they have the same nonzero eigenvalues) since the SVD can be seen from the orthogonal diagonalisation of the symmetric matrix $\mathbf{A}^T\mathbf{A}$ (also from $\mathbf{A}\mathbf{A}^T$). For details, consult Ref. [2].

Consequently, if d is a singular value of $\mathbf{A}$, $\lambda = d^2$ is an eigenvalue of both $\mathbf{A}^T\mathbf{A}$ and $\mathbf{A}\mathbf{A}^T$, and vice versa, whereas if λ is a nonzero eigenvalue of $\mathbf{A}^T\mathbf{A}$ (or $\mathbf{A}\mathbf{A}^T$), then there is a singular value d of $\mathbf{A}$ such that $d^2 = \lambda$.

For that reason (or simply by transposing in Equation (2.16)), it is clear that the singular values of $\mathbf{A}^T$ and $\mathbf{A}$ are identical. Also, if $\mathbf{A}$ is symmetric itself, then its singular values are the absolute values of its eigenvalues.

For example, the SVD of the 3×2 matrix $\mathbf{A}$, $\mathbf{A}_{32} = \begin{pmatrix} -1.8 & 2.4 \\ 0 & 0 \\ 4.8 & 3.6 \end{pmatrix}$ is

$$\mathbf{A} = \begin{pmatrix} 0 & -1 & 0 \\ 0 & 0 & -1 \\ 1 & 0 & 0 \end{pmatrix} \begin{pmatrix} 6 & 0 \\ 0 & 3 \\ 0 & 0 \end{pmatrix} \begin{pmatrix} 0.8 & 0.6 \\ 0.6 & -0.8 \end{pmatrix} = \mathbf{U}\mathbf{D}\mathbf{V}^T.$$

It is left to the reader to verify that $\mathbf{U}$ and $\mathbf{V}$ are orthogonal. (Note that without the need for obtaining their inverses, it can be verified "manually" by computing the inner products between each two columns, which must be zero since they are orthogonal to one another. Clearly, the norm of all columns in $\mathbf{U}$ is one, whereas for $\mathbf{V}$, again, it is a matter of a few calculation steps.)

The minimum between the number of rows and columns is 2, so there are two singular values of $\mathbf{A}$, which are 6 and 3, the diagonal elements of $\mathbf{D}$. In particular, rank $(\mathbf{A}) = 2$, which was already clear because it has a null row.

On the other hand, $\mathbf{A}^T\mathbf{A} = \begin{pmatrix} 26.28 & 12.96 \\ 12.96 & 18.72 \end{pmatrix}$ is indeed a square matrix of rank 2, which contains the sum of squares of the corresponding column in the diagonal and the inner product between columns in the off-diagonal terms (which are, of course, equal).

The other square matrix will be $\mathbf{A}\mathbf{A}^T = \begin{pmatrix} 9 & 0 & 0 \\ 0 & 0 & 0 \\ 0 & 0 & 36 \end{pmatrix}$, with the sum of squares of each row in the diagonal and the two-by-two inner products of the rows of $\mathbf{A}$ in the off-diagonal terms. We see that, in this case, the rows are orthogonal to one another because the matrix is diagonal. This fact makes it easy to see that, indeed, its rank is 2 and it has two nonzero eigenvalues, namely 9 and 36, that are the squares of the singular values of $\mathbf{A}$.

Another exercise for the reader is to verify that 9 and 36 are the two eigenvalues of $\mathbf{A}^T\mathbf{A}$. *Hints:* with the definitions given in this chapter, it is necessary to look for the eigenvectors $\mathbf{v}$ and $\mathbf{w}$, related to 9 and 36, respectively, by solving the respective homogeneous systems of equations

$$\begin{pmatrix} 26.28 - 9 & 12.96 \\ 12.96 & 18.72 - 9 \end{pmatrix} \begin{pmatrix} v_1 \\ v_2 \end{pmatrix} = \begin{pmatrix} 0 \\ 0 \end{pmatrix} \quad \text{and}$$

$$\begin{pmatrix} 26.28 - 36 & 12.96 \\ 12.96 & 18.72 - 36 \end{pmatrix} \begin{pmatrix} w_1 \\ w_2 \end{pmatrix} = \begin{pmatrix} 0 \\ 0 \end{pmatrix}.$$

Alternatively, for the readers familiar with determinants, it will suffice to see that 9 and 36 are the unique roots of the characteristic polynomial, that is, solutions of the following equation:

$$\begin{vmatrix} 26.28 - \lambda & 12.96 \\ 12.96 & 18.72 - \lambda \end{vmatrix} = 324 - 45\lambda + \lambda^2 = 0.$$

2.4. Higher-Dimensional Arrays

In the context of the current book, a matrix is a two-way array. Similarly, a three-way array (or a data cube) can be obtained by stacking matrices of the same size, for example, one on top of another. In that case, we will have an $m \times n \times k$ tensor – a data cube.

Likewise, four-way, five-way, and, in general, n-way arrays can be defined. Usually, they are denoted as **X**. The particularities of operating with such arrays will be explained in the chapters where they are used.

References

[1]   Adachi, K. (2020). *Matrix-Based Introduction to Multivariate Data Analysis*, 2nd edn. Singapore: Springer.
[2]   Gentle, J. E. (2017). *Matrix Algebra. Theory, Computations and Applications in Statistics*, 2nd edn. Switzerland: Springer.
[3]   Larson, R. and Falvo, D. C. (2009). *Elementary Linear Algebra*. Boston: Houghton Mifflin Harcourt Publishing Company, pp. 73–74.

Chapter 3

The Basics of Univariate and Multivariate Calibration

José Andrade-Garda

Objectives and Scope

This chapter summarises the main concepts associated with traditional calibration (or standardisation, as it is preferred to call it here) and relates them to multivariate classical linear regression. This serves as an introductory background to fully understand the advantages of more advanced regression methods presented in subsequent chapters.

Some insights are given on how to use the correct jargon (despite many traditional terms being deeply rooted in our common language) and the importance of validating the models (which is a fundamental issue in obtaining fit-for-purpose results).

Much effort is also devoted to showing the limitations of the chemometric techniques presented here (namely ordinary and inverse multivariate linear regression).

Finally, a guided example using the CAT software is included to illustrate the ideas presented hereinafter.

3.1.   Introduction: Epistemological Principles

Every analytical methodology that is used to solve practical problems must be validated in advance. This is not an option but a mandatory requirement nowadays. Indeed, the most up-to-date epistemological definitions of analytical chemistry describe it as a branch of science (in particular, chemistry) that gathers chemical information about material systems from experimental measurements of quantities of any nature [1–3]. Laboratories exist to acquire information about a "system", which can be any collection of circumstances appearing in daily life where chemical (biochemical or physical) information is needed for decision-making (economic, process control, quality acceptance, health-related, etc.).

This means that people (hereinafter, in this chapter, customers) request help from scientists working in facilities equipped with – usually – complex and expensive devices (such as those found in a laboratory). The scientists then perform some "esoteric" work and, in the end, offer their customers the "keystone" necessary to make sound decisions. Of course, the key concept here is that decisions must be "sound", as good decisions can only be achieved when their foundations are strong and trustworthy. Ultimately, their decisions will be based on the data (transformed into information) that we (or laboratories) offer them. This seems fine and quite trivial. However, every scientist knows that completely perfect data do not exist. At the very minimum, uncertainty exists in any measurement, gross errors may appear, and systematic procedural biases are also common. As an example of the last situation, in clinical chemistry, proprietary method-defined analytes (e.g., genes, proteins, and DNA/RNA fragments) may prevent the existence of reference standards to assess trueness. (You can find more on this challenging issue by searching for "commutability in clinical chemistry" using any web search engine.) Usually, we face many more problems. Then, how can we be sincere with our customers whenever they ask us about the quality of the data? And, for certain, this is a question that invariably appears in any laboratory–customer relationship.

Full explanations of the uncertainty concept are not presented here, as they fall outside the scope of this book. For a comprehensive exposition, with some practical examples, one can consult – for free – one of the most renowned and widely used international guidelines on uncertainty [4]. An interesting guide was also developed to tackle the issue of target measurement uncertainty [5]. For the purposes of this chapter, it suffices to say that uncertainty is "a parameter associated with the result of a measurement, that characterises the dispersion of the values that could reasonably be attributed to the measurand" (i.e., the property you are interested in) [4].

The key is to make customers understand that, despite scientific data always having uncertainty attached, we are "quite sure" that the conclusions we offer them are, to a certain extent, confident given the circumstances. And this is where the fitness-for-purpose concept appears. Admittedly, no experiment is perfect, *but* we should be able to demonstrate that we absolutely made all our best efforts to ensure accuracy and that the resulting information we deliver is good enough for the intended use by the customer. Curiously, this requires our customer to explain to us in advance how they intend to use the information/data that they will receive. In many cases, this does not happen despite its importance. The intended use is important because in one situation, our work may be perfect; however, in another, it might be almost useless or even detrimental. For instance, we can determine the concentration of sulphates in a solution using a gravimetric method (if the amount is large), but the same procedure is not appropriate for quantifying them within a cell.

So, fitness-for-purpose requires "... that the tests carried out are appropriate for the analytical part of the problem that the customer wishes solved, and that the final report presents the analytical data in such a way that the customer can readily understand it and draw appropriate conclusions" [6]. Analogously, procedure (or method) validation is "the process of defining an analytical requirement, and confirming that the method under consideration has capabilities consistent with what the application requires. Inherent in this is the need to evaluate the method's performance". Note also that verification is "provision of objective evidence that a given item fulfils specified requirements" [6] and that validation is a "verification, where the specified requirements are adequate for an intended use". More details and related discussions can be found elsewhere [6]. Observe that validation applies to perfectly described methods of analysis (referred to as standard operating procedures) and that it is only possible to validate a methodology, not individual results.

To what extent is this possible? In principle, everything that can influence a result should be validated (staff, reagents, equipment, software, etc.); however, on many occasions, we cannot study, evaluate, or assure everything we can imagine. So, validation is always a balance between costs, risks, knowledge, and technical possibilities.

In this book, we are committed to showing you not only how to develop chemometric models but also how to validate the results that you can obtain from them. All chapters strongly stress that you have to validate models, and this is not always trivial.

In this chapter, we concentrate first on how to carry out and validate a usual "calibration". This is likely an almost universal step in every commonly used analytical method (except for those that are considered absolute, like gravimetry, titrimetry, radioactive methods, and some electrochemical and mass-fragment methods), and the quality of the results will depend on it, as well as on your own working skills. Then, we generalise those ideas to obtain multivariate linear regression (both ordinary and inverse), study its main drawbacks (which will justify the need for more advanced techniques, as addressed in subsequent chapters), and, finally, present an example.

3.2. Understanding the Jargon: *Chemical Calibration* or *Standardisation*

When an instrumental system is going to be used to measure an aliquot of an unknown sample, it is almost always necessary to obtain in advance a relationship between the signal such an instrument offers and the property we are determining (such as the resistance of a material, concentration, viscosity, the extent of weathering, etc.). A common practice is to plot the signals obtained from a set of "samples" whose property value is known (called standards) and calculate a mathematical function that relates these to the property of interest. Before performing that procedure (as you may have done it many times in the laboratory!), let us reflect a little bit on what we will study.

First, regarding the language. It is common to be confused about the use of the terms *correlation* and *regression*. The former is used to explain the relationship between two variables, and it is very closely related to how that *relationship* can be calculated. It can be calculated using the covariance between two variables, termed v and w; for example, a chromatographic peak height and a concentration. The sample covariance is defined by $\text{cov}(v, w) = [\sum_{i=1}^{I}(v_i - \bar{v}) \cdot (w_i - \bar{w})]/(I-1)$, where $i = 1, \ldots, I$ are the experimental data pairs and $\bar{v}$ and $\bar{w}$ are the arithmetic means of the two variables.

The concept of covariance is intuitive; however, a major problem with it is that it depends on the units employed to measure the quantities. To avoid this problem, we can divide by the standard deviation of both variables, so that the units in the numerator and denominator cancel out. This is the Pearson correlation coefficient, $r = \text{cov}(v, w)/S_v \cdot S_w$, where S denotes the sample standard deviation.

It is very important to bear in mind that correlation does not mean dependence: a variable can be dependent on another without being correlated, as exemplified by the equation of a circle. A high correlation implies that the value of a variable changes in the same way as the value of the other variable. It is also worth noting that a correlation does not necessarily imply a cause–effect relation. You can certainly calculate the correlation between hair colour and the weight of inhabitants in a village; however, they do not depend on one another – they exist independently. This is very important in science and a source of gross misunderstandings (!).

> To demonstrate causality (i.e., cause-effect relation), it is required to interpret deterministic relationships and explain the underlying reason(s) why variables are correlated [7].

Similarly, there is no justification to state that a high correlation coefficient denotes a straight line, as is widely mentioned in the literature. You can have a correlation coefficient of around 0.99 if there is a curvilinear relationship between two variables. One must be careful with the use of this term. Readers are kindly encouraged to search the internet for a case study called Anscombe's datasets and admire the overall equal results that appear with the vastly differing values. Some situations are reasonable, whereas others are clearly wrong.

When the dependency of a variable (the signal of an instrument, or response, r) on an analytical parameter (the property under study; usually a concentration, c) is investigated, the term "regression" must be used. However, even the use of this term is not statistically robust according to its historical development, as Draper and Smith [8] pointed out. The modern use of this term was introduced by Fischer in his studies on the distribution theory of errors [9] in the 1920s. Hence, calibration is a form of regression [10].

Another problem is the strict use of the term calibration. For physical metrology, it is "the verification of the response of an instrument to a material of known properties and, if necessary, the correction by a factor to take the instrument to the corresponding mark" [11]. Most of us will never set marks on the scales of the instruments, so the term "standardisation" seems much more appropriate. This term refers to the "characterisation of the response of an instrument according to the known properties of the material". The standardisation of the response of an instrumental system is done by means of a series of samples whose concentrations/properties under study are known (standards), and this is usually done using the "calibration curve" (which should be called the "standardisation curve").

Hence, when we talk about regression, we usually mean the construction of an empirical mathematical relation (function or model) by which a "dependent" variable (predicted or "r"-variable) is predicted using one or several "independent" variable(s) (predictors or "c"-variables). The quotes indicate that these terms are no longer highly recommended nowadays because either they may be ambiguous or independence does not truly exist in the context of working with vector signals (which will be discussed later in this chapter).

> *Note*: As discussed in the following sections, there is both classical and inverse calibration, so you can predict c from r or r from c. For example, you may want to predict a concentration from the height of a peak or predict the height of a peak from concentration.
>
> This is why, likely, the best way to avoid being ambiguous in our writing is to explain that the property(ies) we are interested in for a given problem (which we want to predict) is(are) predicted using a set of predictors.

This is what we usually do, and so it is recommended to use the term standardisation [11] in the same way as we all talk about standards or, at least, refine the term calibration into "chemical calibration". Accordingly, both terms refer to the process that determines the functional relationship between measured values (instrumental outputs, signal intensities, or instrumental response, r) and the magnitude of the chemical property(ies) that characterise a set of samples (octane number of gasoline, analytes, hardness, biological activity, toxicity level, etc.). In analytical chemistry quite commonly the property of interest is concentration, c. I personally like this definition because it reflects what we do in practice since chemical calibration/standardisation is more a process (a series of tasks) than a unique step. Logically, prediction is the process of using the model to ascertain properties of a sample given an instrumental output, as stated in a classic textbook published in 1998 [12].

In this book, we use the terms "model development", "standardisation", and "chemical calibration" synonymously since the last term is so rooted in normal language that changes will only occur slowly (unless you are involved in an ISO17025 accreditation scheme, in which this ambiguity should not exist and standardisation is preferred for the issue we are discussing here).

Note: Standardisation is often also used as a means of data transformation, i.e., mean centring and dividing by the standard deviation (as will be studied in Chapter 5). In line with this, by the end of this chapter, we also refer to standardised residuals, which are defined as the residuals of the model divided by their standard deviation.

These are different uses of the same word. Doubts regarding the sense in which the term is used can be avoided easily by taking into account the context in which it is applied.

Model development involves the selection of the model, the estimation of the parameters and their errors and subsequent model validation. The idea of a regression triplet (data, model, and method) for performing satisfactory standardisation is thus appropriate [13]. Such an idea is also present in the International Vocabulary of Metrology as "the operation that, under specified conditions, in a first step, establishes a relation between the quantity values, with measurement uncertainties, provided by measurement standards and corresponding indications, with associated measurement uncertainties, and, in a second step, uses this information to establish a relation for obtaining a measurement result from an indication" [14].

It is worth noting that the inherent empirical nature of a chemical calibration/standardisation is reflected by the term *specified conditions*. So, in my laboratory, a working system may result in linear behaviour, while in your lab, the behaviour of a similar system might be curvilinear. For other applications, imagine that you obtain a linear relation between a measured signal and the toxicity of a compound, whereas I can only find a curvilinear trend. These situations might lead to different conclusions and/or results in decision-making. This reasoning extends to the fact that any model deemed useful for a given application is purely empirical, and its particular behaviour can vary with the working session, batch of reagents, day, or personnel, for example.

Thus, a question that arises immediately for any laboratory worker is how to proceed experimentally. The basic concept is quite well known to almost every student; however, quite surprisingly, it is often overlooked, even by senior workers. So, let us recall the basic points:

1. Create an adequate setup for your instrument, ensuring that the system is stable. Or, to put it more accurately, it should be in a state of statistical control such that there exists a range of values – contained between upper and lower control limits – that define the acceptable performance of your system.
2. Prepare a suite of standards, wherein the property of interest varies according to your control (e.g., a set of solutions with differing levels of one or more properties). It is critical that the matrix of these standards matches the matrix of unknown samples, whose properties you are going to predict. To achieve this, you may have to separate the analytes from the bulk matrix, add acids, modifiers, etc. If possible, treat standards as you would treat samples, including the sample preparation steps.
3. Measure the standards exactly as you will do with future sample aliquots.
4. Obtain an *ad hoc* mathematical function to relate the instrumental responses and the property of interest. This is the chemical calibration line (or standardisation function).
5. Finally, measure the unknowns and obtain the predicted values of their properties.

Never extrapolate outside the experimental region because you never know what happens outside your experimental working conditions!

3.3. The Linear Model in the Univariate Case

It is clear that the quality of an analytical result depends on the quality of every step of the working procedure; therefore, it is obvious that the establishment of a valid (useful) mathematical model is as important as performing adequately other steps of the analytical procedure, such as sampling and sample pretreatment. For this reason, it is of interest to learn about the diverse types of chemical calibration, together with their mathematical/statistical assumptions and limitations, the methods for validating these models, and the possibilities of outlier detection. The objective is to select the calibration method that will be most suited for the type of study one is carrying out.

Why do we (almost) always select a linear function for standardising our measurement systems? In essence, it is a question of convenience, as explained in the following:

(i) The linear model is the most simple and easy-to-understand one. It has an immediate meaning: something depends proportionally on something else (cause–effect) and it is easy to explain.

(ii) Many physical and chemical systems yield linear responses in some specific working ranges. It is a correct approach to many situations within the inherent limitations of the analytical techniques (for instance, remember that the Lambert–Beer–Bouguer law becomes curved at high concentrations of the absorbing species).

(iii) It is simple with straightforward mathematics, and associated statistics are quite well known, broadly available in the literature, and interpretable.

Bear in mind that the linear model is a working hypothesis that must be verified/validated. A model cannot be observed directly; we estimate it empirically by studying how the experimental values adhere to the systematic and random parts of a function (Equation (3.1)) – in our case, a linear model:

$$R = f(c) + \varepsilon \rightarrow r_i = \alpha + \beta \cdot c_i + \varepsilon_i (i = 1, \ldots, I)$$
$$\rightarrow r_i = a + b \cdot c_i + e_i (i = 1, \ldots, I), \qquad (3.1)$$

where, r, c, α, β, and ε represent the measured response (or signal), the concentration of the analyte under study (or any property of interest), an underlying intercept term for the model, the model slope, and the model error, respectively. The subscript i corresponds to the number of pairs of data. The Greek letters denote that this is a theoretical model, whereas the Roman letters denote the experimentally derived model. The parameters α and β are called regression coefficients (or a and b, with correspondingly equivalent meanings for the calculated model) and determine the relationship between the two types of data. Note that the model error (ε) is inherent to any model and not observable.

The parameters a and b are the estimates of the underlying parameters α and β. They are obtained from the experimental signals and known concentrations (or whatever property is to be studied) by applying, currently, the least-squares (LS) method to adjust a predefined function to a set of experimental points. There, e_i is called a residual, and it is calculated as the difference between the observed and estimated r values for each standard. Incidentally, note that setting the relation between r and c is not the same as setting the relation between c and r. So, be careful when entering data into any software.

The LS criterion is a universal procedure to adjust functions, and it is based on three major assumptions, which are often quite overlooked:

(i) There is no error in the magnitudes considered "independent" (or predictor) variables (for example, the concentration c, drawn at the abscissa). This might have been true at the beginning of the 20th century when the principles of linear regression were established (in ca. 1920) and relatively high concentrations were measured in laboratories. But nowadays, it is highly difficult to accept that there are no errors when preparing ppb, ppt, or even lower concentrations. Mathematicians have demonstrated that this principle may hold when the error in the abscissa is much lower than the error in the signal. But even this might not be true nowadays, so be absolutely careful when preparing standards.

(ii) There is a random error in the r-direction (i.e., the experimentally measured response or signal), and it is normally distributed. In general, we can accept this postulate because we *know* (or we consider it as given) that our instruments should not drastically vary their functioning over a short period; therefore, the errors in the signal would primarily be random. Strictly, we can only verify this point by replicating the measurements a number of times, which is not frequently conducted due to time, resource, and economic constraints. There is usually a balance between the number of replicates and the number of unique experimental conditions.

(iii) The magnitude of the (random) error in the r-signals (ε) does not depend on the level of the parameter (c) – this is called homoscedasticity. This point is difficult to ascertain unless we prepare true replicates of the standards and measure them independently, which is not common practice (again due to practical limitations).

Recall also that, by definition, the LS criterion minimises the sum of *all* the (squared) residuals, and this is the origin of many problems because when there is a mistake in an experimental measurement, it can have a major influence on the calibration line (which may even become invalid). Note that we differentiate "errors", which are inevitably a part of any measurement or model (and are usually considered random but may also be of a different nature), and "mistakes", which correspond to wrong values/operations whose root cause might be avoided (e.g., typesetting mistakes, wrong constants, wrong calculations, wrong solutions, degraded reagents, and uncontrolled experimental conditions). Errors are nowadays very much associated with the concept of uncertainty [4].

So, with the LS fit, we calculate the standardisation (or calibration) line, i.e., we calculate coefficients a and b as the best estimators of α and β, the regression coefficients of the underlying, true, unknown model. Is our task complete? Unfortunately, no. First, the LS fit is highly dependent on the correctness of the experimental points; are they always fine? Disappointingly, mistakes appear frequently in our daily work, and usually a single point is often insufficient. Second, recall that in the LS criterion, random errors are expected in the measured signals (the r's); will this error propagate to subsequent calculations? Clearly, it will. Let us reflect, briefly, on these two issues:

1. To check for wrong data points (outliers, as they are termed usually), which do not adhere to the general trend shown by the majority of points, the most convenient and fruitful approach is to study their graphical representations. The first option, undoubtedly, should be the typical plot where the experimental data points are represented in the Cartesian r (ordinate) and c (abscissa) coordinates. In many cases, this is not enough to detect small inconsistencies; therefore, another plot showing the residuals is required. This is just a representation of the difference between the observed and the estimated values of the signal ($r_{predicted} - r_{observed}$) for all pairs of data, where $r_{predicted}$ is the response value calculated by the model for each standard and $r_{observed}$ is the signal measured for the same standard, both at each c value. The distribution of the residuals should be random without definite patterns. To get straightforward decisions, the use of standardised residuals is recommended. They are defined by the residuals divided by their standard deviation. Thus, standardised residuals exceeding $+3$ or -3 are clearly outliers and should be discarded (and the calculations should then be repeated). These plots are very useful for validating the models (as seen in Section 3.6).

   Another unfortunate situation (and indeed very frequent) is that a parabolic pattern is visualised for the residuals rather than a random one. This is an obvious demonstration that we are trying to fit a straight line to data which indeed are not straight-line oriented. So, the function must be modified (i.e., a curve should be considered instead; for instance, a quadratic function that accounts for parabolic behaviour), the experimental work must be reviewed, or the working range of the standardisation line must be reduced. More details and examples can be found elsewhere [15].

   Although the graphical representation of the residuals is a good way to ascertain this behaviour, it can be addressed more formally using Mandel's test and some spreadsheet work (see, e.g., Ref. [15] for a simplified presentation), but it is unfortunately uncommon to find this quite simple test in routine applications.

2. The second issue is somewhat more complex, as it involves error propagation. We will not go into mathematical details; instead,

we will concentrate on the repercussions for us in the laboratory. In science, there are always errors (or uncertainties – they do not disappear, even if they are random), which propagate to subsequent calculations. This means that the random errors that occur when measuring the signals propagate to the ordinate and to the intercept; i.e., the regression line you calculate is an estimation of the *true*, unknown, one. Therefore, the values that we predict using the regression line after measuring the response of unknown laboratory sample aliquots (which is termed interpolation if the response is within the experimental working range, as it should always be) must have confidence intervals around them, which we have to report when offering our results. These intervals appear as shown in Equation (3.2):

$$a \pm t \cdot S_a; \quad b \pm t \cdot S_b; \quad r = a \pm (t \cdot S_a) + b \pm (t \cdot S_b) \cdot c;$$

$$c_0 \pm t \cdot S_{co} \quad \text{must be reported.} \tag{3.2}$$

Here, t corresponds to the Student's t-statistic (usually at a 95% confidence level, $I - 2$ degrees of freedom, and two tails); S_a, S_b, and S_{co} are the standard errors of the ordinate, intercept, and interpolated value, respectively. The former two are calculated straightforwardly (as they are given immediately by any software) using Equations (3.3–3.5). In contrast, Equation (3.6) is not typically provided by common software or spreadsheets, although it can be programmed easily (if you use specialised chemometric software, they often offer it). The derivation of Equation (3.6) can be seen in Ref. [15] and the literature sources cited therein:

$$S_a^2 \cong S_{r/c}^2 \left(\frac{1}{I} + \frac{\bar{c}^2}{\sum_{i=1}^{I} (c_i - \bar{c})^2} \right), \tag{3.3}$$

$$S_b^2 \cong \frac{S_{r/c}^2}{\sum_{i=1}^{I} (c_i - \bar{c})^2}, \tag{3.4}$$

$$S_{r/c}^2 = \frac{\sum_{i=1}^{I} e_i^2}{I - 2} = \frac{\sum_{i=1}^{I} (\hat{r}_i - r_i)^2}{I - 2}, \tag{3.5}$$

$$c_0 \pm t \cdot \frac{S_{r/c}}{b} \cdot \sqrt{\frac{1}{q} + \frac{1}{I} + \frac{(r_0 - \bar{r})^2}{b^2 \sum_{i=1}^{I} (c_i - \bar{c})^2}}. \tag{3.6}$$

In these equations, I is the total number of standards, r_0 is the signal obtained for the measured aliquot from which c_0 is interpolated, the "hat" ($\wedge$) denotes the signal predicted (calculated) by the model for each standard, q is the number of ("true") replicates measured per standard and sample aliquot (in most cases, $q = 1$, but if $q > 1$, r_0 should be the average of the q signals), and $S_{r/c}$ is the standard error of the regression of the response (ordinate) against the concentration (abscissa). Here, "replicate" means preparing standards from scratch and measuring them, which differs from taking several aliquots from a final solution and measuring them. Of course, you can consider this latter option, but in that case, you will include only instrumental repeatability (short-term) errors into your model. However, when "complete" or "true" replicates are considered (i.e., replicates from scratch), you will take into account reproducibility (long-term, personnel, volumetric materials, etc.) error. Regardless of your preference, be consistent; if you measure three aliquots of a solution for standards, do the same with the samples and consider $q = 3$. Do not mix "true" replicates (e.g., duplicated aliquots of a soil sample treated from the beginning) with instrumental replicates measured for the standards (e.g., aliquots of a given standard injected into a chromatograph).

An interesting issue here is that, as soon as you calculate a confidence interval for a regression coefficient, you can check whether it is statistically different from zero, and therefore, it is possible to evaluate its relevance for the model. In current regression applications, it is common to check whether the intercept is zero using a Student's t-test. If it is statistically zero some people eliminate the intercept from the equation in order to "simplify" it, but this is not good practice because it affects the accuracy of the standardisation equation [15].

It is also common to apply a one-way analysis of variance (ANOVA) to assess the statistical significance of the regression. ANOVA tests whether the variance modelled by the standardisation function is significantly larger than the random variance. In almost every common situation in analytical chemistry, the standardisation line is likely to be statistically significant since it was designed that way. Therefore, this tool is not very relevant for our usual applications. In other sciences, where the relation between the signal and the property of interest is not as clear, there may be more reasons to apply this approach; however, replicates are usually needed.

If you are working with different multivariate experiments, you will have to determine whether there are squared terms, interactions between them, etc. This will be discussed in Chapter 4, when studying the design of experiments and how to obtain information from them.

> You can check the statistical significance of a regression coefficient by simply applying the Student's t-test to it. If a regression coefficient is not different from zero, in principle it is irrelevant to the model. This point may not seem important here, but it will be in the following sections of this chapter.

Another critical point to bear in mind always when using the kind of models presented in this section, which is the classical way of working, is that they are reliable only when you are sure there are no interferent effects caused either by the sample matrix or other concomitant(s). These models can account for backgrounds (or undue effects) if and only if they remain constant between standards and samples or the samples do not have different concomitants/interferents that cause different backgrounds/effects. This is exemplified in Figure 3.1. If your experiments determine that the opposite situation occurs, you have to modify the laboratory procedure to avoid them; for instance, during sample preparation, you can apply some kind of analyte separation, filtration through active cartridges, add some sort of ionisation suppressors, etc. In other words, you need to improve the selectivity of your methodology at the workbench.

For the purposes of this chapter, the term "concomitant" is used here to denote a species that is present in the sample and

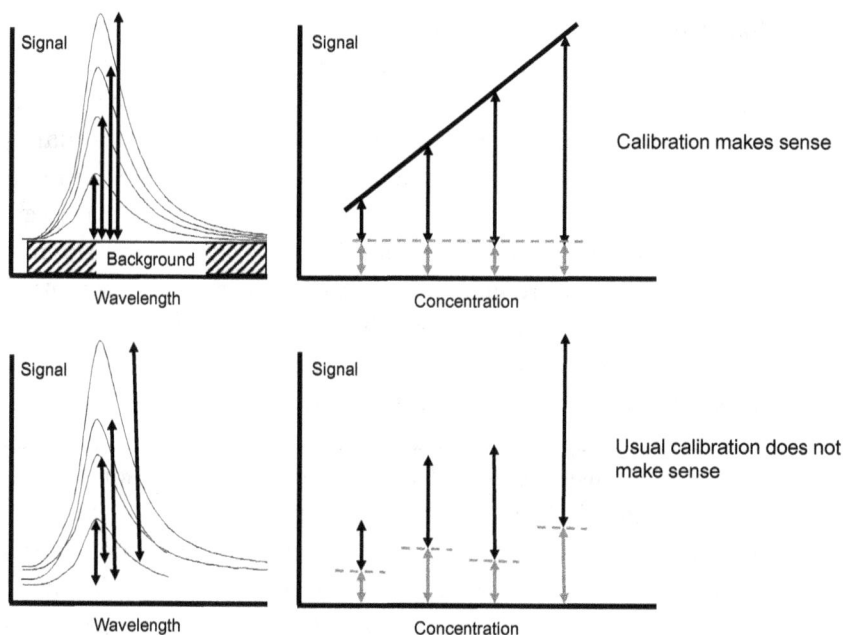

Figure 3.1. Classical standardisation can be applied when a constant background is present (upper plots), but it is not reasonable to apply it whenever different non-constant effects affect the standards or the sample aliquots (lower plots).

that incidentally also has a response on its own that adds to or subtracts from that of the analyte of interest (e.g., it absorbs at a given wavelength while the analyte also absorbs). Interferent means a substance present in a sample that somehow reacts with an analyte (or competes with it in some reaction or process) so that the "original" signal associated with the analyte becomes modified.

It is also worth recalling that the most relevant law of spectroscopy, the Lambert–Beer–Bouguer law, determines that the total signal (absorbance) recorded at a given wavelength is the linear sum of the absorbances caused by all components of the aliquot under measurement. In fact, this basic law of spectroscopy must be considered carefully; it is only true within a certain experimental range of concentrations due to physicochemical interactions among particles in concentrated solutions. Nevertheless, let us consider here that this problem does not occur in our system because, for instance, we work with diluted solutions, which is a common practice. Therefore, be sure to know accurately the composition of your samples. This is almost trivial whenever you prepare standards in the laboratory, but what happens with your sample aliquots (or those aliquots obtained after some sample treatment) (e.g., environmental samples)? If they contain something new, do not expect the calibration to remain valid because, currently, new matrix effects or interferences do not agree with those from standards, so it makes little sense to use the previously prepared calibration function. This point is of paramount importance, but it is often overlooked.

> Note that, when using classical univariate standardisation, you will not have statistical tools at your disposal to detect the presence of unknown species, simply because the model you are using does not allow for such tools [12]. It is too simple. So, whether the results are robust or not, is a purely subjective decision. Hence, the best you can do is try not to violate the basic assumptions of the LS method and the requisites of the physico-chemical law underlying your studies (e.g., the Lambert–Beer–Bouguer law) and hope that the true samples do not contain or produce something unexpected.

3.4. Multivariate Linear Regression by the Ordinary Multiple Linear Regression Method

Going a bit further and based on basic studies on spectroscopy, you may know that it is possible to determine the concentration of two or three species in a mixture using a generalisation of the Lambert–Beer–Bouguer law. However, for this to be possible, you need to measure the absorbances of two (or three) standards at two (or three) wavelengths, where each species is characterised by a specific peak (and hopefully, the other species do not result in interferents at these wavelengths). Following the measurements, you calculate factors that include the molar absorptivities of each chemical species and the path length. Then, you measure your unknown mixture and set a system of two (or three) equations with two (or three) variables; this can be solved to determine the concentrations of each species. To get acceptable results there should be no interference occurring, either at the standards or at the wavelengths characterising the species (but for some constant participation, that may be included as a blank or background).

It may seem quite intuitive that this idea can be generalised to more analytes, although in real life, unfortunately, this is not a practical approach because the equations become too complex and it also becomes increasingly difficult to find characteristic (in theory, specific) wavelengths for each species. Nonetheless, can we somehow use this idea?

Many explanations given in this chapter are illustrated using spectrometric measurements, not only because of their almost countless applications but because most students already apply them during their laboratory practices and possess some – quite useful – practical skills. However, other analytical techniques should not be excluded. Indeed, the exact same reasons, advantages, and difficulties would be encountered in chromatography, electrochemistry, NMR, etc. In other words, any measurement technique that yields a vector of experimental values can benefit from multivariate calibration and the chemometric techniques studied in this and subsequent chapters.

Scientific measurements performed in many fields face highly complex sample matrices, far exceeding those of laboratory-prepared standards. Think about food products, petrochemical goods, clinical and environmental samples, etc. Besides, analysts are often interested in determining various properties for the same samples (concentration of different species, physicochemical parameters, etc.). In spectroscopy, the use of the generalised Lambert–Beer–Bouguer law might be a reasonable starting point for establishing a framework similar to that seen in the previous section.

However, to achieve those objectives, it is much more convenient to use the whole spectrum of a sample/standard (or at least a portion of it) than measuring absorbance at only a few isolated wavelengths. This will not require much additional time, money, or staff workload, but, in turn, it will retrieve a lot of information. Furthermore, it is quite intuitive that if the overall spectrum is a function of all the absorbing species in that spectral region, then such a spectrum should contain information about them. Not only that, it is also quite logical to think that if we register the absorbance (signal) at many wavelengths (or elution times, NMR shifts, etc.), we will not be restricted to studying only one species. In addition, among the wealth of signals that constitute the overall response vector (e.g., absorbances at a given range of wavelengths), each species should have its own characteristic (or, at least, quite specific) signal, and it should even be possible to ascertain whether some variables are more related to noise, interference, and so forth. Although this is not possible by visual means or intuitively, so we often need multivariate chemometric methods.

This is the first-order advantage (where "first order" refers to the order of the data used: vectors). So the concentration(s) of the analyte(s) – or the properties we are looking for – can be determined without the need for highly selective analytical methods. Technically, it is said that the signals of the analytes and the interferents/concomitants become separate mathematically. This is undoubtedly an important advantage of multivariate calibration over its univariate counterpart. More details can be found elsewhere [16].

It is obvious that we are abandoning the use of scalars (single values, scalar or zero-order data, such as pH, absorbance at a given wavelength, signal intensity at a specific retention time, and voltage at a point in time) and moving towards the use of spectra or vectors (sometimes called first-order tensors). This requires mathematical formulations – matrix notation – that are capable of handling vectors easily. Calculations will not be done by hand but using computers. Nonetheless, recall that you are in command of the computer – it will do exactly what you request it to do. So, you need to know the principles of the methodologies and critically evaluate the outputs in order to obtain reliable conclusions and reports.

How to extract the information from the response vector that we record is a matter of choosing from the many available multivariate chemometric methodologies. In this section, we present the oldest and simplest one: ordinary multivariate linear regression (OMLR). Probably it is not the most commonly used method nowadays, but it remains the fundamental basis for understanding and visualising the concepts presented in the following chapters of this book (and many others that will not be considered as they exceed our introductory scope).

The procedure is very similar to what you do currently in a laboratory when you calibrate/standardise your usual instruments:

1. Define the species or property of interest, and search for concomitants and/or interferents known or suspected to be present in the true samples.
2. Record separately the pure response (a spectrum or any other complex signal) for each species and interferent/concomitant (e.g., using pure commercial products); alternatively (depending on the regression method to be used), prepare/collect true samples or standard mixtures and measure them.
3. Develop a regression model.
4. Record the whole response for each sample under the same experimental conditions as those fixed for the standards, and use the model to predict the concentrations (properties) of interest.

Figure 3.2 depicts a situation where signals are measured sequentially along a continuous range of variables; when plotted, they yield a band shape. It could be a band in the UV-vis region (or any other spectral region), an HPLC peak, a voltamperometric signal, etc. Note that the overall measured signal contains – non-constant – participations of various concomitants and interferents in addition to the signal of interest and noise (which is always present). You can think of as many complicated participations as you wish (after all, this is an understanding exercise). Note that, in practical situations, these concomitants and interferents are not constant throughout all the standards because they usually depend on the concentration levels. The additive nature of the signals can be seen in the inset

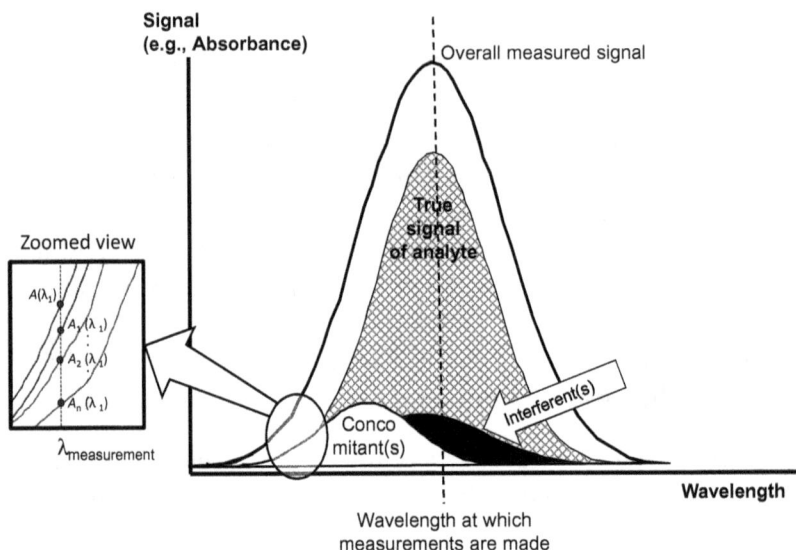

Figure 3.2. Overall measured signal which is intrinsically comprised of the true signal of the analyte of interest, a background, and other varying signals attributed to concomitants and/or interferents.

of Figure 3.2. Note that, in this exemplary plot, we have only recorded a reduced range of variables (e.g., the visible region where coloured solutions absorb) and that there is only one analyte of interest.

Considering UV-vis spectroscopy – a paradigmatic example that is well known to every student – for the simplest cases, we know how to apply the Lambert–Beer–Bouguer law. In this situation, for each wavelength, we can formulate Equation (3.7). Here, $a_{\lambda j}$ is the absorbance at any j wavelength ($j = 1, \ldots, J$); the c's are the concentrations of the species of interest and/or the concomitants/interferents ($m = 1, \ldots, M$); $S_{m(\lambda j)}$ represents the well-known multiplicative term (molar absorptivity times path length) of the Lambert–Beer–Bouguer law (which is specific for each analyte and wavelength; sometimes this is called the chemical sensitivity or, simply, sensitivity); and λ_1 means that we are considering only the first wavelength. The error term (ε) is used here to be coherent with Equation (3.1).

$$a_{(\lambda 1)} = c_1 \cdot S_{1(\lambda 1)} + c_2 \cdot S_{2(\lambda 1)} + \cdots + c_m \cdot S_{m(\lambda 1)} + \varepsilon. \qquad (3.7)$$

Now, for each measured wavelength, you can deploy a similar equation by simply changing λ_1 to λ_2, etc. You can also visualise immediately that you will construct as many equations as there are wavelengths, which can be very many. Note that this collection of equations is only for one standard/sample; for other standards/samples, you will get corresponding collections of equations. Clearly, they are not easy to handle, and we should use matrix algebra instead of the fully developed equations. Indeed, this is an overdetermined system of equations (i.e., you have more equations than strictly required to obtain a solution).

For those samples that have to be measured, the unknowns of the system of the equations are the c's, for which we need to know all the $S_{m(\lambda i)}$ regression coefficients, exactly as when calculating the regression line in Section 3.3. Therefore, during calibration, we need to prepare as many standards as needed to evaluate the regression coefficients (because, of course, for the standards, all the c's are known), so that the equations are solved and the regression coefficients are calculated to define the standardisation model.

In matrix notation, Equation (3.8) is applied to set the model after measuring each spectrum (or any other signal) of each standard mixture (which is equivalent to the modelling/calibration step). Here, matrix $\mathbf{A}$ contains I rows and J columns, associated with the I standards measured at the J wavelengths; $\mathbf{C}$ is the matrix containing the values of the concentrations of each of the M absorbing species (analytes of interest plus interferents/concomitants) in the I standard mixtures; and $\mathbf{S}$ is the matrix of the multiplicative terms (sensitivities) for each M absorbing species at each jth wavelength. The random error terms are avoided hereinafter for simplicity:

$$\mathbf{A}_{(I \times J)} = \mathbf{C}_{(I \times M)} \cdot \mathbf{S}_{(M \times J)}. \tag{3.8}$$

To calculate matrix $\mathbf{S}$, it is required to perform a calculation similar to the "$\mathbf{A}/\mathbf{C}$" division, but division is not defined as such for matrices! Circumventing this is not straightforward, and the so-called Moore–Penrose pseudo inverse (also termed a pseudoinverse matrix, denoted as $\mathbf{C}^+$) needs to be calculated instead. For OMLR, the ordinary least-squares solution is $\mathbf{C}^+ = (\mathbf{C}^T \cdot \mathbf{C})^{-1} \cdot \mathbf{C}^T)$ [16], where $(^T)$ means transposed and $(^{-1})$ denotes the inverse. (Review Chapter 2 for a more detailed treatment of this issue.) Therefore, Equation (3.9) holds:

$$\mathbf{S}_{(M \times J)} = \mathbf{C}^{+}_{(M \times I)} \cdot \mathbf{A}_{(I \times J)} = (\mathbf{C}^T \cdot \mathbf{C})^{-1} \cdot \mathbf{C}^T)_{(M \times I)} \cdot \mathbf{A}_{(I \times J)}. \quad (3.9)$$

Next, for an unknown sample to be predicted, whose spectrum denoted by the row vector $\mathbf{a}_{\text{unkown}}$ was measured, the concentrations of the analyte and other concomitant species are given by Equation (3.10a) (if there is more than one unknown sample, use matrices $\mathbf{C}_{\text{unknown}}$ and $\mathbf{A}_{\text{unknown}}$ instead):

$$\mathbf{c}_{\text{unknown}(M \times 1)} = \mathbf{S}_{(M \times J)} \cdot (\mathbf{a}_{\text{unknown}(1 \times J)})^T$$
$$= [(\mathbf{C}^T \cdot \mathbf{C})^{-1} \cdot \mathbf{C}^T)_{(M \times I)} \cdot \mathbf{A}_{(I \times J)}] \cdot \mathbf{a}^T_{\text{unknown}(J \times 1)}.$$
$$(3.10a)$$

In general, we will not be interested in the concentrations of the concomitants/interferents, and so we can focus only on the mth element (or position) of vector $\mathbf{c}_{\text{unknown}}$, where we know the analyte is located (because we designed the matrices and organised the values at the very beginning). For example, let us imagine that the compound we want to quantify is located in the third row. Analogously, we need the same row of matrix $\mathbf{S}_{(M \times J)}$, say, $\mathbf{s}_{(3 \times J)}$ (in general, if we are interested in the mth row, then it is $\mathbf{s}_{(m \times J)}$). Thus, for practical use, we can simplify Equation (3.10a) to Equation (3.10b), from which the scalars and vectors are obtained [16]:

$$c_{m(\text{unknown},1 \times 1)} = \mathbf{s}_{(m th \text{ row} \times J)} \cdot (\mathbf{a}_{\text{unknown}(1 \times J)})^T. \quad (3.10b)$$

Two general remarks are in order before continuing:

(i) When presenting the models and explaining their calculations, many textbooks and papers add a first column of ones in the response matrix to account for the intercept of the model. However, since in almost all cases, some kind of data preprocessing is applied to the signals (and almost always, mean centring is implicit there), this offset is cancelled out, and thus, this step is not mandatory. In many cases, the values of the property to be predicted are also mean centred, which is why this issue is not included in the explanations here.

(ii) Recall that the preprocessing parameters (mean, standard deviation, etc.) should be the same for standards (modelling step) and unknown samples (prediction). This means that the processing parameters of the standards should be stored within the computer and used to preprocess the signals of the unknowns. This is done automatically by any validated software; however, if you use your own, it's important to be cautious as this issue is a common source of mistakes since some individuals preprocess the two datasets independently.

3.5. Inverse Multivariate Linear Regression for Solving Some Ordinary Multivariate Linear Regression Problems

The formulation above and its mathematical implications pose some restrictions and involve some problems that might impede successful calculations. They will be described briefly here and constitute the reasons why this methodology is not widely applied nowadays: mostly because the other typical methods, including PCR and PLS, which will be studied in subsequent chapters, overcome all these limitations.

The first point is obvious and was already mentioned: to set the model (calibration step), not only the analytes of interest but also the concomitants/interferents need to be known(!), otherwise, you will not solve the equations for all the species that contribute to the final signal, and your model will be flawed. This also means that you need to know not only the concentrations of the concomitants/interferents (in the calibration standards) but also their pure signals (spectra) because you need to calculate the multiplicative factors (in spectroscopy, the absorptivities or the absorptivity times the path length factors). And these concomitants must be the same as those present in your unknown samples. This is a major, usually irresolvable, problem when applying the OMLR method.

A way to avoid this problem is to formulate the regression from another viewpoint, which is called inverse multivariate linear regression (IMLR), or, simply, MLR because linear regression is typically applied in this manner.

It does not require knowing all the concentrations and signals (spectra) of all the constituents of the standards. Similarly, we can model a physico-chemical property (of any type, including taste and odour), without caring for other undesired compounds or properties, as long as the standards used to set the model contain those compounds and properties in the same form, quantity, etc. [17]. (As stated in previous sections, standards should emulate unknowns closely!)

Using some classical words: "It is an oversimplification to say that one can be completely ignorant of the additional sources of variation in a system and still produce good models. In fact it is very important to ensure that all the significant sources of variation are present when the calibration models are estimated. However, these other sources of variance are not included as additional variables in the model (explicitly), but are implicitly modelled [...], the ideal approach is to manipulate all of the sources of variance using an experimental design [...] if it is not feasible, another approach is to measure many samples and assume that the variance that is relevant to the prediction of future samples is present" [12].

Further, instead of knowing and evaluating each and every sample constituent, the overall signal has to be measured and the concentrations or properties of the species of interest must be known; therefore, unknown concomitants/interferents can be present in the standards. The formulation is as shown in Equation (3.11).

Observe that we estimate the concentrations of the analyte (property of interest) by means of regression coefficients instead of applying a form of the Lambert–Beer–Bouguer law, to which nevertheless they can be related. That is why it is called "inverse". As only one concentration or property is studied, we can use vectors instead of full matrices:

$$\mathbf{c}_{(I \times 1)} = \mathbf{A}_{(I \times J)} \cdot \mathbf{b}_{(J \times 1)}. \tag{3.11}$$

As indicated above, I is the number of standards used to develop the model, J is the number of wavelengths, $\mathbf{A}$ is the matrix of the corresponding vectors of responses, and $\mathbf{b}$ is the vector of regression coefficients (which can be interpreted as related to the inverse of the chemical sensitivity, as discussed in Section 3.4).

To predict the concentrations of the analytes of interest in the unknown samples (let us denote them by the subindex $K(k = 1, \ldots, K)$), vector $\mathbf{b}$ must be calculated previously considering the set of standards and using the Moore–Penrose pseudoinverse ($\mathbf{b} = [(\mathbf{A}^T \cdot \mathbf{A})^{-1} \cdot \mathbf{A}^T] \cdot \mathbf{c}$). Then, Equation (3.12) is applied to predict the concentration or property of the analyte in the unknown samples after recording their signals ($\mathbf{a}_{unknown}$):

$$\text{For one sample, } \mathbf{c}_{unknown} = (\mathbf{a}_{unknown(1 \times J)}) \cdot \mathbf{b}^T_{(J \times 1)};$$

$$\text{For various samples } (1, \ldots, K), \mathbf{c}_{unknown(K \times 1)}$$

$$= (\mathbf{A}_{unknown(K \times J)}) \cdot \mathbf{b}^T_{(J \times 1)}. \tag{3.12}$$

A second problem, common to both the OMLR and IMLR approaches, is that the inverse of a matrix must be calculated to obtain the Moore–Penrose pseudoinverse. This is only possible if the columns of the matrix to be inverted are independent of one another, i.e., they should not be correlated. Unfortunately, in chemistry and many other sciences, the signals we register from any measuring system are strongly correlated (e.g., think about spectra, chromatograms, voltamperometric signals, etc.); therefore, it might not be possible to invert the matrix. Some mathematical methods avoid this problem by using a reduced set of new, independent variables, and in fact, they determine other multivariate regression techniques, such as PCR and PLS (studied in next chapters).

However, a simple option to ameliorate this problem would be to select a set of experimental variables under the condition that their correlation is low. That would require special efforts to look for a set of wavelengths (variables) where the analytes still yield signals with low noise and with no or – very low mutual correlation.

Nowadays, we have tools capable of performing such extensive searches; however, these techniques, including genetic algorithms, are not described in detail in this text as they are not frequently used with MLR. The other techniques (e.g., PCR and PLS) are simpler to apply and more computationally efficient.

A classical option to obtain a reduced set of variables, and not so time- or computation-intensive solution, consists of constructing successive models with an increasing number of predictors. First, a predictor – the best possible one – is considered; then the second-best predictor is added, followed by the addition of the third-best predictor to yield another model, and so forth. This is called the forward MLR approach. The opposite possibility exists: we start with a model with all predictors and reduce them by deleting the worst one (in the first step), then the next worst (in the second step), and so forth. In theory, both approaches should end up with the same solution; however, this is not usually the case due to the ill-conditioning of the data. Details on how the predictors are introduced into or removed from a model, according to a Fisher's F statistic, can be found elsewhere [18]. Other methods are discussed in Ref. [17].

A third problem is related to the number of variables (λ_j) we are considering. For each λ_j, we need to calculate the sensitivity of the analytes and concomitants/interferents, as done in Equations (3.7) and (3.8) (OMLS), or, at least, the regression coefficient (IMLR). As most spectra consist of a large number of variables, we need at least as many standards as there are variables(!), and that is not usually feasible in our work.

Further, for multivariate regression, the preparation of the standards is not as simple as that for classical univariate standardisation. For the latter, serial dilutions are applied quite often to obtain simple solution standards (almost no interferents at all), but this is not a good strategy here as we need to evaluate the concomitants and, inherently, interactions and interferents. So, the standards should contain different concentration ratios of the species. This involves additional work, and it will significantly increase when considering that you would need at least as many samples as predictors [7].

As you will prepare, currently, much fewer standards than predictors, your model will suffer from ill-conditioning. This means that it will be difficult to calculate the Moore–Penrose pseudoinverse, and so, the regression coefficients can be unreasonably big and have wrong signs [18]. Moreover, the predictions obtained for the unknown samples will not be stable and will have large associated variances (very large confidence intervals) [7]. A related problem is that many regression coefficients in MLR have no chemical meaning or, even worse, are statistically not significant because they also have large confidence intervals. This is caused by the way in which the information is distributed among all the variables, the distribution of errors, and ill-conditioning. In MLR, all variables contribute to the model, regardless of whether they add some meaning to it or not. This is one of the problems that PCR and PLS (in the following chapters) try to avoid.

A fourth problem, although not a minor one, is that any new interferent(s) (concomitants) in the unknown samples that contribute to the signal or modify the signal of the analyte(s) will invalidate any previous calibration, and you would need to recalibrate the model again. This is less problematic in IMLR.

Hence, MLR should be applied only to variables which are independent, to predict a reduced number of analytes/property of interest, and to systems where the composition of future samples may be reasonably anticipated. That could be possible, for instance, in stable industrial production systems where a sound subset of variables can be selected to characterise a product. Note that various properties can be predicted "simultaneously" without the need for additional laboratory work. If you set an MLR model for each property of interest, the experimental work has already been done previously, and you "only" have to apply each model to the signals associated with the unknown samples.

As a final remark, I refer to a comprehensive multidisciplinary book written by Professor Michele Forina, a pioneer of chemometrics in Europe, in which different options are provided to circumvent the typical problems that arise in MLR [19]:

1. Use inverse MLR instead of ordinary MLR.
2. Select a subset of predictors to obtain more standards than variables; this will help avoid ill-conditioning. Bear in mind that

sensible models require the number of compounds under study to be less than or equal to the smaller of the number of samples or variables [10].

3. Compress your variables so that you have multivariate factors (which is what we describe in the following chapters, when dealing with PCR and PLS).

4. Use "biased" regression methods instead of the ordinary least-squares one (also performed in PCR and PLS).

5. Combine the options above, i.e., regression techniques that combine variable selection plus data compression plus biased regression methods.

A concluding interesting issue is that inverse calibration is appropriate for handling situations where the errors associated with the predictands (the property of interest, e.g., concentration) are (much?) higher than the errors on the signals [10]. This might sound strange, but remember that weighing, volumetric flasks, make-up of volumes, etc., have not changed greatly in the past two centuries, but the quality of instruments has! Random errors in most instruments nowadays could be lower than the errors linked to the preparation of standard solutions, solid mixtures, etc. Remember the three principles of the least-squares fit given in Section 3.3.

> *Note*: OLS is considered an *unbiased* method in the sense that it assumes all predictors are needed and relevant for the model. As we have seen above, this is not always true for our experimental, real systems, and we would like to get rid of noisy variables or potentially overlook effects and minor compounds that contribute only slightly to the signal (and, hopefully, to the prediction). As soon as we avoid them, it is said that we "bias" the model (it becomes a *biased regression*), although this can indeed be quite satisfactory because the model and its predictions can substantially improve. Therefore, in regression models, the term "biased" does not imply "wrong", rather it means that we are avoiding the use of a part of the initial information. Keep this in mind in subsequent chapters!

3.6. Validating the Model

Similar to well-known univariate models, the MLR models also need to be evaluated critically. This issue is independent of whether MLR is applied infrequently to practical applications. If you apply it, you need to validate it. Several key criteria are as follows:

- First, in close analogy to the definition of the correlation coefficient for a univariate regression, the *coefficient of multiple determination, R* (given as a percentage), can be calculated as shown in Equation (3.13) [18]. It is interpreted as the proportion of variance in the parameter of interest (concentration, property, etc.) that is explained by the regression model. As already explained, the classical Pearson's correlation coefficient should be used with caution:

$$R^2 = 1 - \left(\frac{\sum_{i=1}^{I} (c_i - \hat{c}_i)^2}{\sum_{i=1}^{I} (c_i - \bar{c})^2} \right) \quad \rightarrow \quad R = 100 \cdot \sqrt{R^2}, \qquad (3.13)$$

where $\hat{c}_i$ are the predicted concentrations (property) of each of the I standards applying the regression model and $\bar{c}$ is the average of their true, or reference, concentration values (property) set when modelling (which are also termed reference, target, known, etc., values and are denoted by c_i).
- Second, examine *the extent of collinearity*. As we mentioned in the previous section, one of the consequences of ill-conditioned data is that the standard errors of the regression coefficients become very large. Somehow, collinearity inflates them. To what extent does this occur?

    A relatively simple way to measure the strength of the correlation between a variable and all other variables consists of calculating the so-called variance inflation factor (VIF), as described in Equation (3.14). This checks the level of (multi)collinearity between the predictors [18] (where the prefix "multi" refers to the fact that we study many variables at the same time):

$$\text{VIF}(\lambda_j) = 1/(1 - R^2(\lambda_j)), \qquad (3.14)$$

where λ_j is the jth variable under consideration (here, a wavelength) and $R^2(\lambda_j)$ is the coefficient of multiple determination for the regression between λ_j (which acts as the parameter to be predicted) and all other predictor variables.

Since a VIF must be calculated for each variable, computations are performed using dedicated software rather than manually. One must examine the values: VIF values higher than 5 (or 10, depending on the literature sources) indicate that this predictor is highly correlated to the others, and it should not be considered in the regression [19].

- Third, calculate *confidence intervals for the regression coefficients*. Of course, this means that you can check the statistical significance of the regression coefficients. But you have to be careful even if they are not significant, and you should not delete them immediately. Recall the problems discussed above regarding the functioning of MLR due to ill-conditioning. The mathematical foundations for calculating these intervals are somewhat complex and will not be included here. They can be consulted in Chapter 10 of Part A of Ref. [18], an outstanding classical textbook. The final solution is given in Equation (3.15a) (in which the terms have the same meaning as in previous equations):

$$b_j \pm t_{(2 \text{ tails},95\%,\text{I}-\text{J})} \cdot S_{bj}, \qquad (3.15a)$$

where S_{bj} is the standard error associated with the jth regression coefficient, which can be obtained from the calculated variance-covariance matrix of the regression coefficients, defined by $\mathbf{V}(b) = S^2_{\text{regress}} \cdot (\mathbf{A}^T \cdot \mathbf{A})^{-1}$, and so, S_{bj} is the jth element of the main diagonal of that matrix.

Matrix $(\mathbf{A}^T \cdot \mathbf{A})^{-1}$ is called the dispersion matrix and is useful for visualising the covariances among the regression coefficients. Recall that $\mathbf{A}_{(I \times J)}$ is the matrix containing the response vector (spectra, chromatograms, electrical signals, etc., composed of J variables) for I samples. If the off-diagonal values of this matrix are close to zero, the predictors tend to be independent of each other. In Chapter 4, we will find that the construction of this matrix determines the quality of the regression model. Thus, the number of variables (J) and samples (I) and how they are distributed in the working space (the so-called experimental design) are relevant to arrive at a good model.

Due to the fact that the standard errors of all regression coefficients are obtained simultaneously from matrix $\mathbf{V}$, it is recommended to modify Equation (3.15a) to Equation (3.15b) [17], where the F statistic is used instead of the t statistic; this takes into account the correlation between the regression coefficients and defines a multidimensional joint confidence region, which is not frequently used in MLR [18] because its interpretation is not straightforward:

$$b_{\mathrm{j}} \pm F_{(95\%,(J+1,I-J-1))} \cdot S_{bj}. \qquad (3.15b)$$

- Fourth, calculate *confidence intervals for the interpolated values* using Equation (3.16):

$$c_0 \pm t_{(2 \text{ tails},95\%,I-J)} \cdot S_{\text{regress}} \cdot \sqrt{\mathbf{a}_0 \cdot (\mathbf{A}^{\mathrm{T}} \cdot \mathbf{A})^{-1} \cdot \mathbf{a}_0}. \qquad (3.16)$$

Recall that c_0 and $\mathbf{a}_0$ are the concentrations predicted for the unknown from its corresponding absorbance vector, respectively; S_{regress} is the standard error of the regression, equivalent to Equation (3.5), although upon changing the degrees of freedom, it now converts to $S_{\text{regress}}^2 = \frac{\sum_{i=1}^{I} e_i^2}{I-J} = \frac{\sum_{i=1}^{I} (\hat{c}_i - c_i)^2}{I-J}$.

Typically, these calculations will be provided by dedicated software because they are not as straightforward as in univariate calibration (Section 3.3). A step-by-step example of the simplest possible situation (involving two predictors) can be found in Ref. [18].

- Fifth, use *diagnostic plots*. Decisions during validation can be a bit easier if graphical outputs are used. This key idea is extensively exploited when using complex models, such as PLS (as will be shown in the following chapters), for which a suite of different kinds of plots offer diverse viewpoints and acceptance and/or diagnostic criteria can be posed. In MLR, most plots focus on the residuals. They are the differences between the predicted values of the properties under study and the reference values for either the standards and/or a small collection of samples that should be kept separate for this task (the validation dataset).

When considering validation, it is always of paramount importance to have a small set of samples/standards for which the property/concentration of interest is known in advance (e.g., using a reference method). This is called the validation dataset. Once the model is developed and deemed acceptable, it is time to check whether it can predict the properties of unknown samples, which constitute the validation set.

If the differences between the predicted and experimental values for the validation samples are smaller than a given (preset) limit, the model is considered adequate. No other method is better than this one to demonstrate that the model is fit for purpose. Keep in mind that this collection of samples should never be used to develop the model.

A collection of different diagnostic plots was presented elsewhere [12]. Here, two typical examples are described. First, the widely used plot where the predicted concentrations are confronted to the reference values ("predicted vs. reference" plots) for all the samples. Second, it is also very common to plot the residuals against the predicted values.

Both diagnostic plots are useful to evaluate whether the regression model accounts for most of the variation of the property/concentration in the calibration and/or validation datasets. The second type of plot is usually a little easier to visualise than the former. The ideal figure obtained for the first type of plots should be a line with zero intercept and unit slope. Otherwise, the line can be used to estimate the average systematic bias (which is the calculated intercept). If a curvilinear pattern is observed in the residuals plot, there is a general modelling problem. Outliers can also be identified as the points that clearly depart from the general straight-line trend.

The plot of the so-called standardised residuals (see Section 3.3) is also frequently used to help visualise trends. When standards closely adhere to a linear model, the residuals should look like a random pattern. If they do not, check for some problems, such as a too-large working range, problems with the standards, and measurement problems.

- Sixth, use *the leverage plot*. This is a relevant, common diagnostic plot that helps researchers identify outliers. Each sample is characterised by its signal (e.g., spectrum), and this is compared to the average signal of the calibration dataset. This serves to identify signals (samples) which are different from the average of those used in the calibration stage. The leverage for the ith sample is the ith position of the diagonal of the hat matrix, $\mathbf{H}$ (Equation (3.17)). Typically, the leverages of samples are plotted in a leverage versus sample number plot:

$$\mathbf{H} = \mathbf{A} \cdot (\mathbf{A}^T \cdot \mathbf{A})^{-1} \cdot \mathbf{A}^T. \qquad (3.17)$$

This statistic has a relevant meaning whenever multivariate regression and experimental designs are used. It will be mentioned and studied in more detail in the following chapters. Indeed, it has very strong similarities with the Mahalanobis distance, as defined in Chapter 2, and is employed in subsequent chapters. It ranges from 0 to 1, with values close to 1 denoting very influential points in the model (potential outliers, although not necessarily). Leverage values higher than $2J/I$ or $3J/I$ should be studied carefully as possible influential outliers [17]. A leverage close to 0 indicates that the spectrum of the sample is very close to the mean of the spectra of the calibration set.

3.7. A Practical Example

In the following chapter, which is dedicated to the design of experiments, you will find various examples where MLR constitutes an excellent methodology for developing suitable and satisfactory regression models that yield relevant information. This is because the predictor variables will be independent, and the system will be correctly planned (or designed). It is, therefore, instructive to work here with an example that can illustrate some of the problems that you might find in other complex – and not so uncommon – situations. Most of them were commented on in the previous sections. You can consider it as a worst-case, exemplary scenario. And, no doubt, such a problematic situation requires alternative regression methods that will be studied in subsequent chapters.

In Chapter 5, you will work with the same dataset as the one used here, and it is highly recommended that you apply what you learn in Chapters 7 and 8 to it as well.

Imagine that you are working in a laboratory of a food processing plant where NIR spectroscopy is being implemented to replace a tedious reference method. In this case, the property under study is the moisture present in soy wheat, but it might also refer to any other typical property (such as the total nitrogen using the Kjeldahl method). You would like to implement a rapid and straightforward operational procedure that avoids most human mistakes and allows measuring multiple samples, but of course you need to demonstrate that it yields results which are totally comparable to the classical ones.

Consider the file "Moisture" already present in the CAT software dataset. Load it as indicated in Chapter 1, observing that it is an Excel-type file, with the property of interest (the moisture of soy wheat) in the first column and the NIR spectra in columns 2–176. Rows (standards) 1–40 correspond to the training set, while rows (standards) 41–54 constitute the validation set.

3.7.1. *Model development (standardisation/calibration)*

Figure 3.3 shows how to load the data using the Data Handling menu. Once you select "Moisture" in the file selection window, disable the "Header" option (because the original data do not have a first row with labels). Then, the very first task that one should always do is to visualise the spectral data. So, follow Figure 3.3 (the Univariate menu). In the figure, you will see that the spectra are not very clear (although they are very typical of NIR spectrometry) as they have proportional baseline effects, offsets, possible reflectance effects (such as diffuse reflectance and more), etc.

1ST LOAD FILE

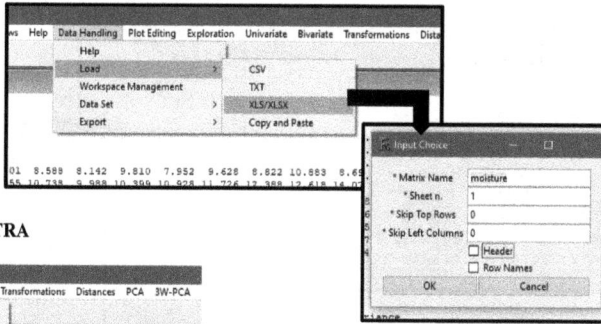

2ND REVIEW SPECTRA

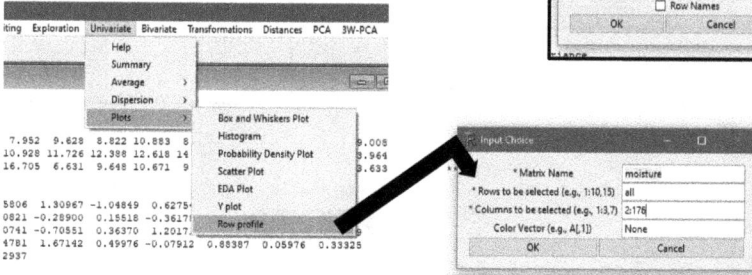

Figure 3.3. Loading files and visualising the raw (spectral) data using CAT.

This situation is not ideal, and a reasonable option to try to get rid of these issues is to preprocess the signals. Chapter 5 is devoted to this task, and subsequent chapters will show how the models are affected by the selections made at this modelling stage. So, we will not anticipate those issues here. For the sake of coherence with Chapter 5, let us suppose that a suitable treatment for the spectra is the standard normal variate (SNV) (whose meaning and fundamentals are found in that chapter). Apply it using the Transformations menu, as shown in Figure 3.4.

Note: The transformed/preprocessed data are stored in the computer's memory with a default name (which you will see at the CAT main desk or on the RGui screen), so when you wish to plot them, remember this fact. In this case, to plot the preprocessed data, you have to type "moisture.snv" in the Univariate menu. This is how you can reproduce Figure 3.4 and see that the preprocessing makes considerable sense (although the sloping trend can be avoided as well, see Chapter 5).

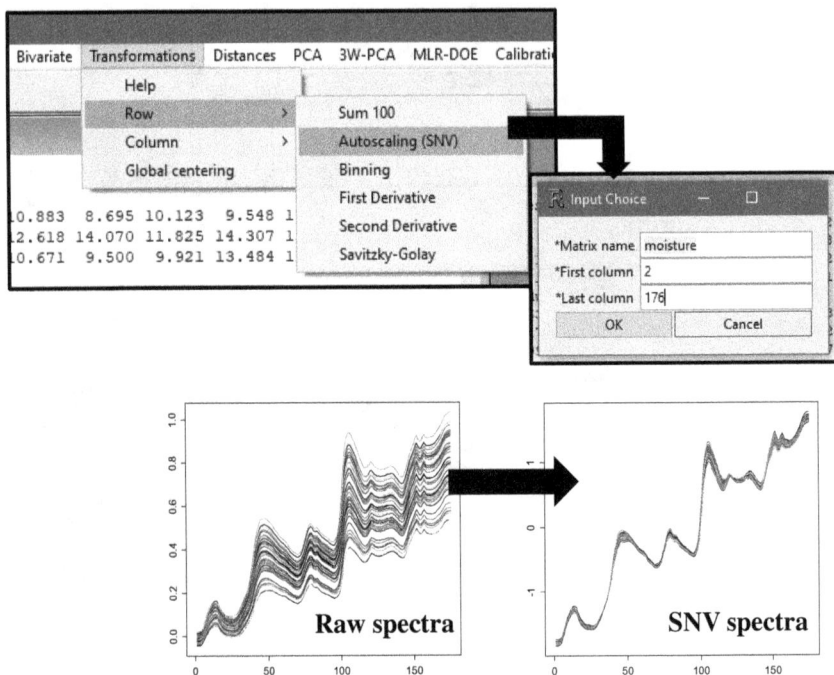

Figure 3.4. Processing signals with CAT and visualising the final result.

Now, let us develop a few trials to get some tentative MLR models. We need the MLR-DOE module (Figure 3.5).

Using all the spectral variables to obtain an MLR model is intuitive. So, select that in the menu. The variable to be predicted (moisture) is in the first column of the file, whereas all others are spectral variables. Recall that we have to use the SNV-transformed data (which is why you must type "moisture.snv" to denote the data file under use). Mark only the box indicating the "Intercept" and unselect "Higher Terms" (which correspond to other types of regression).

The error message obtained is due to huge multicollinearity, which impedes the calculation of the pseudoinverse; consequently, the model cannot be calculated at all. This is reasonable because in spectroscopy, there are too many highly correlated variables. As a trivial example, think of the adjacent variables to a maximum

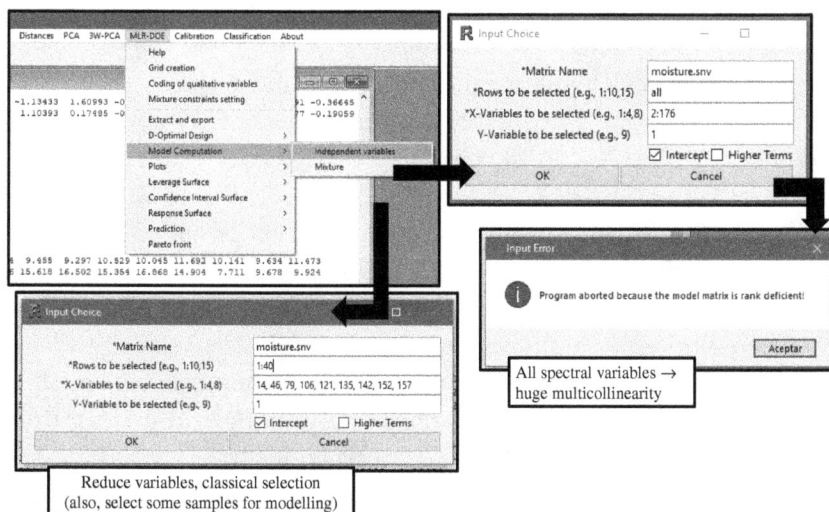

Figure 3.5. Setting up a model, while showing how strong multicollinearity can impede its calculation.

(or a minimum): they have almost the same values. So, MLR cannot work in this case; however, it's not the worst-case scenario, as here we "only" have 175 NIR spectral variables. In current mid-IR spectroscopy, we could easily have 3600 variables or more (depending on the spectral working range).

To obtain a reduced suite of variables, we can resort to traditional analytical chemistry techniques and look for those maxima or minima that clearly show variability across the samples. If we select variables which are quite separated, we expect them not to be too correlated. Although this is not really true because, likely, they will exhibit similar trends for the samples. But at least they are not adjacent. If you observe the SNV spectra (Figure 3.4; develop it yourself to maximise the plot on your screen), you can deduce that the most relevant maxima occur at predictors #14, 46, 79, 106, 121, 135, 152, and 157, plus an additional interesting minimum at variable #142.

At the same time, select rows 1–40; these are the standards we will consider for developing the model. Select an MLR model using those predictors and those samples (Figure 3.5). Include the intercept term, but no higher terms. CAT will ask you how to estimate the variance of the model; select residuals.

A model is calculated (which can be found towards the end of the tables that appear at the RGui main screen of CAT): Moisture $= -256.98 - 28.58 \cdot X14 + 18.16 \cdot X46 + 28.22 \cdot X79 + 13.21 \cdot X106 + 218.92 \cdot X121 - 177.61 \cdot X135 + 85.85 \cdot X142 - 59.94 \cdot X152 + 164.75 \cdot X157$. The coefficient of multiple determination is quite good (percentage explained variance $= 94.4\%$), so the model explains quite well the amount of moisture in the standards.

Their associated standard errors are seen in the table "Std.Dev. of coefficients", and they are quite large, sometimes higher than the corresponding regression coefficient itself! It is also clear from Figure 3.6 that the confidence intervals include zero (i.e., the coefficients are not statistically significant). These are bad news. Only

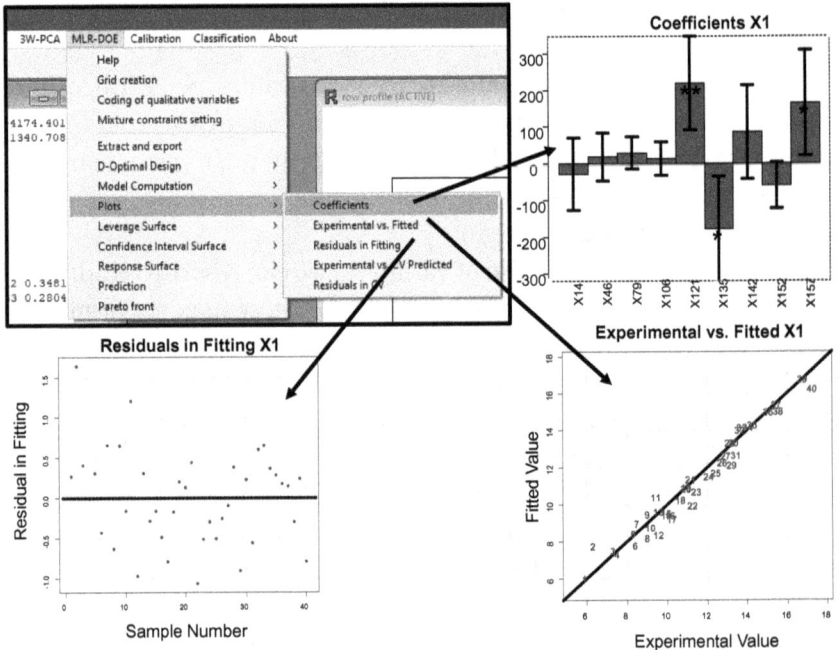

Figure 3.6. Studying the model by means of some typical diagnostic plots in CAT.

variables #121, 135, and 157 are significant at a 95% confidence level (variable #152 is in a borderline position).

The correlation between the variables remains important, as you can observe from the table headed "Dispersion Matrix" (look at the CAT main desk). Many off-diagonal terms, which denote correlations among the regression coefficients of the model, are large (in an absolute sense), but others are not so. The VIFs for the regression coefficients are clearly much higher than 5 or 10 (see the VIFs in the RGui main screen of CAT), indicating that all the variables are highly correlated with one another.

Despite these problems, the model seems useful to predict moisture, as the predicted values are very close to the true ones (see Figure 3.6), and they fit very nicely to the "perfect agreement line" (the diagonal of the plot representing predicted vs. reference, or observed, values) with a standard error of prediction (the standard deviation of the residuals) of about 0.66 (units of humidity, %), which seems good. The average of the absolute errors (residuals) is 0.66%, and the root mean square of the predictions for the calibration set (RMSEC = sum of the squared residuals$/40 - 10 - 1$) is 0.58%, which represents the level of error we would like to obtain when predicting new unknown samples.

Further, the standardised residuals do not show any relevant trend or serious suspicious samples since all are below ± 2 (see Figure 3.6). This is reinforced by the leverages of the standards, which are lower than the maximum recommended values (0.5 or 0.75).

3.7.2. *Model validation*

To evaluate how well this model predicts future samples, additional standards are needed. So, let us proceed to predict them (follow Figure 3.7, and remember that we have to use the same predictors as those used for the model). CAT offers you, automatically, the plots of the residuals and the predicted versus reference values. The results are excellent; perhaps the last sample could be examined in more detail, as its standardised residual barely passes the ± 2 line,

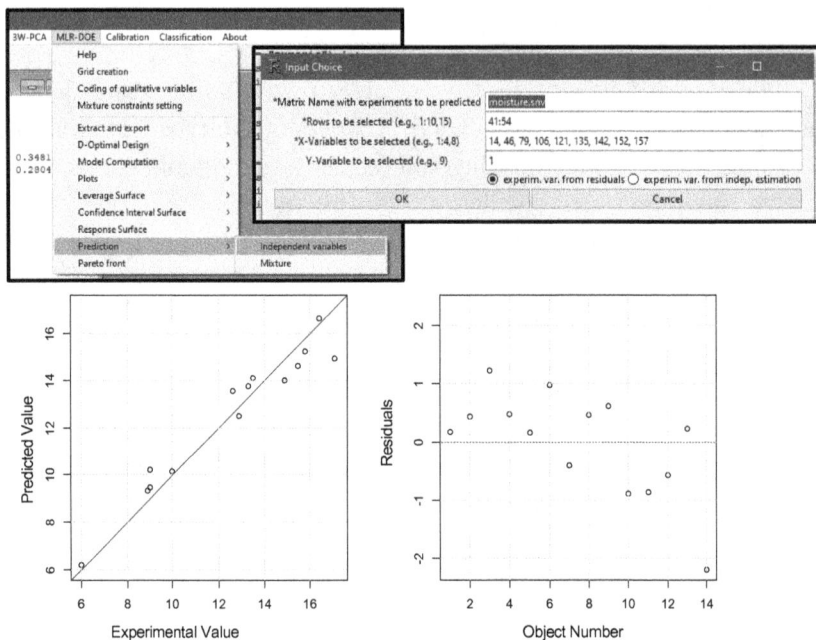

Figure 3.7. Validating the model by means of a validation set.

but it remains far from passing ±3, which indicates outliers. Also, the leverages are lower than the limits we mentioned in the previous paragraph, and the confidence intervals are reasonable. The average absolute error is 1.22%; however, if we discard the last standard, it is 0.48%. The root-mean-square error of prediction (RMSEP = sum of squared residuals/14 samples) is 0.86%, which is slightly higher than RMSEC; however, if the last sample is removed, it becomes 0.66%, showing excellent agreement with RMSEC.

Further, if you calculate the "predicted vs. reference or experimental" regression line for both the calibration and validation steps, you will find the following:

Calibration: Predicted value = 0.49(±0.76) + 0.96(±0.07). Experimental value. The slope agrees statistically with unity (t-test: 95% confidence), although the F test, performed to check simultaneously whether the intercept and the slope are zero and one, respectively, rejects this hypothesis.

Validation: Predicted value $= 1.98(\pm 1.72) + 0.84(\pm 0.13)$. Experimental value. The slope does not include unity (classical t-test), and the joint F test indicates that the intercept and slope are not simultaneously equal to zero and one, respectively. These are bad news: for the validation samples, both systematic and proportional errors exist in the predictions. It might be because either the validation samples are not representative or the model is probably not good enough. It fits the calibration standards quite well, but it does not make correct predictions for new samples. It is said that the model overfits the calibration standards. This can be a consequence of the correlation between the regression coefficients and the instability of the model and its predictions.

> *Note*: The calculations not displayed by CAT can be done easily in a spreadsheet by simply copying and pasting the outputs provided by CAT (e.g., the sample residuals or the predicted and real values can be used to calculate different statistics).

Let us try to refine the model. First, delete the variables whose regression coefficients are statistically zero, repeat the model, and compare it with the previous study. After a preliminary trial, you will find that variable #157 has a regression coefficient that is statistically not different from zero, so delete the variable and proceed to obtain a new model using only two predictor variables. This is left for the reader. We can anticipate that the model does not exhibit statistical problems in accordance with the criteria we have seen, with calibration errors similar to those presented above.

Using this new model, slightly higher prediction errors are obtained for the validation set than for the calibration samples again. The predicted versus experimental values plot leads to a regression in which the slope is not statistically equal to unity (even after deleting the outlier, sample #41).

An interesting exercise is left for the reader: use second-derivative preprocessed data to eliminate the sloping trend in the spectra and check if the model can be improved.

Acknowledgements

I am indebted to Richard Brereton for his enlightening discussions and suggestions that have improved the readability and pedagogical approach of this chapter.

References

[1]   Cammann, K. (1992). Analytical Chemistry-today's definition and interpretation. *Fresenius' Journal of Analytical Chemistry*, 343, 812–813.
[2]   Kellner, M. (1994). Education of analytical chemists in Europe. The WPAC Eurocurriculum on analytical chemistry. *Analytical Chemistry*, 66, 9A–101A.
[3]   Valcárcel, M., López Llorente, A. I., and López Jiménez, M. A. (2016). *Foundations of Analytical Chemistry*. Springer International Publishing.
[4]   Ellison, S. L. R. and Williams, A. (2012). *EURACHEM/CITAC Guide. Quantifying Uncertainty in Analytical Measurement* (3rd edn.). Available at: www.eurachem.org.
[5]   Bettencourt da Silva, R. and Williams, A. (eds.) (2015). *EURACHEM/CITAC Guide: Setting and Using Target Uncertainty in Chemical Measurement* (1st edn.). Available at: www.eurachem.org.
[6]   Magnusson, B. and Ornemark, U. (eds.) (2014). *Eurachem Guide: The Fitness for Purpose of Analytical Methods — A Laboratory Guide to Method Validation and Related Topics* (2nd edn.). Available at: www.eurachem.org.
[7]   Westad, F. B. and Marini, F. (2013). Regression. In Marini, F. (ed.), *Chemometrics in Food Science*. Amsterdam, The Netherlands: Elsevier.
[8]   Draper N. R. and Smith, H. (1998). *Applied Regression Analysis*. New York, USA: John Wiley & Sons.
[9]   Aldrich, J. (2005). Fisher and regression. *Statistical Science*, 20, 401–417.
[10] Brereton, R. (2018). *Chemometrics, Data Driven Extraction for Science* (2nd edn.). Chichester, UK: John Wiley & Sons.
[11] Ortiz, M. C., Sánchez, S., and Sarabia, L. (2009). Quality of analytical measurements: Univariate regression. In Brown, S. D., Tauler, R., and Walczack, B. (ed.), Comprehensive Chemometrics: Chemical and Biochemical Data Analysis, Vol. 1. Amsterdam, The Netherlands: Elsevier.

[12] Beebe, K. R., Pell, R. J., and Seasholtz, M. B. (1998). *Chemometrics, A Practical Guide*. New York, USA: John Willey and Sons.

[13] Meloun, M., Militky, J., Kupka, K., and Brereton, R. G. (2002). The effect of influential data, model and method on the precision of univariate calibration. *Talanta*, 57, 721–740.

[14] ISO/IEC Guide 99:2007. (2008). International Vocabulary of Metrology: Basic and General Concepts and Associated Terms (VIM). Geneva, Switzerland: International Organization for Standardisation.

[15] Andrade-Garda, J. M. and Gómez-Carracedo, M. P. (2017). Basic data analysis. In Andrade-Garda, J. M., Carlosena-Zubieta, A., Gómez-Carracedo, M. P., Maestro-Saavedra, M., Prieto-Blanco, M. C., and Soto-Ferreiro, R. M. (eds.), *Problems of Instrumental Analytical Chemistry*. New Jersey, USA: World Scientific.

[16] Ferré-Baldrich, J. and Boqué-Martí, R. (2013). Ordinary multiple linear regression and principal components regression. In Andrade-Garda, J. M. (ed.), *Basic Chemometric Techniques in Atomic Spectroscopy*. Cambridge, UK: RSC Publishing.

[17] Kalivas, J. H. and Gemperline, P. J. (2006). Calibration. In Gemperline, P. J. (ed.), *Practical Guide to Chemometrics* (2nd edn.). Boca Raton, USA: CRC.

[18] Massart, D. L., Vandeginste, B. G. M., Buydens, L. M. C., De Jong, S., Lewi, P. J., and Smeyers-Verbeke, J. (1997). *Handbook of Chemometrics and Qualimetrics, Part A*. Amsterdam, The Netherlands: Elsevier.

[19] Forina, M. (2014). Fondamenta per la chimica analítica. (written in Italian language). Available by free at different websites. In particular the Società Italiana di Spettroscopia NIR. http://www.sisnir.org/sisnir/download/fondamenta-per-la-chimica-analitica; consulted, April 2024.

Chapter 4

Statistical Experimental Design

Riccardo Leardi and Emanuele Farinini

Objectives and Scope

In this chapter, we aim to address the following major issues. At the end of this chapter, readers should be able to understand:

- the importance of rationalising a chemical problem (including complexity and experimental error);
- why to apply a multivariate instead of a univariate approach;
- the basic mathematics behind the design of experiments (DoE), together with their interpretation;
- the different DoE strategies according to the specific objectives (for each strategy, a real case study is reported along with the statistical elaboration conducted using the CAT software).

4.1. Introduction

Chemometrics is a scientific discipline that emerged with the purpose of interpreting and solving problems in chemistry using multivariate approaches. Within chemometrics, according to its definition, there are two primary areas of major interest. The first is the methodology of experimental research, known as experimental design (i.e., design of experiments, or DoE). This branch is concerned with designing or selecting optimal experimental procedures to obtain high-quality data by means of a series of planned experiments. Such a plan allows the construction of a mathematical model for quantifying the effects of various factors on the system and, when necessary, enables one to make predictions. The second focus deals with pattern recognition techniques, which are employed to extract useful information from analysed data and recognise patterns or trends within chemical systems. However, it is crucial to establish a suitable plan before applying various statistical methods to extract valuable insights from collected data. Without adequate preliminary planning and a comprehensive examination of its variability for accurate sampling, no statistical methods would yield meaningful results.

Broadly speaking, experimental design can be defined as a systematic approach used to plan experiments and investigate how changes in the settings of experimental factors influence the outcome of the process [1, 2].

4.2. The Chemical Problem

When facing real problems, it is fundamental to recognise and appreciate their distinctive characteristics before implementing effective methods to address them. Rushing in without proper understanding can lead to mistakes. To ensure a thorough understanding is obtained, it is important to gather as much information about the problem as possible. Asking questions and obtaining answers regarding the investigation's objectives, historical data, data collection methods, measurement protocols, equipment, and existing knowledge about the phenomenon is essential. Defining the goal of experiments in advance is also crucial, as many researchers often start experiments without a clear goal in mind. Taking the time to gather information and set goals lays the foundation for effective problem-solving and reliable results. The researcher faces two main challenges: complexity and experimental error.

4.2.1. *Complexity*

Complexity arises when studying multiple controlled input variables (or factors) and their effects on various output variables (or responses) that may be (a) maximised, (b) minimised, (c) acceptable beyond a threshold, or (d) have a specific target. Understanding the relationships between factors and responses becomes crucial for process improvement. Experimental design allows for simultaneous experimentation with multiple factors, minimising experimental error and providing a clear understanding of their individual and combined effects.

> Before performing any experiment, always define the aim, detecting (by critical thinking) all factors that may have an effect on the responses measured to describe the system under investigation.

Sometimes, the complexity of the system cannot be attributed only to the large number of factors but also to the intricate nature of the experimental process, which can involve multiple sequential steps, requiring careful consideration and potentially forcing the problem to be divided into several blocks. Conducting the whole study using a single experimental design could unduly perturb the system, leading to excessive variability in the results. In those cases, it may be advisable to divide the variables into different blocks, allowing for the identification of local optima through distinct experimental designs without excessively perturbing the system. This iterative approach between blocks facilitates arriving at a final solution. For instance, this approach could be used in a production process which can be separated into working blocks (e.g., involving reaction and filtration) or in the study of analytical methods with distinct blocks (e.g., a sample pretreatment phase and an analytical chromatographic phase).

The final goal of a general study is not always to identify an optimal solution. In industrial settings, reaching a qualitative target that minimises resource consumption (e.g., raw materials and operating costs) is often sufficient. This is where experimental design plays a fundamental role, providing operators with a clear map that indicates the direction to pursue and the associated risks in terms of uncertainty.

The complexity can arise from various aspects of experiment execution. Running the experiments in random order is fundamental to mitigate any time-related effects. Furthermore, all the experiments should be independent. However, a conscious trade-off with these "golden rules" can always be considered. An illustrative example is the "set and reset" of the temperature of industrial equipment; this operation may result in a significant loss of product, which might be a problem for the company. The complexity can also stem from the raw material itself, including both chemical reagents produced from different batches and natural organic products characterised by high seasonal variability.

4.2.2. *Experimental error*

Experimental error refers to variability in results that cannot be explained by known factors. While some degree of experimental error is unavoidable, knowing how to manage it effectively is crucial. Commonly, only a small fraction of the experimental error can be attributed to analytical measurement errors. Variations in raw materials, sampling methods, and experimental conditions often contribute more significantly to the overall variability. For these reasons, devising good experimental designs plays a crucial role in minimising the impact of experimental errors, allowing real effects to be observed accurately. However, as the complexity of a process increases, so does the experimental error. It becomes essential to evaluate and understand experimental variability, especially by identifying the critical step that contributes to this variability the most.

> The reliability of the results depends on a reliable estimate of the experimental error, preferably considering the global process.

In general, typical sources of experimental error are (a) sampling, (b) sample pretreatment, (c) "process" (e.g., the chemical reaction), and (d) analytical measurements. By identifying the main sources of variability, researchers can focus their efforts on optimising and controlling these factors to minimise experimental errors and improve the reliability of results. Simple experiment planning and data processing techniques, such as variance decomposition, can help estimate the variance of each step in the experimental process.

4.3. Why Multivariate?

To fully comprehend the potential of the multivariate approach, a comparison with the conventional "one variable at a time" (OVAT or "one factor at a time", OFAT) approach is essential [1, 2]. Thus, a simulated case study is shown with a chemical reaction involving two variables: temperature (°C) and time (min), with the aim of maximising the reaction yield (%) (Figure 4.1).

The contour plot in Figure 4.1 shows the presence of a significant interaction between the two variables, as well as a quadratic effect. If there were no interaction effects, the contour plot would have ellipses with axes parallel to the graph axes. The OVAT approach

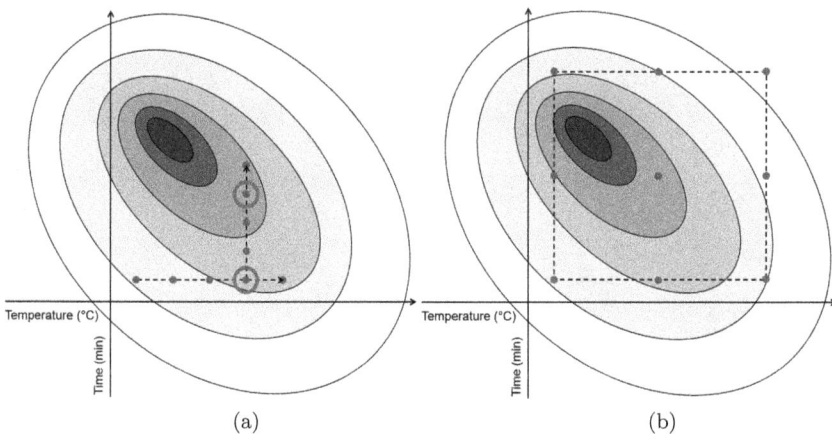

(a)   (b)

Figure 4.1. (a) Experiments conducted using the OVAT approach overlook interactions between factors, potentially missing the true optimum of the system. (b) In contrast, DoE considers various dependencies and exploits a mathematical model that leads to the identification of the "true" optimum.

may not be able to detect the actual optimum of the process. On the contrary, employing a DoE approach, a set of experiments is designed to discover the real effects of the factors on the response variable.

There is a fundamental shift in concept: experimental points are projected into a researcher-defined space, tailored to the specific problem. This projection results in a diverse range of geometric figures within the subsequent experimental domain. Figure 4.2 illustrates typical experimental domains for quantitative and continuous factors, encompassing squares, cubes, hypercubes, spheres, hexagons, and constrained shapes, depending on the related constraints between studied factors. Additionally, the domain may take the form of a triangle or a tetrahedron. On the other hand, for quantitative discrete variables, the experimental domain becomes layered based on the number of levels, while for qualitative variables with up to two levels, there is a similar layered structure (with just two layers).

> The experiments should not be considered a mere list of tasks but a series of points within a predefined space or experimental domain.

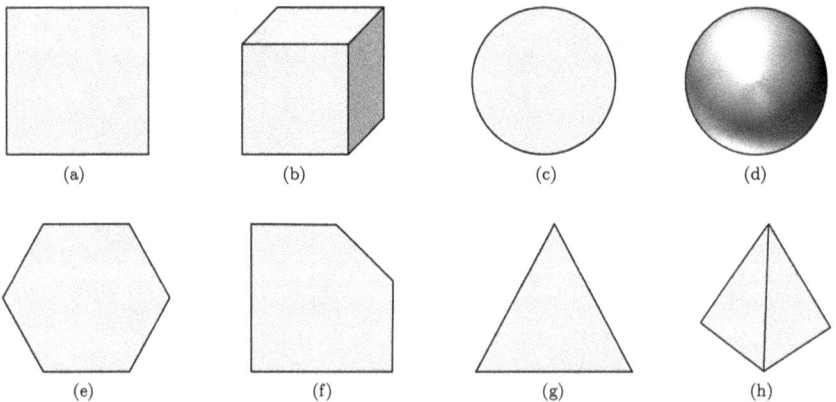

Figure 4.2. Typical experimental domains: (a) square; (b) cubic; (c) circular; (d) spherical; (e) hexagonal; (f) constrained; (g) triangular; (h) tetrahedral.

The following general but important considerations about the univariate and multivariate approaches can be reported:

(a) Having a design for experiments is always better than having no design at all. Conducting experiments without an objective will lead to biased results. That is why it is crucial to plan and execute experiments in a systematic manner, focusing on *which* experiments to run. Poorly designed experiments cannot provide useful information, regardless of the following computational methods or statistical tricks used.

(b) A DoE approach provides the opportunity for extensive thinking and collaboration among experts before conducting experiments. It emphasises the importance of having a clear goal, identifying all possible influential factors on the final outcome, and finding the proper measurable responses. DoE is a sequential process that must be applied whenever multiple factors are involved.

(c) DoE offers clear advantages over OVAT. With the latter, the variables are examined separately (Figure 4.1), which limits the ability to explore the entire experimental domain and thus to detect potential interactions between factors. In other words, OVAT does not consider whether the effect of a factor depends on the levels of one or more other factors, indicating that an interaction is present. OVAT can lead to completely wrong conclusions, and it would only be valid if no interaction among the variables were present.

> In real cases, interactions are likely to occur: the effect of a factor is not always the same but varies depending on the level of another factor involved in the system.

(d) With a limited number of experiments to perform, the univariate sequential approach very rarely leads to the detection of the real optimum, as the final result strictly depends on the previous results and initial choices.

(e) By employing DoE, it becomes feasible to construct a mathematical model based on the planned experiments. This enables the quantification of the effects (linear, interaction, or quadratic) of factors on the response variable. After validation, the model can be used to make predictions within the entire experimental domain at a known level of uncertainty.

(f) The univariate approach is intrinsically biased. *A priori* knowledge can often lead chemists to begin with the most cost-effective experiments, which may not necessarily be the best starting point and can result in reaching only a suboptimal solution. This logical concept is often overlooked within the organic synthesis field.

(g) A screening step is always recommended. This is because subjective decisions based on limited information can lead to a biased and suboptimal design; furthermore, the univariate approach tends to oversimplify and often overlook the specific "details" of a chemical problem.

(h) Despite the benefits of DoE, many scientists hesitate to use it, offering excuses such as the requirement for too many experiments. However, this is totally wrong, as the number of experiments needed using the OVAT approach is unpredictable and usually much higher than those required by DoE.

Traditionally, the primary objective of DoE has been to maximise knowledge gained while minimising the number of experiments conducted. In the past, conducting many experiments was not practical due to limited technology. Nowadays, certain fields still face constraints when it comes to the number of experiments that can be carried out. However, in most cases, the progress of technology, including faster analytical techniques and the integration of robotic equipment, has enhanced the capacity to conduct a significantly larger number of experiments within shorter time frames. Given these circumstances, there is an emerging trend of exploring complex experimental systems, making DoE even more crucial for efficiently selecting the right experiments to perform.

With proper planning, statistics can help chemists with useful tools:

• to determine the effect of changing experimental conditions on several responses (yield, selectivity, product purity, etc.);

• to predict optimal conditions;

- to get the best out of historical data and run fewer experiments, in addition to those already performed to obtain complementary information and useful models;
- to verify the robustness of a specific process against variations in external factors.

It is worth noting that its broad applicability spans a wide spectrum of fields, ranging from everyday life problems to intricate challenges in engineering, economics, sociology, marketing, computational science, and psychology.

The ideal design for any planned investigation depends on the study's objectives. The first step is to clearly specify the goals of the investigation and translate them into measurable responses. It is important to identify all the factors (k) that can potentially influence the selected responses (y) and, for each factor, determine the appropriate range of variability, considering any constraints between the different levels. It is crucial to identify all potential sources of variability that impact the system, which often results in a large number of factors to consider. Researchers must choose an appropriate experimental design based on available resources and limitations, taking into account the maximum number of experiments that can be conducted. The number of experiments inherently limits the number of parameters that can be investigated. The available resources should be carefully considered in order to effectively manage the workload throughout the different phases of the experimental campaign. It is also crucial to assess experimental variability through the use of replicates within the same time block and between different time blocks, especially in the case of multi-step studies.

Experimental designs can be classified into different steps based on their objectives:

- *Screening*: When researchers face a large number of factors, fractional factorial or, even better, Plackett–Burman designs are often considered the best choices. They help researchers understand the key factors and guide them in planning subsequent experimental designs. These screening designs offer effective alternatives to the commonly used OVAT approach in preliminary experiments, enabling more accurate insights into the key factors influencing the responses.

- *Improvement*: After a first screening, full factorial designs can be used as the next step. By using simple linear models with interactions, specific characteristics can be enhanced, and the interactions between input variables can be interpreted. These models, once properly validated, can be used to predict outcomes within the studied experimental domain.

- *Optimisation*: Similar to the previous ones, central composite and Doehlert designs are used to establish relationships between a selected set of factors and the response(s). In addition, these designs involve estimating quadratic terms in the model. As a result, they enable the identification of either maximum or minimum outcomes, thereby facilitating the discovery of optimal conditions.

- *Robustness*: This is an integral part of method validation, commonly observed in analytical chemistry (i.e., chromatography). However, it can also find application in various other fields, including production processes. Its primary objective is to evaluate the impact of various factors on performance. Specifically, it aims to demonstrate the statistical non-significance of certain factors within a narrower range, as the system performance should remain unaffected by small changes in these factors. To assess robustness, different experimental designs can be employed depending on the number of factors involved, such as the above-mentioned Plackett–Burman, fractional factorial, and full factorial designs. Typically, the robustness assessment occurs towards the end of the process, following model validation.

Additionally, mixture designs are used when dealing with mixtures. Here, it is important to consider the implicit constraint that the sum of all components must be 1 (or 100%). This means that the components of a mixture cannot be varied independently, as changing the percentage of one component will also affect the percentages of the other components.

In addition to traditional designs that adhere to symmetrical geometric figures, there are "optimal designs" that offer the flexibility

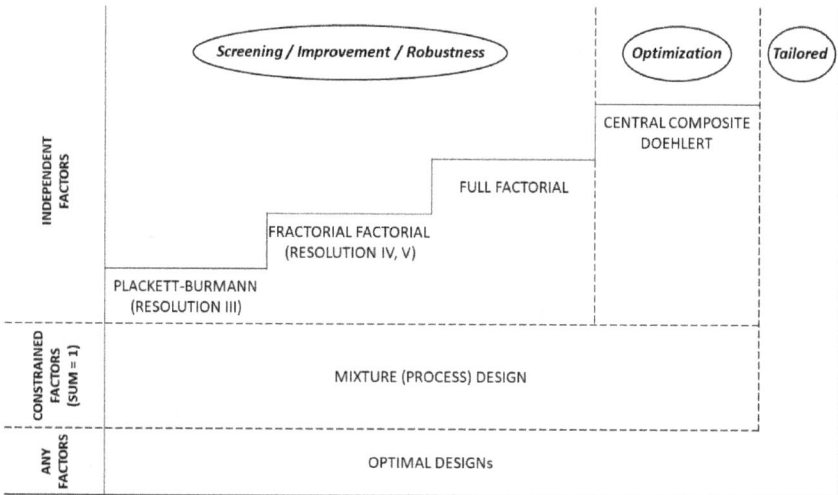

Figure 4.3. The "complexity" steps of experimental design.

to plan and create tailor-made solutions. These designs maximise the information obtained by optimising the experimental effort; therefore, they are suitable for pursuing any goal (from screening to optimisation). They allow the study of quantitative and qualitative factors at more than two levels and can accommodate non-regular experimental domains that may arise from incompatibilities when setting certain levels between factors. In addition, optimal designs are particularly useful in mixture-process designs, where different formulations are tested under various process conditions. These "complexity steps" are reported in Figure 4.3.

4.4. Mathematics Behind the Experimental Designs

After defining the goal(s), a model is postulated *a priori*, i.e., before conducting any experiments. Considering the previous example about optimising a chemical reaction with two factors, temperature (40–$60°$C) and time (20–60 min), where the response is the yield (%). In this case, the second-order full quadratic model with an intercept is as follows:

$$y = b_0 + b_1 x_1 + b_2 x_2 + b_{12} x_1 x_2 + b_{11} x_1^2 + b_{22} x_2^2 + \varepsilon, \qquad (4.1)$$

where y is the response variable and x_1 and x_2 are the two factors. With just nine experiments, it is possible to estimate a constant term (b_0), two linear terms (b_1 and b_2), the two-term interaction (b_{12}), and two quadratic terms (b_{11} and b_{22}); ε is the error term.

Once the model and its complexity have been determined, the experimental matrix can be obtained. It will contain as many rows as there are experiments and as many columns as there are factors. The experimental matrix contains the coded levels for each factor. Usually, the range is coded between -1 and $+1$; however, in some designs (e.g., central composite circumscribed designs), it can be different. Working with coded values is critical because it allows for interpretation and direct comparison of model coefficients. The experimental plan is derived from the experimental matrix by converting the coded values of each factor into their corresponding actual values.

The coded levels for each factor can be determined using the following equation:

$$\frac{(\text{actual level} - \text{centre value})}{\text{half range}}. \qquad (4.2)$$

> Experimental matrix = coded input values
>
> Experimental plan = actual input values

For quantitative factors (both continuous and discrete) and qualitative factors with two levels, the values in the matrix are normally assigned as -1 and $+1$. This allows for estimating the coefficients of the corresponding linear equation, including an intercept and all possible interactions between the factors. To account for nonlinear effects and estimate more complex models for quantitative factors, additional intermediate levels are used. However, note that when working with historical datasets, they usually contain multiple additional levels for the considered factors, which must be coded within the -1 and $+1$ range. The software, CAT, suggested throughout this book allows converting any matrix from actual values to coded values within the -1 to $+1$ range, as shown in Figure 4.4.

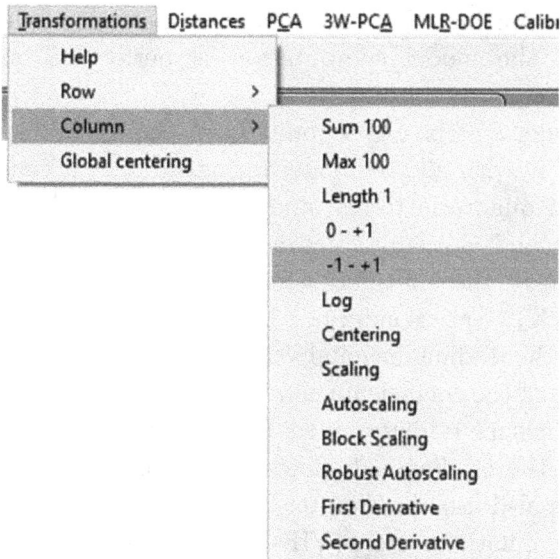

Figure 4.4. Column transformation: coding of factors within the range −1 and +1.

Qualitative variables with more than two levels are handled differently in the experimental matrix, as described in Section 4.7. Unlike quantitative variables that can be coded as −1 and +1, qualitative variables do not have a numerical scale; therefore, there are no intermediate levels between different categories or classes. Examples of categorical factors include suppliers, raw materials, catalysts, production plants, instruments, and operators. These variables are typically represented using different methods, such as dummy coding, to capture their categorical nature in the analysis. Instead, mixture designs involve factors ranging between 0 and 1, with the combined components or ingredients always summing up to 1 (or 100%), as explained in Section 4.8.

Once the experiments have been conducted, their results (y) are recorded in the same table. As explained above, it is mandatory to run the experiments in a random order. By doing this, possible time-related effects, such as worn equipment or changes in raw materials, contribute to the overall variability of the model rather than covarying with certain factors.

After importing the experimental matrix, along with the response(s), the model computation is performed, as shown in Figure 4.5.

The model matrix ($\mathbf{X}$) is built from the experimental matrix, augmenting it with all the model terms, such as a constant, interactions, and quadratic terms, and has as many rows as there are experiments and as many columns as there are coefficients in the model. From it, the information matrix ($\mathbf{X}^{\mathrm{T}}\mathbf{X}$) and the dispersion matrix ($\mathbf{X}^{\mathrm{T}}\mathbf{X}$)$^{-1}$ are computed. By multiplying the experimental variance by the leading diagonal values (c_{ii}) of the dispersion matrix, the variance of the coefficients can be obtained, while the covariance of the coefficients is determined by multiplying the experimental variance by the off-diagonal values of the dispersion matrix. These matrices are also used in Chapter 3.

Variance inflation factors (VIFs) are computed using the information matrix and provide an estimation of the level of multicollinearity among variables. They serve as useful indicators to assess the quality of the model, often referred to as "tolerance". For each coefficient in the model, a VIF value is calculated, indicating the degree of covariance between terms (multicollinearity). A VIF value of 1 indicates no correlation; as a rule of thumb, VIF values below 4

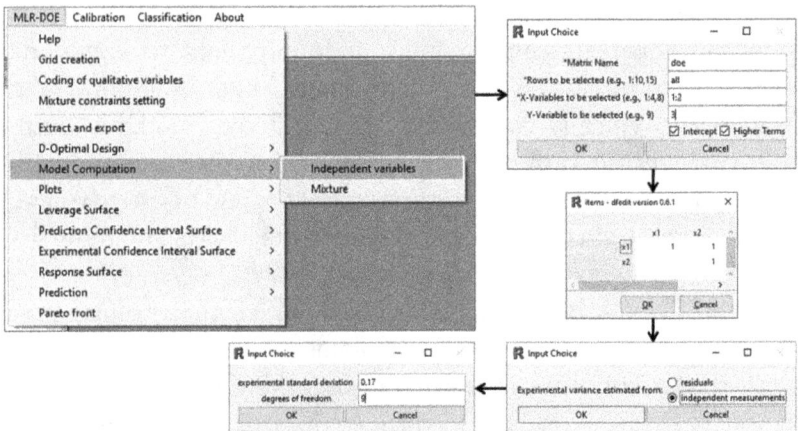

Figure 4.5. Model computation for independent variables, with the model postulation and the definition of the experimental variability.

are considered acceptable, while a coefficient with a VIF > 8 is too strongly affected by covariance and therefore should not be estimated. In the case of classical experimental designs, VIFs are very close to 1, whereas historical datasets may exhibit covariances between factors (high VIFs), resulting in unstable multiple linear regression (MLR) models (see Chapter 3) (when covariances occur between terms, it becomes impossible to isolate and quantify the independent contributions of each factor).

The leverage ($H = x[\mathbf{X}^{\mathrm{T}}\mathbf{X}]^{-1}x^{\mathrm{T}}$) is calculated using the dispersion matrix. It can be determined for any point within the experimental domain, and when multiplied by the experimental variance, it yields the variance of the estimate. A leverage value of 1 implies that the response can be predicted with the same precision as the actual experiment. A leverage value lower than 1 means that the confidence interval for predicting the response is narrower than that of an actual experiment conducted at the same point. This statistic will be used again in most chapters. With a proper design, the leverage can be brought to <1 at each point of the experimental domain; therefore, the quality of the information (in terms of precision) is higher than that obtained through the OVAT approach. For the latter, as no model is present, the knowledge of the system is limited to the points where the experiments have been performed, with a leverage equal to 1.

> The quality of the information, in terms of precision, obtained through the experimental design is higher (leverage <1 for any point of the experimental domain) than for the information obtained through the OVAT approach (leverage $= 1$, limited to the points where the experiments have been performed).

It is possible to generate a leverage surface plot as well as a plot representing the semiamplitude of the confidence interval of the response (the latter concept will be explained in the subsequent "Model validation" section). It is worth noting that, if the experimental variance is known, this information can be obtained before running the experiments of the design.

The vector of the regression coefficients (**b**) can be obtained by applying the equation expressed in matrix notation (which is the same approach used in the following chapters of the book):

$$b = (\mathbf{X}^T\mathbf{X})^{-1}\mathbf{X}^T y. \tag{4.3}$$

The statistical significance of the model coefficients can be assessed easily using a Student's t-test under the null hypothesis that the coefficient equals zero. Confidence intervals for each coefficient are computed by multiplying the coefficient's variance by the corresponding Student's t-value for the chosen significance level (which is typically indicated as * for $p < 0.05$, ** for $p < 0.01$, and *** for $p < 0.001$).

With regard to model interpretation, some notes are important:

- The constant term (b_0) will estimate the response when all factors are set to zero, corresponding to the centre of the domain (this does not hold when dealing with qualitative variables with more than two levels, as described in Section 4.7).
- The linear coefficients are measures of the linear dependence of the corresponding factors, representing the slope of the tangent plane. A variable that has a large influence on the response will have a large coefficient. If the variables are coded in the range of -1 to $+1$, the coefficients of the linear terms, being equal to a one-unit variation in the response, represent half the effect of the corresponding factor. Here, the effect of a factor is considered as the difference in response between the two extremes of the studied range, as shown in Figure 4.6.
- The cross-product coefficients constitute an estimation of the interaction effect between the variables. Geometrically, a linear model with interactions corresponds to a twisted plane.
- Quadratic (square) coefficients are an estimation of nonlinearity. Geometrically, a model with quadratic terms corresponds to a curved surface (convex in the case of negative coefficients, while concave in the case of positive coefficients). If the experimental domain contains the maximum (or the minimum), then it can be said that the optimum conditions have been detected.

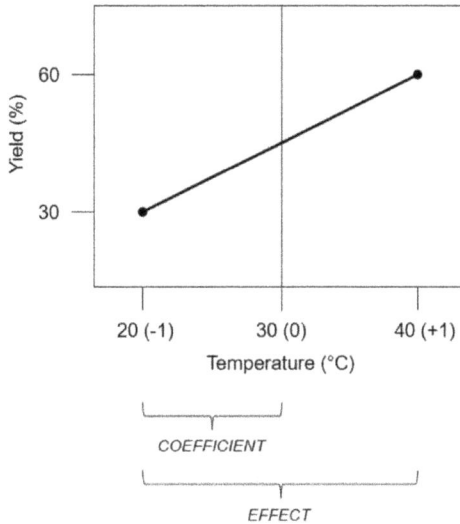

Figure 4.6. Linear effect of an independent factor (temperature in °C) on a response (yield in %), which is double the corresponding coefficient.

The resulting regression model can be applied to generate a contour plot that visualises how the response behaves across different regions of the experimental domain, enabling the interpretation of interactions between factors, as illustrated in Figure 4.7. After validating the model, it becomes possible to make predictions about the outcome of potential future experiments within the investigated domain.

4.4.1. *Model validation*

Validation, a crucial step in modelling, ensures the model is effective in addressing the problem at hand. The validation process involves statistical methods, allowing for the use of the mathematical model. However, it is imperative to first assess the model's fitting ability to experimental data to ensure that it adequately approximates the system. There are various figures of merit for assessing the model's quality obtained through MLR (some of which are reviewed in Chapter 3). The most compact and informative one is the adjusted R-square (also known as explained variance or determination coefficient, adjusted

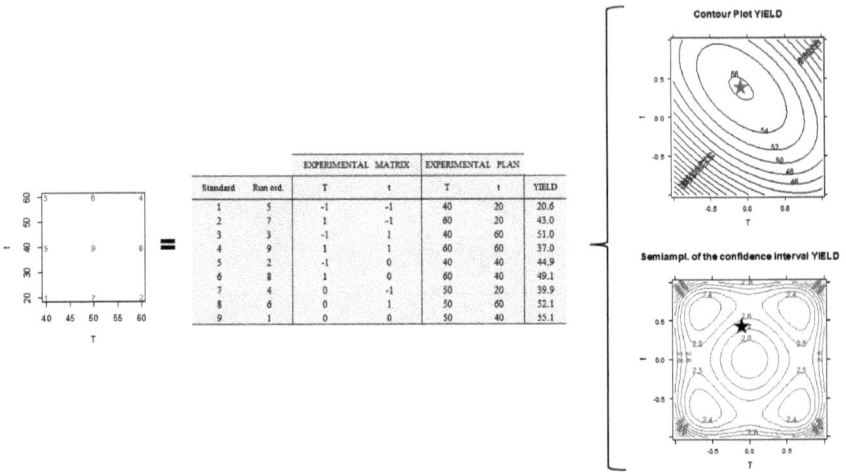

Figure 4.7. Set of experiments considering two factors: temperature (40–60°C) and time (20–60 min). The yield (%) from these experiments was estimated and used to construct a multiple linear regression model. The results include a contour plot and the semiamplitude of the confidence interval plot for the model (experimental standard deviation = 1.5; degrees of freedom = 5).

according to the degrees of freedom), which provides insights into the model's capacity to explain variation, as compared to the total variation in the original data. This indicator is obtained through the following equation:

$$R_A^2 = 1 - \frac{\sum_{i=1}^{N} (y_i - \widehat{y}_i)^2}{\sum_{i=1}^{N} (y_i - \overline{y}_i)^2} * \frac{N-1}{N-p-1}, \quad (4.4)$$

where $\widehat{y}_i$ is the fitted response, $\overline{y}_i$ is the average of the experimental responses, y_i is the experimental response, N is the number of observations, and p is the number of regressors.

The standard deviation (SD) of the residuals indicates the error of the model in fitting, where the residual is the difference between the fitted value and the experimental value:

$$s_{\text{res}} = \sqrt{\frac{\sum_{i=1}^{N} (y_i - \widehat{y}_i)^2}{N-p}}, \quad (4.5)$$

where $\widehat{y}_i$ is the fitted response, N is the number of observations, and p is the number of regressors, with $N-p$ being the degrees of freedom

(DoF) of the regression model. The analysis of residuals according to the order of execution of the experiments provides valuable insights into systematic effects caused by external factors, such as time-related block effects, learning improvements, or system degradation; it is considered acceptable when residuals exhibit normal random distribution around zero. An example concerning a clear block effect is illustrated in Figure 4.8.

A crucial aspect involves estimating the experimental variability, which serves as a reference when compared to the SD of the residuals. If the latter is significantly larger than the former, a so-called "lack of fit" (LoF) occurs. The experimental variability can be estimated by performing replicates of specific experimental points, test points, or historical data if replicates are accessible. In DoE, it is advisable to perform the replicates on different points (twice for each point is sufficient) and not several times on a single point as usually suggested. This allows measuring variability across the entire domain, thus improving the quality of the model. In this scenario, variability is calculated using the pooled SD.

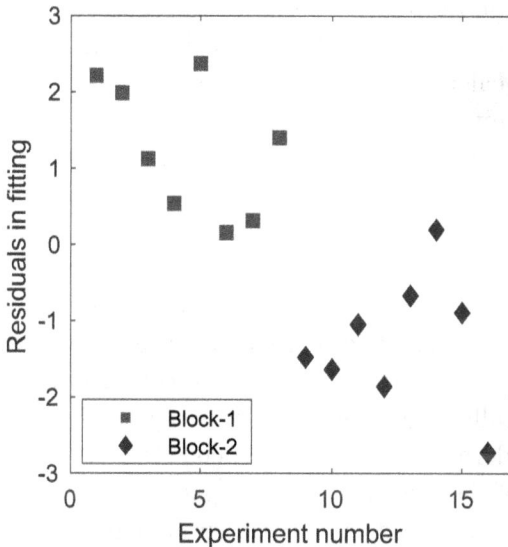

Figure 4.8. The residuals in fitting highlight the systematic block effect based on the experimental session, with the model tending to overestimate in the first session and underestimate in the second session.

> It is always advisable to run in duplicate different points of the experimental matrix instead of performing several replicates on a unique point. Then, to get a compact statistical indicator, a pooled standard deviation will be computed. This strategy applies for all types of DoEs.

It is worth noting that the SD of the residuals in fitting can serve as an indirect estimate of experimental variability, most notably when working with a limited number of replicates, as this results in a low number of DoF and in the absence of LoF. Consequently, all model diagnostics are computed based on this estimate, with its DoF usually being larger (resulting from the number of observations minus the number of regressors) than the ones used to estimate the experimental variability [3].

An extensively emphasised internal validation strategy in commercial software is Q^2, which is the explained variance offering insights into predictive performance. This approach utilises cross-validation (CV) with a "leave-one-out" (LOO) scheme, involving as many cancellation groups as samples (more details on this strategy are given in Chapters 6 and 7). It works by excluding one observation at a time and calculating the model based on the remaining data to predict the omitted point. In the context of experimental design, where the number of observations is often limited and essential for covering the entire experimental domain, this approach can significantly impact the model. Consequently, the resulting diagnostics may be overly pessimistic.

Another model validation tool to prove its effectiveness involves comparing the predicted responses with the experimental responses from selected test points, which will be conducted at a later stage. If no statistically significant differences are observed, the model can be considered valid. It is crucial to acknowledge that the construction of the model relies on experimental observations, which are inevitably influenced by experimental errors. High experimental variability can make model validation easier, but it often renders the model inadequate. On the contrary, when experimental variability is minimal, for example, in chromatographic responses, the likelihood that the model will not be validated is quite frequent. In these cases, the need for statistical rigour may be reduced; instead, the SD of the residuals and the error in prediction should be taken into account.

For factorial designs, typically associated with linear models, the test point is commonly placed at the centre of the domain, which must be replicated. If the model does not pass validation, the linear model may be insufficient for accurately describing the response.

In the case of quadratic models, the validation step primarily aims to assess their predictive ability. In this case, two scenarios may appear according to the specific aim:

(a) Validation at a single test point, typically defined afterwards and corresponding to the model-detected optimum. This necessitates a direct comparison between predicted and experimental values. The focus is not on building a model that has good predictive ability for every point within the experimental domain. The interval within which the experimental value (or the average of the experimental values) is expected to fall in order for the model to be validated must take into account both the model error and the experimental error (Figures 4.9(a) and 4.9(b)).

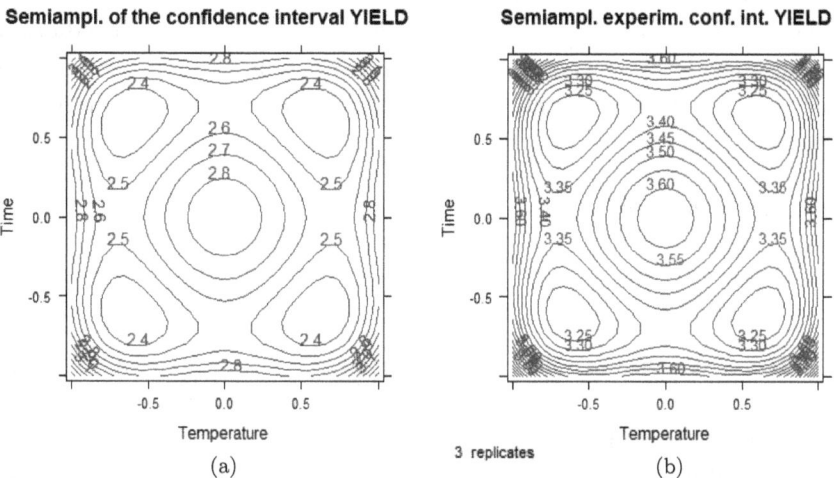

Figure 4.9. Semiamplitude of the confidence interval of the (a) prediction and (b) experimental response, with the latter calculated for the case of three replicates (experimental standard deviation $= 1.5$; degrees of freedom $= 5$).

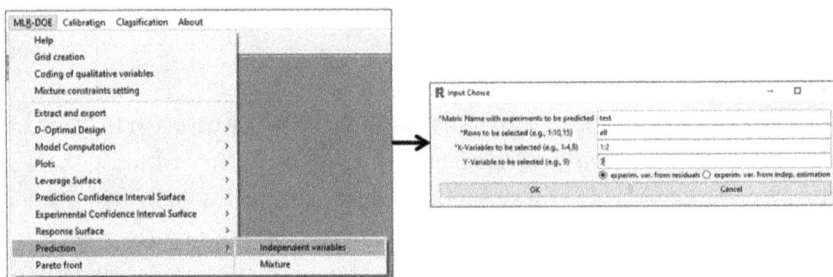

Figure 4.10. Prediction of test points.

(b) Validation of multiple test points, encompassing the entire experimental domain, where the root mean square error of predictions (RMSEP) should not be significantly larger than the SD of the residuals in fitting. Various strategies can be employed to select test points, as outlined in Section 4.9, including the D-optimal design by addition. This step is crucial for stress-testing the model in areas where it lacks information and, if validated, enhancing its quality by incorporating these additional points.

Figure 4.10 shows how to perform the predictions using CAT. The final output will include the predicted value, leverage, prediction confidence interval, and residual value for each point, along with a graphical output of the experimental values versus the predicted values.

4.4.2.   *Questions before statistics*

Before continuing with the application of statistical tools, it is critical to address the following questions:

(a) Have all sources of variability been identified accurately, or in other words, have all variables that could potentially influence the responses been correctly identified?
(b) If dealing with independent variables, are they genuinely independent?

(c) Are the selected variable ranges *logically* broad enough to ensure sufficient variability in the selected responses while avoiding excessively wide experimental ranges that might yield unusual results?

(d) Is it possible to perform the experiments while taking into account the relative range of variability for each identified variable?

(e) Is it feasible to execute all the experiments within the same time frame?

(f) Last but not least, has the experimental variability of the system been quantified?

If satisfactory answers cannot be offered to the above questions, it becomes necessary to reconsider the formulation of the problem and engage in discussions with the team experts.

4.5. Screening/Improvement/Robustness Designs

The first step, after defining the goal of the experimental campaign, is to detect all the factors that may affect the response(s) to be measured. Screening designs provide a valuable tool for studying a large number of factors within a relatively small number of experiments. The primary objectives of this stage are to obtain an initial overview of the system under investigation, to identify the factors that have the greatest influence on the response (thus not considering those with a low influence), and eventually to determine suitable levels for further exploration of these factors. Screening designs may also help evaluate interactions between the studied factors, although they require a larger number of experiments [1].

The types of designs described in this section can handle *quantitative, discrete quantitative,* and *qualitative factors,* simultaneously considering only two levels (coded as −1 and +1) for each factor.

4.5.1. *Plackett–Burman designs*

Plackett–Burman designs can be useful when the number of factors is quite high (above 6) [4] and offer effective alternatives to the commonly used OVAT approach in preliminary experiments.

> Having a design for experiments is always better than having no design at all; therefore, Plackett-Burman designs should replace the so-called preliminary experiments.

The Plackett–Burman design is named after the two statisticians, Plackett and Burman, who developed it. The historical context surrounding the origin of this design is noteworthy. It emerged during the Second World War, when the United Kingdom was facing attacks from the German Luftwaffe. In response, the British needed to enhance their anti-aircraft artillery and develop more efficient shells to counter the threat [5]. Therefore, a design capable of coping with many variables and necessitating a low number of experiments was required. Table 4.1 displays the experimental matrix, designed to study 11 independent factors through 12 planned experiments.

Table 4.1. Experimental matrix of a Plackett–Burman design with 12 experiments (maximum number of factors is 11).

#	x_1	x_2	x_3	x_4	x_5	x_6	x_7	x_8	x_9	x_{10}	x_{11}
1	1	1	−1	1	1	1	−1	−1	−1	1	−1
2	−1	1	1	−1	1	1	1	−1	−1	−1	1
3	1	−1	1	1	−1	1	1	1	−1	−1	−1
4	−1	1	−1	1	1	−1	1	1	1	−1	−1
5	−1	−1	1	−1	1	1	−1	1	1	1	−1
6	−1	−1	−1	1	−1	1	1	−1	1	1	1
7	1	−1	−1	−1	1	−1	1	1	−1	1	1
8	1	1	−1	−1	−1	1	−1	1	1	−1	1
9	1	1	1	−1	−1	−1	1	−1	1	1	−1
10	−1	1	1	1	−1	−1	−1	1	−1	1	1
11	1	−1	1	1	1	−1	−1	−1	1	−1	1
12	−1	−1	−1	−1	−1	−1	−1	−1	−1	−1	−1

In the experimental matrix, shown in Table 4.1, each row corresponds to a specific experiment in coded values, and each column represents a factor.

The number of experiments (N) required is the first multiple of 4 greater than the number of factors. The experimental matrices can be constructed without any software. The construction of the different designs starts from the first row of the experimental matrix, as listed in Table 4.2. (For up to 23 factors, the other matrices can be found in the literature.) The next row begins by copying the rightmost value of the previous row and pasting it into the leftmost position and shifting the following values by one position to the right. Subsequent rows are constructed in the same way until the $(N - 1)$th row, with the final row consisting entirely of -1. This construction method can be deduced from Table 4.1.

These designs can estimate up to $N - 1$ terms, resulting in what is known as a Resolution III design. Therefore, linear terms can be estimated, but they are confounded with the two-term interactions. More specifically, the design always estimates the effect of $N - 1$ terms, but when the actual number of factors is lower, the remaining factors are treated as dummy variables (and those terms can be used to estimate the incertitude of the coefficients since their real effect should be equal to zero).

The model with $k = 11$ is therefore the following:

$$y = b_0 + b_1 x_1 + b_2 x_2 + b_3 x_3 + b_4 x_4 + b_5 x_5 + b_6 x_6 + b_7 x_7$$
$$+ b_8 x_8 + b_9 x_9 + b_{10} x_{10} + b_{11} x_{11} + \varepsilon. \tag{4.6}$$

As mentioned, this strategy should replace the preliminary experiments. Even in the context of preliminary experiments, conducting replicates remains crucial to assess experimental variability. However, replicates at the central point (with all the levels of the variables set to zero) in Plackett–Burman designs are not informative, as these models serve primarily for screening purposes rather than making predictions or inferences about the coefficients, particularly concerning the risks associated with disregarding interactions. Instead, it is advisable to perform different replicates of the experiments based

102 Basic Chemometrics for Analytical Chemists

Table 4.2. First rows of the experimental matrices of the first five Plackett–Burman designs, requiring a number of experiments equal to the first multiple of 4 greater than the number of factors.

N. exp.	x_1	x_2	x_3	x_4	x_5	x_6	x_7	x_8	x_9	x_{10}	x_{11}	x_{12}	x_{13}	x_{14}	x_{15}	x_{16}	x_{17}	x_{18}	x_{19}	x_{20}	x_{21}	x_{22}	x_{23}
8	1	−1	−1	1	−1	1	1																
12	1	1	−1	1	1	1	−1	−1	−1	1	−1												
16	1	1	1	1	−1	1	−1	1	1	−1	−1	1	−1	−1	−1								
20	1	1	−1	−1	1	1	1	1	−1	1	−1	1	−1	−1	−1	−1	1	1	−1				
24	1	1	1	1	1	−1	1	−1	1	1	−1	−1	1	1	−1	−1	1	−1	1	−1	−1	−1	−1

on the Plackett–Burman design. This approach allows for a more robust model construction by incorporating a greater number of experiments and enabling the simultaneous presence of DoF, thereby providing more accurate estimations of the statistical significance of coefficients.

Case study: Disc brake pads (I)

This case study concerns a company which operates in the automotive sector, specifically in the production of brake pads. The study aims to determine the extent to which the production process influences the quality of the braking pads. Eleven process variables, listed in Table 4.3, both qualitative and quantitative, are identified at two levels (for reasons of industrial secrecy, the true names of the raw materials and their original values cannot be reported). The response variable used to assess product quality is the compressibility of the pad, which should be minimised. The goal is to identify which of the 11 process variables significantly impact the response and which do not. With a total of 11 variables, 12 experiments are needed according to the Plackett–Burman design, as shown in Table 4.4. These experiments are useful to estimate 12 coefficients (a constant and 11 linear terms), and therefore no DoF are available.

Table 4.3. Process independent factors involved in the study.

Variable ID	Variable name	−1	1
x_1	Resin type	Slow	Fast
x_2	Press type	Old	New
x_3	Press time	Short	Long
x_4	Press pressure	Low	High
x_5	Press temperature	Low	High
x_6	Oven temperature	Low	High
x_7	Oven time	Low	High
x_8	Scorching time	Low	High
x_9	Scorching temperature	Low	High
x_{10}	Pressure at high temperature	Low	High
x_{11}	Pressure at low temperature	Low	High

Table 4.4. Experimental matrix together with the response.

#	Resin type	Press type	Press time	Press pressure	Press temperature	Oven temperature	Oven time	Scorching time	Scorching temperature	Pressure at high temperature	Pressure at low temperature	Compressibility
1	1	1	-1	1	1	1	-1	-1	-1	1	-1	163
2	-1	1	1	-1	1	1	1	-1	-1	-1	1	121
3	1	-1	1	1	-1	1	1	1	-1	-1	-1	152
4	-1	1	-1	1	1	-1	1	1	1	-1	-1	100
5	-1	-1	1	-1	1	1	-1	1	1	1	-1	93
6	-1	-1	-1	1	-1	1	1	-1	1	1	1	173
7	1	-1	-1	-1	1	-1	1	1	-1	1	1	133
8	1	1	-1	-1	-1	1	-1	1	1	-1	1	131
9	1	1	1	-1	-1	-1	1	-1	1	1	-1	157
10	-1	1	1	1	-1	-1	-1	1	-1	1	1	157
11	1	-1	1	1	1	-1	-1	-1	1	-1	1	101
12	-1	-1	-1	-1	-1	-1	-1	-1	-1	-1	-1	236

Coefficients compressibility

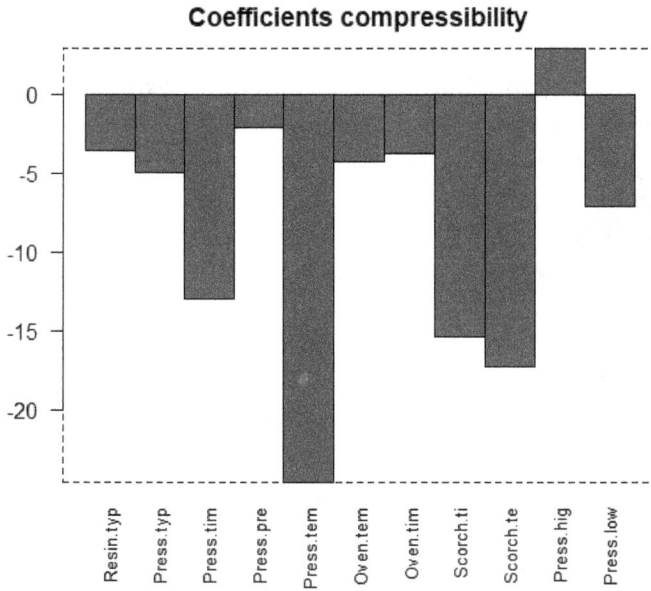

Figure 4.11. Bar plot of the coefficients.

Following MLR, the equation relating the factors to the response is obtained:

$$y = 143.1 - 3.6x_1 - 4.9x_2 - 12.9x_3 - 2.1x_4 - 24.6x_5 - 4.3x_6$$
$$- 3.8x_7 - 15.4x_8 - 17.3x_9 + 2.9x_{10} - 7.1x_{11}. \tag{4.7}$$

From Figure 4.11, it is clear that the relevant variables are press time, press temperature, scorching time, and scorching temperature (respectively, $b3 = -12.9$, $b5 = -24.6$, $b8 = -15.4$, and $b9 = -17.2$). In addition to identifying the variables that impact the response the most, this design allows for the identification of a preferred direction within the experimental domain: since compressibility should be minimised, it is better to set the four variables at their higher levels.

4.5.2. *Full factorial designs*

The full factorial designs are the simplest designs that can be employed, necessitating a total of 2^k experiments, where k represents the number of variables being investigated. In this design, as mentioned, each factor has two levels, and all possible combinations are tested. For full factorial designs, the number of experiments required increases exponentially with the number of factors. Consequently, this design is only suitable for situations involving a relatively small number of potential factors and, therefore, when there are no limitations imposed by the number of possible experiments. Table 4.5 displays the experimental matrix designed for the study of three independent factors through eight planned experiments.

The experimental matrix can be constructed easily without any software. The matrix consists of 2^k rows, each corresponding to a specific experiment, and k columns, each representing a variable. In the first column, the values alternate between -1 and $+1$ for each row. In the second column, the values alternate every second row, while in the third column, they alternate every fourth row, and so on. The reported experimental matrix is within the standard order; however, to minimise time-related effects, it is mandatory to run the experiments in a random order.

The experiments must be performed in a random order to mitigate any time-related effect!

Table 4.5. Two-level full factorial design experimental matrix reported within the standard order.

#	x_1	x_2	x_3
1	-1	-1	-1
2	1	-1	-1
3	-1	1	-1
4	1	1	-1
5	-1	-1	1
6	1	-1	1
7	-1	1	1
8	1	1	1

Geometrically, in a 2^3 full factorial design, the set of coded points forms a cube, with each point located at one of the eight corners (Figure 4.13(a)). Unlike the OVAT approach, where variable 1 is changing while variables 2 and 3 remain constant, factorial designs involve changes in variable 1 while variables 2 and 3 assume various values. This principle applies to all the variables. Consequently, factorial designs are suitable for evaluating interactions between variables, allowing for the examination of changes when variable 1 changes simultaneously with variable 2 at its upper or lower levels, and so on. The data obtained using experiments planned according to a full factorial design allow for model estimation and therefore enable the estimation of the effects of the single variables on the response and the effects of the interactions between the variables.

The number of terms in a two-factor interaction model is

$$1 + k + \frac{k(k-1)}{2}, \tag{4.8}$$

where 1 is the constant and k is the number of factors.

The model with $k = 3$ is therefore the following:

$$y = b_0 + b_1 x_1 + b_2 x_2 + b_3 x_3 + b_{12} x_1 x_2 + b_{13} x_1 x_3 + b_{23} x_2 x_3 + \varepsilon. \tag{4.9}$$

In CAT, a dialogue box shows the available higher terms (Figure 4.12); in the case of a full factorial design, 2^3, there are three

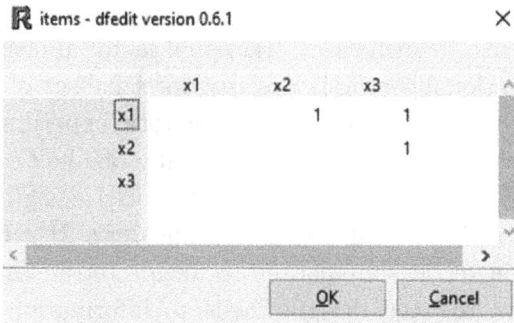

Figure 4.12. Dialogue box for higher terms definition; within the current case, only cross-terms are available.

cross terms. The user can discard one of the current coefficients by replacing 1 with 0. It has to be noted that this removal must be done *a priori* (when postulating the model, i.e., before performing the experiments) and not *a posteriori* (after computing the coefficients).

Each term in this model consumes one DoF (i.e., eight experiments to estimate seven coefficients; one DoF available). The three-term interaction term ($b_{123}x_1x_2x_3$) is not reported since these terms are both unlikely to be statistically significant and their interpretation is also not straightforward.

In the context of full factorial designs, the model validation step is quite relevant. To achieve this, it is good practice to include a series of centre-point replicates during this phase, with quantitative variables (coded levels) in this case being [0, 0, 0]. By comparing the experimental results with the predicted values at the central point, if no significant differences are observed, the model can be considered validated. These replicates also offer the advantage of estimating experimental variability. Precisely, this variability can then be compared with the overall design-induced variability and the SD of the model residuals, enabling an evaluation of any potential LoF, as described in Section 4.4.1. If the linear model is not validated, there is a possibility of observing quadratic behaviour within the experimental domain being investigated. Therefore, a more complex model is required.

However, when the number of factors exceeds five ($2^5 = 32$ experiments), a two-level full factorial design requires significantly more experiments than what is needed to estimate linear and interaction terms. In such cases, fractional factorial designs are often employed. Fractional factorial designs are a subset of full factorial designs, characterised by a fraction of the total experiments required. They are denoted as 2^{k-p}, which is typically one-half or one-quarter of the experiments required by a full factorial design. Figure 4.13 shows a fractional factorial design, specifically a 2^{3-1} design, where half of the cube vertices are chosen to form a tetrahedron. However, executing these reduced designs leads to information loss. In this specific example, interactions between factors cannot be estimated

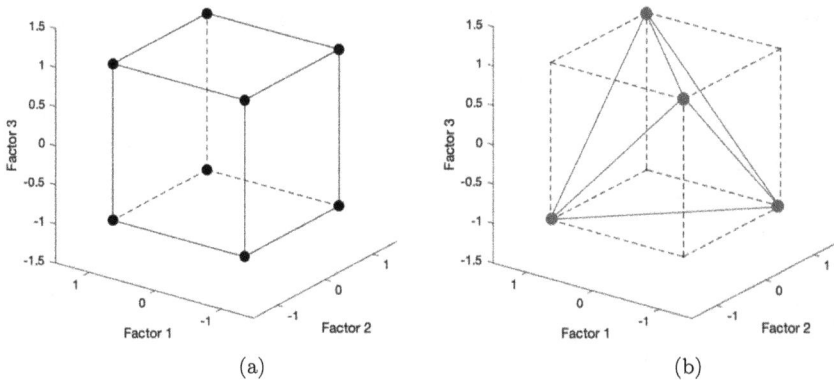

Figure 4.13. Graphical representation of (a) a three-factor full factorial design showing a cubic domain and (b) a three-factor fractional factorial design 2^{3-1} showing a tetrahedral experimental domain.

independently from the main factor effects, resulting in confounded effects.

Case study: Lysozyme nucleation at NASA

The aim of the study was to investigate the effects of precipitant, supersaturation, and impurities on improving the crystallisation conditions of a protein. Detailed results are presented in Ref. [6]. Two levels were identified for each variable. The response variable was the number of crystals on a logarithmic scale, which should be minimised (i.e., the lower the count, the larger the crystals). Following the full factorial design strategy, a total of eight experiments, plus central points for linear model validation, were deemed sufficient for three factors. However, considering the high experimental variability and the low cost, it was decided to perform each experimental point in duplicate, resulting in a total of 18 experiments (16 factorial points plus the central point in duplicate). The experimental matrix, plan, and response are shown in Table 4.6, with the experiments reported in the standard order (i.e., they were performed simultaneously). Otherwise, the order must be randomised to avoid any time-related effects. Replicates were used to estimate experimental variability by means of the pooled SD, which was 0.12, with nine DoF.

Table 4.6. Experimental matrix and experimental plan together with the response.

X	Experimental matrix			Experimental plan			Response
	Precip.	Supers.	Impurity	Precip.	Supers.	Impurity	log(crystals)
1	−1	−1	−1	3	2.4	0	1.16
2	−1	−1	−1	3	2.4	0	1.17
3	1	−1	−1	7	2.4	0	1.14
4	1	−1	−1	7	2.4	0	0.75
5	−1	1	−1	3	3	0	2.28
6	−1	1	−1	3	3	0	2.07
7	1	1	−1	7	3	0	1.66
8	1	1	−1	7	3	0	1.69
9	−1	−1	1	3	2.4	0.9	1.67
10	−1	−1	1	3	2.4	0.9	1.71
11	1	−1	1	7	2.4	0.9	1.36
12	1	−1	1	7	2.4	0.9	1.63
13	−1	1	1	3	3	0.9	2.16
14	−1	1	1	3	3	0.9	2.16
15	1	1	1	7	3	0.9	1.95
16	1	1	1	7	3	0.9	1.86
17	0	0	0	5	2.7	0.45	1.75
18	0	0	0	5	2.7	0.45	1.76

Following MLR, the model (Figure 4.14) was obtained, where stars represent the statistical significance (p-value: $* = 0.05$, $** = 0.01$, and $*** = 0.001$):

$$y = 1.65 - 0.15x_1(**) + 0.33x_2(***) + 0.16x_3(***)$$
$$- 0.04x_1x_2 + 0.03x_1x_3 - 0.11x_2x_3(**). \tag{4.10}$$

SD of the residuals (9 DoF): 0.13; experimental SD (9 DoF): 0.12; explained variance (%): 91.1. Since the SD of the residuals and the experimental SD were not significantly different, there is no LoF.

The isoresponse surface, together with the plot of the semiamplitude of the confidence interval, is reported in Figure 4.15, showing the interaction between the two factors. The predicted value at the

Coefficients log(crystals)

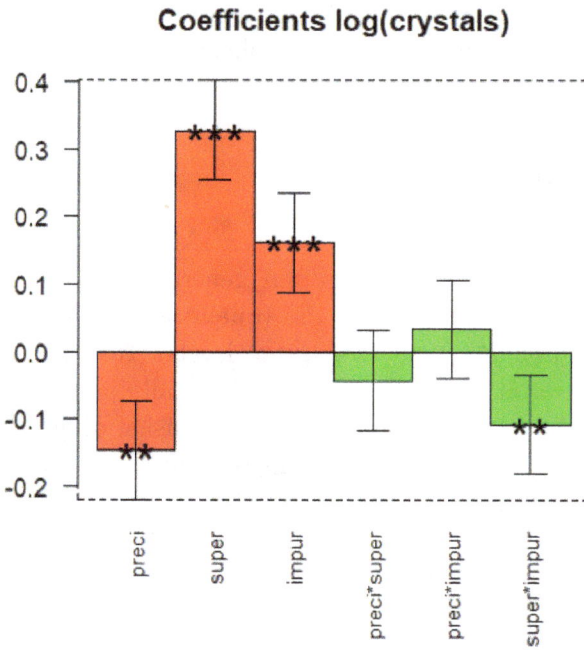

Figure 4.14. Bar plot of the coefficients.

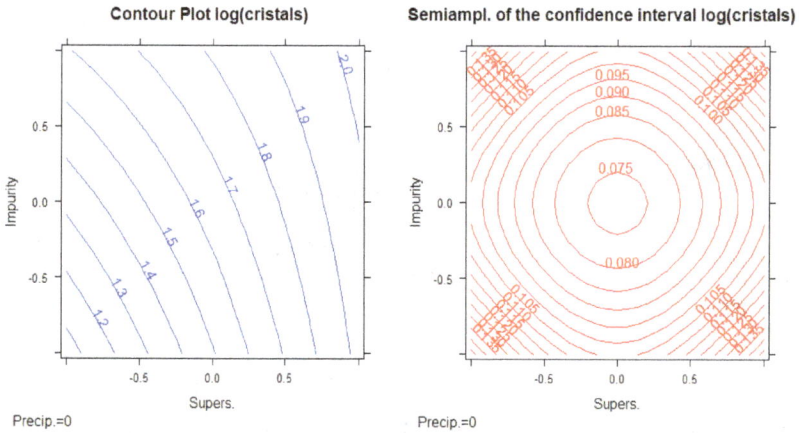

Figure 4.15. Contour plot and plot of the semiamplitude of the confidence interval in the prediction of the log(crystals) on the supersaturation–impurity $(x_2–x_3)$ plane, setting precipitant (x_3) at the central level.

central point is 1.65±0.07, while the experimental value is 1.76±0.20; the model adequately describes the phenomenon and can therefore be applied.

4.6.   Optimisation Designs

When the important factors have been identified and their ranges defined, a subsequent DoE is often generated. More experiments are needed for these steps, as the (quantitative) factors are studied more thoroughly (i.e., with more levels). Real-world problems frequently exhibit maxima, minima, or saddle points within their experimental domains. As a result, the inclusion of quadratic terms is imperative to provide a comprehensive and accurate description of these phenomena. Therefore, only one additional intermediate level for each factor is enough $(-1, 0, +1)$. Three-level full factorial designs could be a solution, requiring nine experiments for two factors (3^2) but with just three factors, they would require 27 experiments (3^3). Some compromise designs allow for performing well-balanced studies with fewer experiments. The domains of these classic designs are symmetrical geometric figures. They are specifically designed to be used only with quantitative variables and are not suitable for incorporating qualitative variables.

These types of designs can handle *quantitative* and *discrete quantitative factors* simultaneously, considering for each factor more than two levels, even if one additional intermediate level for each factor is enough $(-1, 0, +1)$. Qualitative factors cannot be studied.

4.6.1.   *Composite designs*

The commonly employed optimisation designs include composite designs, specifically the central composite face-centred (CCF) and central composite circumscribed (CCC) designs, as depicted in Figure 4.16. These designs are derived from two-level factorial designs (2^k) and are extended by incorporating star points that are placed

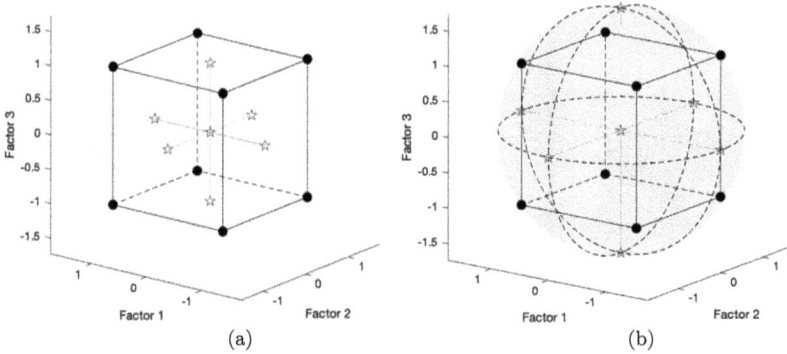

Figure 4.16. Graphical representation of (a) a three-factor central composite face-centred design and (b) a three-factor central composite circumscribed design. The face-centred design explores a cubic domain, while the circumscribed design shows a spherical domain.

either on the surfaces of a k-dimensional cube (CCF) or outside the cube (CCC). More specifically, for each factor, two axial points are included, representing the $+\alpha$ and $-\alpha$ levels of that factor, where α is the distance of the star point from the origin, while the remaining factors are set to zero. The design is complemented with a central point (at which all the variables are set to their central value corresponding to a vector of zeros in coded units), often replicated for estimating experimental variability and improving prediction quality (by reducing the leverage, as illustrated in Figure 4.17). However, it is recommended not to replicate the central point more than three times, considering both logical and orthogonal aspects. This is because it is more beneficial to have distinct experimental conditions, each with two replicates, rather than replicating the same condition six times. Table 4.7 illustrates the experimental matrix of a CCF designed for the study of three independent factors through 15 planned experiments, with one central point.

The number of experiments required by a central composite design is

$$2^k + 2k + N, \tag{4.11}$$

while the number of terms in the model is

$$1 + 2k + \frac{k(k-1)}{2}, \tag{4.12}$$

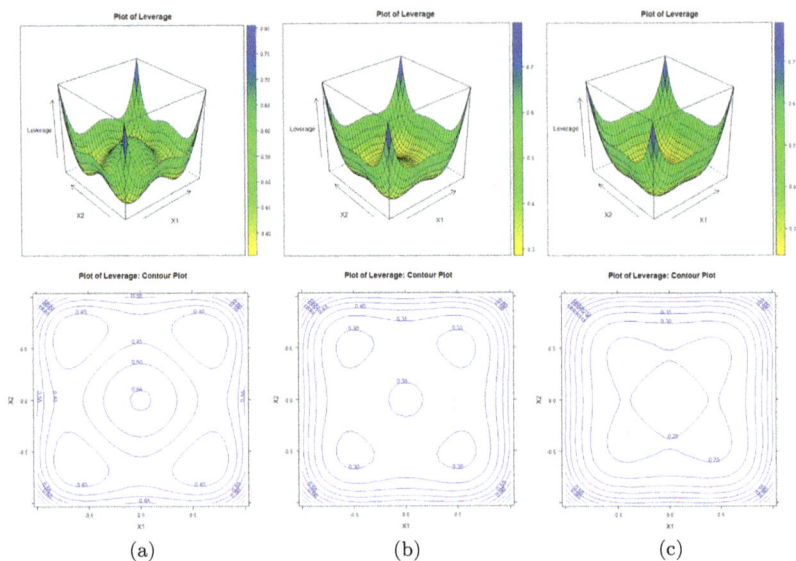

Figure 4.17. The leverage surfaces of a two-factor central composite face-centred design, highlighting the effect of the central point replicates (N) on the leverage: (a) one, (b) two, and (c) three central points.

Table 4.7. Experimental matrix of a three-factor central composite face-centred design with one central point.

#	x_1	x_2	x_3
1	-1	-1	-1
2	1	-1	-1
3	-1	1	-1
4	1	1	-1
5	-1	-1	1
6	1	-1	1
7	-1	1	1
8	1	1	1
9	-1	0	0
10	1	0	0
11	0	-1	0
12	0	1	0
13	0	0	-1
14	0	0	1
15	0	0	0

where k is the number of factors and N is the number of central point replicates.

The full quadratic model with $k = 3$ is therefore the following:

$$y = b_0 + b_1 x_1 + b_2 x_2 + b_3 x_3 + b_{12} x_1 x_2 + b_{13} x_1 x_3 + b_{23} x_2 x_3$$
$$+ b_{11} x_1^2 + b_{22} x_2^2 + b_{33} x_3^2 + \varepsilon. \tag{4.13}$$

In CAT, a dialogue box shows the available higher-order terms (Figure 4.18) in the case of a central composite design with three variables, including both interactions and quadratic terms. Again, only according to the postulated model, the user may discard one of the current coefficients by replacing 1 with 0.

CCF is characterised by α being equal to one, while in CCC α is equal to sqrt(k), as shown in Figure 4.16. In the CCC design, each variable has five levels, and the experimental domain takes the shape of a sphere (k equals to 3). Points on the sphere's surface are equidistant from the centre, and the design gains the property of rotatability, meaning that the leverage of a point is determined by its distance from the centre and not influenced by its orientation, and thus an equal prediction variance lies at a fixed distance from the origin. In contrast, the CCF design assigns three levels to each variable, and the experimental domain can be visualised as a cube (k equals to 3). In this case, not all points are equidistant from the centre; therefore, the leverage of a point is influenced by both its distance from the origin and the direction being considered.

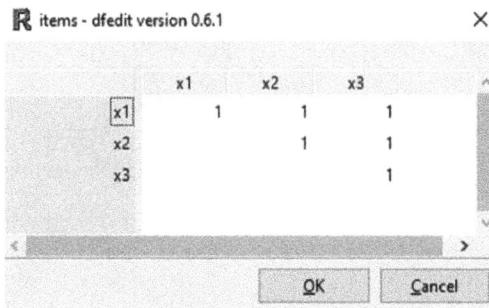

Figure 4.18. Dialogue box of the available higher terms beyond the constant and the linear ones.

Despite the statistical benefits of CCC designs, having five levels for all variables is not always straightforward from a chemical or practical perspective. Additionally, in CCC designs, the levels for each variable are not always equidistant from one another. Since it is the square root of the number of variables, α is 1.41 for two variables, 1.73 for three variables, and 2 for four variables, in which case the levels are equidistant.

Another advantage of the CCF design over CCC is its operational aspect. Figure 4.19 provides a two-dimensional example. While the CCF offers a fully accessible domain, CCC does not allow making predictions for the experimental conditions corresponding to the combinations of the extreme values (e.g., the highest temperature and the lowest pressure) since they fall outside the domain. This limitation becomes more significant when transitioning to a higher-dimensional space.

Central composite face-centred designs can be useful as the subsequent step of two-level full factorial designs when the linear model has not been validated due to the presence of a curvature. By adding the star points, along with some central point replicates, it is possible to initially identify a potential "time block effect" by

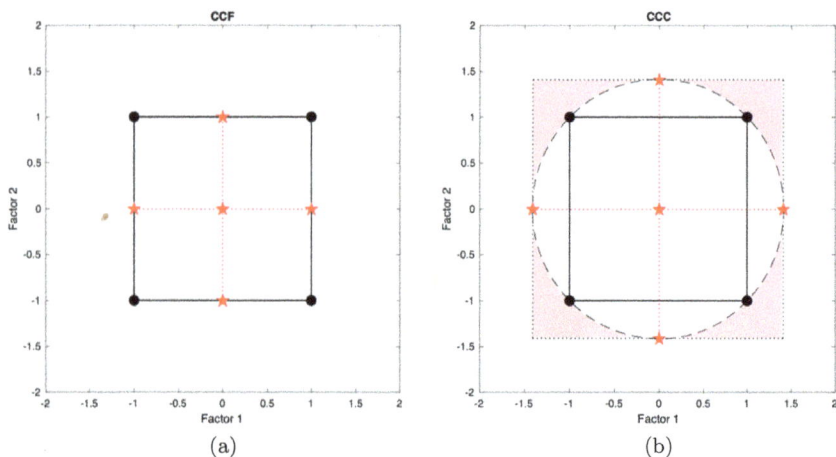

(a)   (b)

Figure 4.19. (a) Two-factor central composite face-centred design and (b) two-factor central composite circumscribed design. The red region shows the combinations of the two factors falling outside the experimental domain.

comparing experiments performed under the same conditions but at different sessions. If the time block effect is found to be non-significant, it becomes feasible to pursue a more complex design involving quadratic terms.

Case study: Coal mill

The aim of the study was to investigate the effects of variables such as coal load (tons per hour) and the setting of a classifier (percentage opening of the mill top, which causes variations in air recirculation) on the quality of ground coal. These variables are both quantitative and studied at three levels. Two responses related to the quality of grinding are considered, concerning the particle diameter as determined using sieves: the percentage of particles with a diameter less than 200 mesh (y_1, fine particles) and the percentage of particles with a diameter greater than 50 mesh (y_2, coarse particles). From a corporate economic perspective, the objective is to produce an acceptable powder by identifying the joint acceptability domain for the two responses based on the following acceptability thresholds: fine particle content greater than 70% and coarse particle content less than 1%. At the same time, it is necessary to maximise mill productivity by operating at the highest possible coal load levels. Nine experiments are required, following a CCF strategy, as reported in Table 4.8.

Following MLR, the equations relating the factors to the response are obtained:

$$\text{Small particles} = 76.90 - 4.84x_1(**) - 5.60x_2(**) + 1.84x_1x_2$$
$$- 1.71x_1^2 - 8.42x_2^2(**). \tag{4.14}$$

(SD of the residuals (3 DoF): 1.18; explained variance (%): 97.8)

$$\text{Large particles} = 0.15 - 0.46x_1(***) - 0.21x_2(**) - 0.23x_1x_2(**)$$
$$+ 0.29x_1^2(**) + 0.08x_2^2. \tag{4.15}$$

(SD of the residuals (3 DoF): 0.06; explained variance (%): 98.4)

According to the coefficients and the response surfaces (Figures 4.20 and 4.21), the quantity of fine particles decreases as

Table 4.8. Experimental matrix and experimental plan in the standard order together with the responses.

#	Experimental matrix		Experimental plan		Responses	
	Coal load	Classifier position	Coal load	Classifier position	Small particles	Large particles
1	−1	−1	5	0	79.00	0.05
2	1	−1	15	0	66.58	1.44
3	−1	1	5	100	63.50	0.03
4	1	1	15	100	58.42	0.50
5	−1	0	5	50	80.75	0.01
6	1	0	15	50	69.22	0.88
7	0	−1	10	0	73.25	0.40
8	0	1	10	100	63.31	0.08
9	0	0	10	50	77.31	0.12

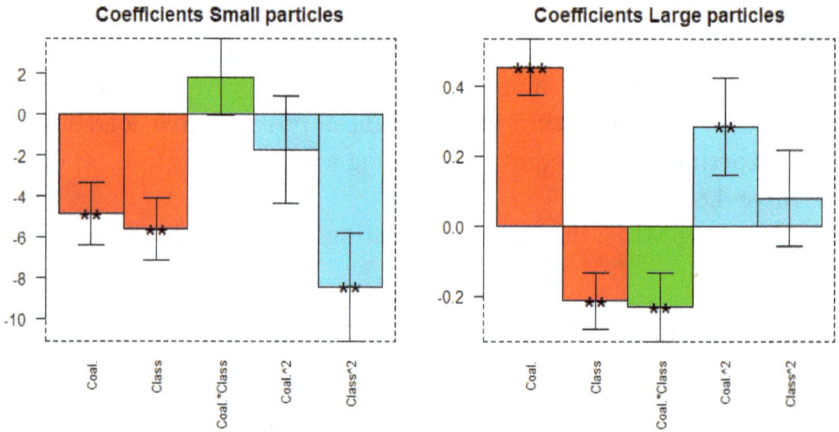

Figure 4.20. Plots of the coefficients of the two responses.

the coal load increases, while the classifier should be set to a half-open position. For coarse particles, an increase in load has a negative effect on quality, resulting in an increase in coarse particles, especially if the mill is operated with the classifier closed. It is then important to observe the joint acceptability domain to achieve maximum

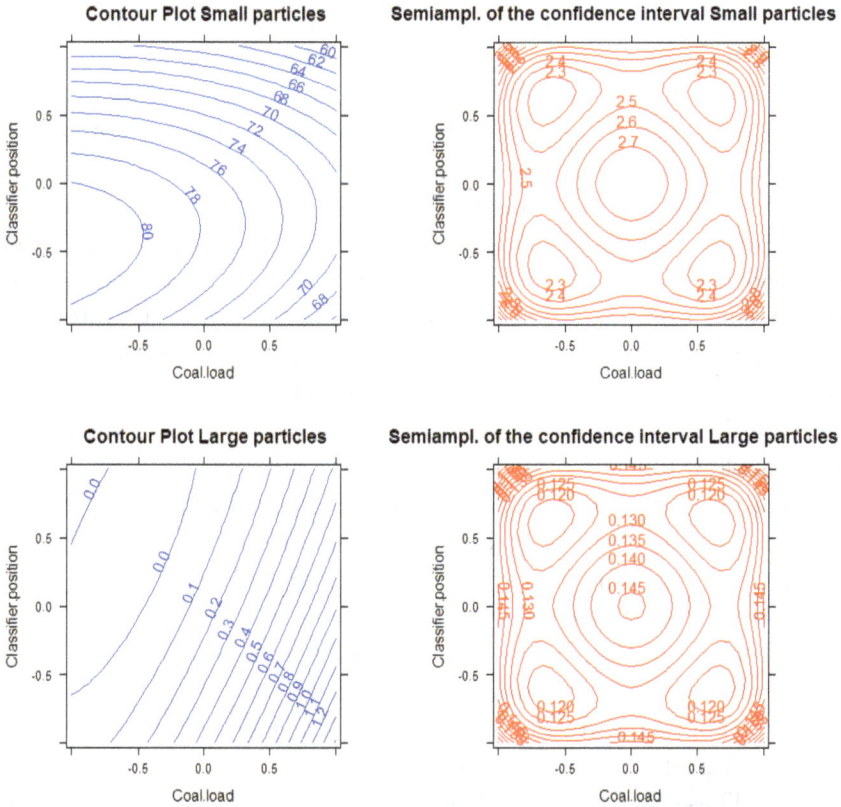

Figure 4.21. Contour plot and semiamplitude of the confidence interval plot for the prediction of both responses.

operational efficiency, specifically operating at the highest possible coal load. However, it is necessary to consider that the boundary is close to the acceptability region for fine particles, necessitating consideration of both model and experimental uncertainties. A more conservative setting will be adopted, taking into account the lower limit of the confidence interval.

4.6.2. *Multicriteria decision-making: Pareto optimality*

When managing multiple responses, it is often necessary to seek the best compromise that leads to a desired outcome. This compromise may not only involve direct responses but also consider indirect factors, such as production costs or input variables. The optimal compromise is determined by estimating and validating a complete quadratic model, which is essential for assessing a local maximum or minimum in the different responses. In this regard, the Pareto optimality approach is a valuable technique that enables simultaneous optimisation of multiple responses, allowing for effective handling of various objectives and identification of optimal solutions based on specific criteria. Another commonly used approach, although potentially misleading, is the Derringer desirability function, which involves transforming individual responses into a mathematical function based on subjective judgements and aggregating them into a single desirability score.

The concept of Pareto optimality, named after the Italian Vilfredo Pareto, is widely applicable in economics, game theory, engineering, and the social sciences. Pareto optimality is achieved when resource allocation reaches a point where further Pareto improvements within the system are no longer possible. This means that enhancing the condition of one individual or aspect cannot occur without a disadvantage to another. The Pareto front represents a collection of optimal solutions, encompassing all non-dominated points. These non-dominated points have the distinctive quality of not being outperformed by any other point, considering all objectives in the optimisation function. Such a framework allows experts to evaluate various optimal solutions. Conversely, a point (i.e., a product or experimental condition) is considered dominated when at least one other point is better (or at least equal) in all responses [7]. This approach is based on a fundamental assumption, namely the assignment of weights (or importance) to each response, but without assigning a number to each response as in the Derringer desirability approach, which is often not so simple. For example, in the case of two responses of equal importance, determining the solution is relatively straightforward, as one aims to maximise the value of one while also maximising the value of the other. Nevertheless, even in such cases, the ultimate decision must be made by the expert. It is essential to acknowledge that the determination of importance, which is often inherently subjective, cannot always be quantified precisely. On the contrary, in scenarios where a response holds greater significance than another, a logical approach involves tracking the Pareto front: this process on the "front" begins with the worst results of the relatively less important response and continues until a compromise is identified between the two, always comparing the actual response values. This compromise, selected based on the expert's knowledge, represents the optimal balance between the two responses, and the decision is made through rational judgement.

One may face a typical situation in which the goal is to maximise quality while minimising costs. The outcome depends on the available budget and the required quality standards. The relative significance of different characteristics must be weighed carefully, and it is worth noting that individual perspectives lead to diverse sets of assigned weights. Consequently, there is no universally applicable solution. Often, the objective is to maximise performance by simultaneously considering process variables as well. Additionally, there are other indirect parameters to consider since performance, quality, or operational costs should be minimised. Therefore, for each point in the experimental domain, it is possible to calculate the operational cost of the system. An increase in quality generally leads to higher operational costs; however, there comes a point where the return on investment in quality improvements decreases, or in other terms, the increase in the overall costs becomes significantly higher than the quality gain. Beyond a certain threshold, further increases in quality may not be cost-effective or competitive in the market. In this context, the Derringer desirability function is misleading, as it enables the software, rather than the expert, to determine the best compromise and make the selection. Moreover, it may produce impractical optimal results, such as temperatures or pH levels that are impossible to operate with. Alternatively, the Pareto optimality approach can be employed, allowing for the setting of constraints and targets to guide the decision-making process.

> Using the Pareto front approach, experts select the best compromise that enables the simultaneous optimisation of multiple responses, according to their knowledge of the problem, by choosing an option within a list of optimal solutions (i.e., non-dominated points).

In the context of experimental design, after establishing models for each response, the procedure is structured as follows (the steps are oriented towards the use of the software CAT):

(1) The user should prepare a list of the possible "candidate points" through the "MLR-DOE" menu and the corresponding "Grid creation" sub-menu. It is essential to ensure that the points to be predicted, within the predefined experimental domain, must be logical, so that the process can operate with each point presented by the expert. To do so, it is necessary to input the number of independent factor and the number of levels for each factor, as well as defining the type (i.e., qualitative or quantitative by using the feature flag on "variable is numeric") and levels for each, as illustrated in Figure 4.22. The final matrix will be displayed in the console and saved in the corresponding object named "cp".

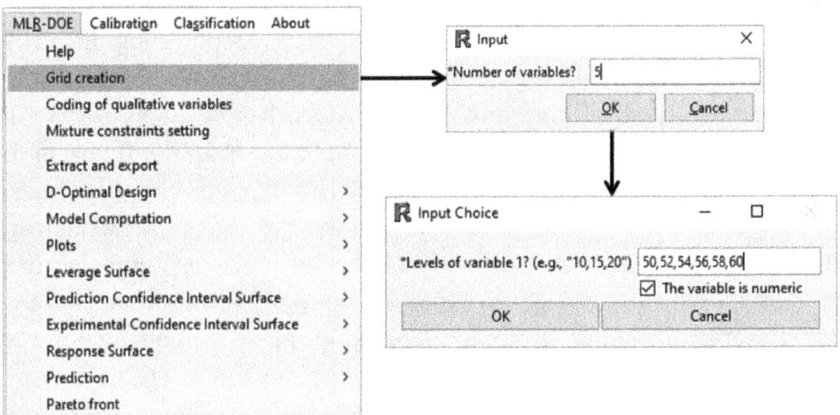

Figure 4.22. MLR-DOE: Grid creation.

(2) If the user has entered the real levels for each factor, before carrying out predictions, it is mandatory to code them within the range of -1 to $+1$, as outlined in Section 4.4.

(3) Predictive models are established for each response, and these models enable predictions across the entire experimental domain. To obtain predictions for each response at each point of the previously generated "candidate points" (stage 1), the user should go to the "Prediction" sub-menu through the "MLR-DOE" menu, as outlined in Section 4.4.1. Once defined, the matrix (name, rows, and columns of the experimental matrix) and the predicted response, along with leverage and confidence interval limits, will be displayed in the main console of the software and stored in "response.pred". The same procedure is applied to each response under study. Eventually, the user should prepare a matrix which includes the N predicted responses (as many rows as candidate points and as many columns as responses).

(4) The list of non-dominated points is then selected based on the objectives using the Pareto front optimality function on the response matrix. To do so, the user should go to the CAT "Pareto front" sub-menu, through the "MLR-DOE" menu and define the response matrix, specifying the three possible aims for each response: maximisation, minimisation, or proximity to a specific target. The matrix of the non-dominated points will be displayed and saved in "nondom"; the whole procedure is depicted in Figure 4.23.

(5) To select the best compromise more adequately, according to the specific threshold or goal, a graphical representation of the non-dominated points should be produced. In CAT, the user can go to the "Bivariate" menu, as shown in Figure 4.24.

4.7. Designs for Qualitative Variables with More Than Two Levels

Qualitative variables with more than two levels are handled differently. Unlike quantitative variables with multiple levels, which can be coded between -1 and $+1$, qualitative variables lack

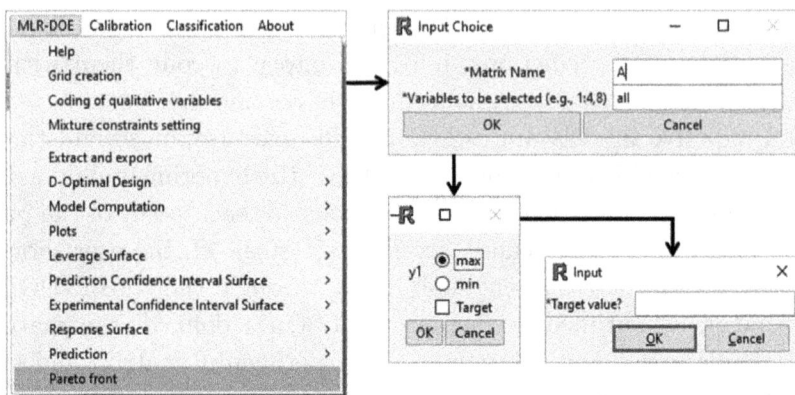

Figure 4.23. MLR-DOE: Pareto front.

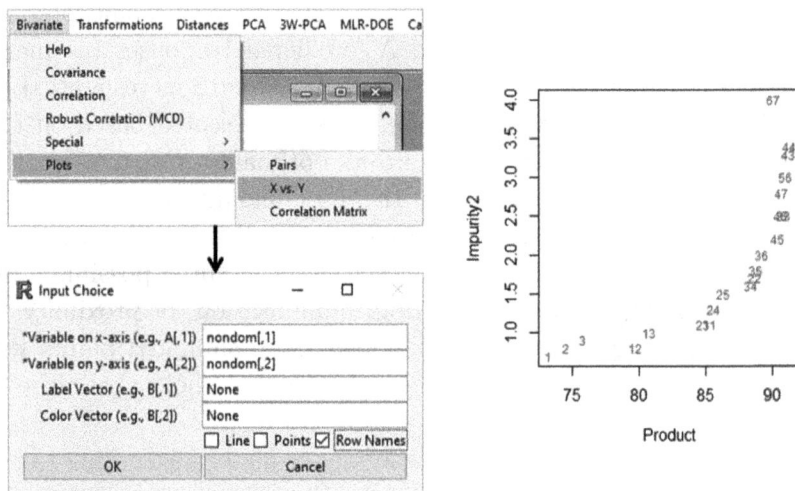

Figure 4.24. Bivariate: X vs. Y plot of the non-dominated points. In the proposed case, the product (x-axis) should be maximised, while the impurity (y-axis) should be minimised. The user, based on their expertise, will select the best compromise settings.

a numerical scale, as there are no intermediate levels between categories. Examples of qualitative factors include suppliers, raw materials, catalysts, production plants, instruments, and operators. To capture their categorical nature in the analysis, these variables are typically represented using different coding strategies, including dummy coding, effects coding, orthogonal coding, and criterion coding. They enable researchers to incorporate categorical predictor variables into multiple regression analysis. Each strategy requires creating one or more variables (i.e., columns of the model matrix) to represent the categories of the predictor variable [8].

Dummy coding is the simplest and most commonly used method for coding qualitative variables with more than two levels, and it is implemented in CAT. It compares levels of the qualitative predictor variable to a specific reference level or an implicit level. For example, when dealing with a qualitative variable with three levels (A, B, and C), two "dummy" variables (i.e., dummy A and dummy B) are created to indicate specific levels (in the example, representing level A and level B, respectively). The experiments at level A will have 1 as dummy A and 0 as dummy B (lines 1, 4, 7, and 10 for variable x_1 in Table 4.9); the experiments at level B will have 0 as dummy A and 1 as dummy B (lines 2, 5, 8, and 11); the experiments at level C will have 0 as dummy A and 0 as dummy B (being neither A nor B, it is obvious that it refers to C; lines 3, 6, 9, and 12). The level not corresponding to any dummy variable is defined as the implicit level. In the model matrix, only $l - 1$ dummy variables, where l represents the number of levels, must be present in order to avoid collinearity issues. The model matrix illustrating this concept is shown in Table 4.9.

The choice of the implicit level is crucial when using dummy coding since it acts as a reference for the other levels. From a mathematical point of view, in the case of "symmetrical" designs (in which all the levels are tested the same number of times), the selection does not affect the quality of the model. In the case of asymmetrical designs, guidelines recommend avoiding selecting the implicit level with few observations, as it plays a crucial role in calculating statistical significance and, consequently, confidence intervals. Such scenarios typically arise when working with non-designed data matrices, such as historical datasets.

In the postulated model, the focus is primarily on the constant term and the linear terms, as studying interactions can be complicated, and they often lack meaningful interpretations and are

Table 4.9. Experimental plan and model matrix to handle qualitative variables with more than two levels. The number of possible experiments is given by the total combinations between the level variables (3×4).

	Experimental plan		Model matrix					
				x_{1A}	x_{1B}	x_{2a}	x_{2b}	x_{2c}
#	x_1	x_2	b_0	b_{1A}	b_{1B}	b_{2a}	b_{2b}	b_{2c}
1	A	A	1	1	0	1	0	0
2	B	A	1	0	1	1	0	0
3	C	A	1	0	0	1	0	0
4	A	B	1	1	0	0	1	0
5	B	B	1	0	1	0	1	0
6	C	B	1	0	0	0	1	0
7	A	C	1	1	0	0	0	1
8	B	C	1	0	1	0	0	1
9	C	C	1	0	0	0	0	1
10	A	D	1	1	0	0	0	0
11	B	D	1	0	1	0	0	0
12	C	D	1	0	0	0	0	0

unlikely to be significant. However, different analytical techniques may exhibit interactions with specific sample pretreatments; therefore, the researcher must pay attention to them. The procedure for estimating interactions with qualitative variables at more than two levels is described in detail in the bibliography [9].

The model for two qualitative variables, with three and four levels, respectively, is the following:

$$y = b_0 + b_{1A}x_{1A} + b_{1B}x_{1B} + b_{1C}x_{1C} + b_{2a}x_{2a}$$
$$+ b_{2b}x_{2b} + b_{2c}x_{2c} + b_{2d}x_{2d} + \varepsilon, \tag{4.16}$$

where the levels of variable 1 are (A, B, C) and the levels of variable 2 are (a, b, c, d).

The number of terms is

$$1 + \sum (\text{number of levels for each factor} - 1), \tag{4.17}$$

where 1 is the constant. It is worth noting that in these designs, it is impossible to obtain VIFs equal to one, even when they are very close to one, because the information matrix in such cases is, by definition, not a centred matrix.

Unlike previous designs, where the constant term corresponds to the response when all the variables are at their central point ($y = b_0$; all the factors have level 0) and the coefficients of the linear terms contribute one-half of the effect of the corresponding factor, in this case the constant term represents the response at the implicit level. (In Table 4.9, it happens when $x_1 = $ C and $x_2 = $ d.) Each regression coefficient indicates the difference in the response between the corresponding level and the implicit level. For instance, the linear term b_{1A} corresponds to the difference between working at level A and working at the implicit level C. Testing the significance of these coefficients is similar to conducting Student's t-tests for comparing group means or using one-way ANOVA procedures.

CAT offers the function of converting a matrix containing factors with qualitative variables with more than two levels within the "MLR-DOE" menu, as depicted in Figure 4.25. In such cases, one

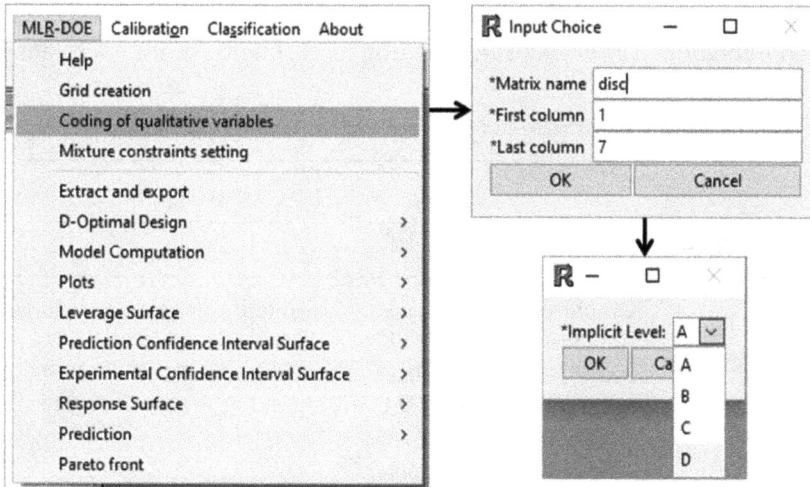

Figure 4.25. MLR-DOE menu: coding of qualitative variables at more than two levels (e.g., A, B, C, and D).

should select the columns containing the interested factors and then choose the implicit level for each, as necessary for dummy coding.

It is quite common to study both qualitative variables (with more than two levels) alongside quantitative variables. In such scenarios, all possible combinations of qualitative and quantitative variables would result in an excessive number of experiments to perform. For this reason, an optimal design approach is preferable to finding the best compromise between chemical information and experimental effort, as outlined in Section 4.9.

Case study: Disc brake pads (II)

The performance of a friction material is influenced by various process variables. An experimental design was implemented to investigate the effects of seven process variables; detailed results are presented elsewhere [10]. The results of the dyno-test were transformed into an indicator of performance at high temperatures through multivariate analysis strategies, which served as the response variable. Table 4.10 lists the independent factors and their respective levels. All factors (x_1, x_2, x_5, x_6, and x_7, which are qualitative, and x_3 and x_4, which are intrinsically quantitative) are coded with the dummy variable in order to be able to interpret the coefficients in the same way (i.e., as a difference with respect to the implicit level). The model includes 14 terms (one constant term, one term for each factor

Table 4.10. Process-independent factors involved in the study.

Variable ID	Variable name	A	B	C	D
		Levels			
x_1	Graphite size	small	medium	large	small/large*
x_2	Sulphide type	Sb	Mo*		
x_3	Graphite/sulphide	0.5:1	1:1	2:1	3:1*
x_4	Lubricants/Abrasives	2:1	3:1	4:1	5:1*
x_5	Abrasives	Al_2O_3	SiO_2*		
x_6	Metal size	small	big*		
x_7	Furnace cycle	scorch	no scorch*		

Note: *implicit levels.

Table 4.11. Experimental plan and model matrix to handle qualitative variables with more than two levels.

#	x_1	x_2	x_3	x_4	x_5	x_6	x_7	b_{1A}	b_{1B}	b_{1C}	b_{2A}	b_{3A}	b_{3B}	b_{3C}	b_{4A}	b_{4B}	b_{4C}	b_{5A}	b_{6A}	b_{7A}	Perf.
			Experimental plan									Model matrix									
1	A	A	A	A	A	A	A	1	0	0	1	1	0	0	1	0	0	1	1	1	54
2	A	A	B	B	A	B	B	1	0	0	1	0	1	0	0	1	0	1	0	0	39
3	A	B	C	C	B	A	A	1	0	0	0	0	0	1	0	0	1	0	1	1	35
4	A	B	D	D	B	B	B	1	0	0	0	0	0	0	0	0	0	0	0	0	44
5	B	B	A	B	B	B	A	0	1	0	0	1	0	0	0	1	0	0	0	1	14
6	B	B	B	A	A	A	B	0	1	0	0	0	1	0	1	0	0	1	1	0	21
7	B	A	C	D	A	B	A	0	1	0	1	0	0	1	0	0	0	1	0	1	49
8	B	A	D	C	A	B	B	0	1	0	1	0	0	0	0	0	1	1	0	0	55
9	C	B	A	C	A	A	B	0	0	1	0	1	0	0	0	0	1	1	1	0	80
10	C	B	B	D	B	B	A	0	0	1	0	0	1	0	0	0	0	0	0	1	79
11	C	A	C	A	B	A	B	0	0	1	1	0	0	1	1	0	0	0	1	0	16
12	C	A	D	B	B	B	A	0	0	1	1	0	0	0	0	1	0	0	0	1	6
13	D	A	A	D	B	A	B	0	0	0	1	1	0	0	0	0	0	0	1	0	10
14	D	A	B	C	A	B	A	0	0	0	1	0	1	0	0	0	1	1	0	1	0
15	D	B	C	B	A	B	B	0	0	0	0	0	0	1	0	1	0	1	0	0	100
16	D	B	D	A	A	A	A	0	0	0	0	0	0	0	1	0	0	1	1	1	82

at two levels, and three terms for each factor at four levels). A total of 1024 combinations are possible (candidate points). A D-optimal solution with 16 experiments was chosen, as shown in Table 4.11, as it offered the best compromise between the number of experiments and the quality of information.

The following model is obtained:

$$y = 46 - 5x_{1A} - 13x_{1B} - 3x_{1C} - 28x_{2A}(*) - 7x_{3A} - 12x_{3B}$$
$$+ 4x_{3C} - 3x_{4A} - 5x_{4B} + 49x_{5A}(*) + 2x_{6A} - 6x_{7A}.$$

(SD of the residuals (2 DoF): 10; explained variance (%): 89)

Analysis of the regression coefficients reveals that factors x_2 and x_5 are significant ($p < 0.05$). The best performance is achieved when x_2 is at level B (implicit level) and x_5 is at level A. A quick check of the data table shows that the four formulations with x_2 at level B and x_5 at level A (experiments 9, 10, 15, and 16) are by far the best ones (see Figure 4.26).

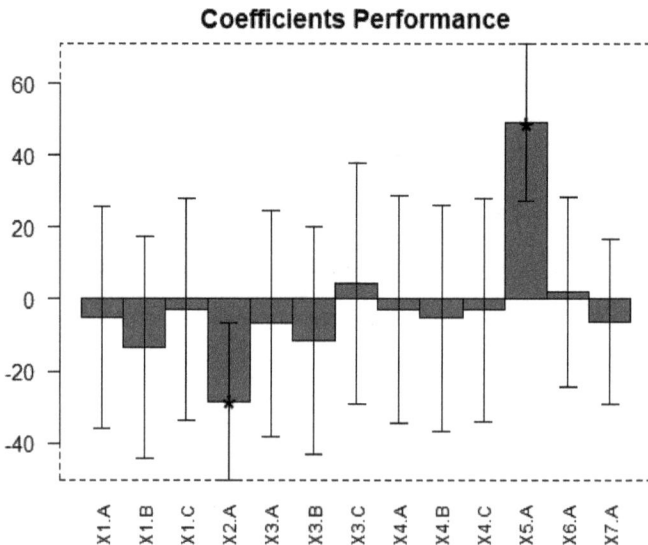

Figure 4.26. Bar plot of the coefficients.

4.8. Mixture Designs

Mixture designs hold particular significance for product development and optimisation of products and processes [11–13]. In the field of chemistry, studies often involve examining the effects of various ingredients and their proportions in different systems. These systems can encompass typical analytical chemistry tasks, such as chromatographic optimisation (e.g., the composition of a mobile-phase solvent). This is crucial in many fields, such as pharmaceuticals, cosmetics, food formulations, culinary recipes, and mortar or fuel compositions. Mixture designs involve factors ranging between 0 and 1, and the combined components or ingredients always sum up to 1 (or 100%). In such scenarios, a constrained mixture analysis approach employing a Scheffé polynomial model is important [14]. It is crucial to emphasise that, in the cases involving mixtures, consideration should focus exclusively on major components. Minor components, typically constituting approximately less than 5%, should not be classified as mixture factors. Instead, they should be treated as process (or independent) factors. This distinction is crucial because, from a compositional standpoint, interpreting and considering minor components together with major components can be challenging, and from a methodological point of view, not entirely correct. The experimental domain that this would yield would be very narrow for the minor components compared to the major ones. However, it is important to note that, despite their lower content, minor components can be essential in the formulation.

A two-component mixture can be visualised as a point located on a segment, whose extremes correspond to a pure component. In this case, a classic coding between −1 and +1 can be applied (e.g., a mixture of A and B, A relative to B, where level −1 corresponds to 0% of A, level +1 to 100% of B, and level 0 to 50% of A and B). A three-component mixture can be visualised as a point within an equilateral triangle, where each vertex represents a pure component and points along the edges represent two-component mixtures. Any other point inside the triangle represents a three-component mixture. When dealing with a four-component mixture, the experimental domain is represented as a tetrahedron. This concept can be generalised for larger numbers of mixing factors (solvents, etc.) in a multidimensional space. The domain of such mixtures forms a regular figure that has as many vertices as pure components. This figure lies in a space with a dimensionality equal to $N-1$ components, although both visualisation and interpretation can become challenging. In these cases, it is suggested that the entire domain be divided into two or more subdomains, each containing a manageable number of components. Subsequently, one can study each subdomain independently, iterating among them.

The mixture designs can be categorised into simplex centroid and simplex lattice designs; however, this manuscript focuses solely on the simplex centroid design (SCD). The experimental matrix SCD for N components comprises $2^N - 1$ points, including all N pure components (the vertex of the geometrical figure), all binary (each 50%) combinations, all ternary (each 33.3%) combinations, all quaternary (each 25%) combinations, and so on (ending with a combination of all components in equal proportion, $100\%/N$). Table 4.12 reports the SCD experimental matrix, while Figure 4.27 shows its graphical representation.

Researchers often face practical limitations and constraints while investigating the factors and their associated ranges of variability. These constraints can be either independent or relational.

Independent constraints
In this case, the constraints are limited to single components. For ternary mixtures, they are represented by straight lines drawn inside

Table 4.12. Experimental matrix of a three-component simplex centroid design.

#	x_1	x_2	x_3
1	1	0	0
2	0	1	0
3	0	0	1
4	0.5	0.5	0
5	0.5	0	0.5
6	0	0.5	0.5
7	0.3333	0.3333	0.3333

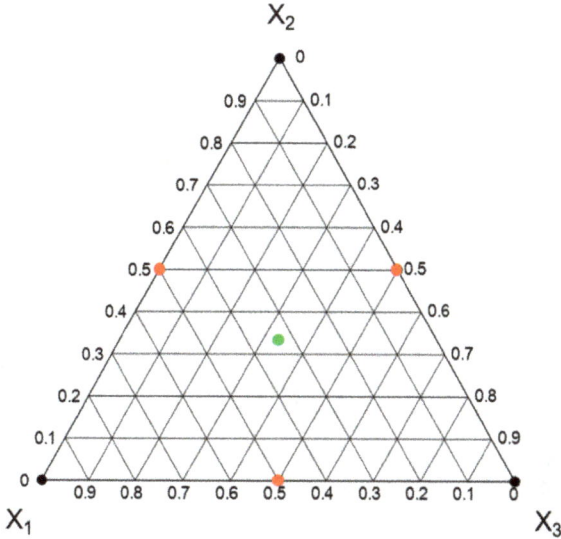

Figure 4.27. Graphical representation of a three-component simplex centroid design showing an equilateral triangle domain. In this representation, the vertices of the triangle correspond to the pure components, the sides to the binary mixtures, and the internal points to the ternary mixtures.

a triangle, parallel to the side opposite the vertex of the constrained component. For example, Figure 4.28 shows the following constraints in red lines: unilateral constraints: $x_3 \leq 0.4$; bilateral constraints: $0.2 \leq x_1 \leq 0.6$ and $0.2 \leq x_2 \leq 0.45$. In the case of a quaternary mixture, the constraint is represented by a plane parallel to the face opposite the vertex of the constrained component.

In some real cases, it may not be possible to perform experiments using pure components. In such situations, researchers can employ pseudocomponents. This approach consists of setting unilateral constraints that define a regular domain within the global domain, where the vertices no longer correspond to pure components but rather represent pseudocomponents. Table 4.13 displays a three-component experimental matrix where each factor exhibits a 20% range of variability, while Figure 4.29 shows a graphical representation of the selected points.

In terms of modelling and interpretation, these pseudocomponents are treated as if they were pure components. Subsequently,

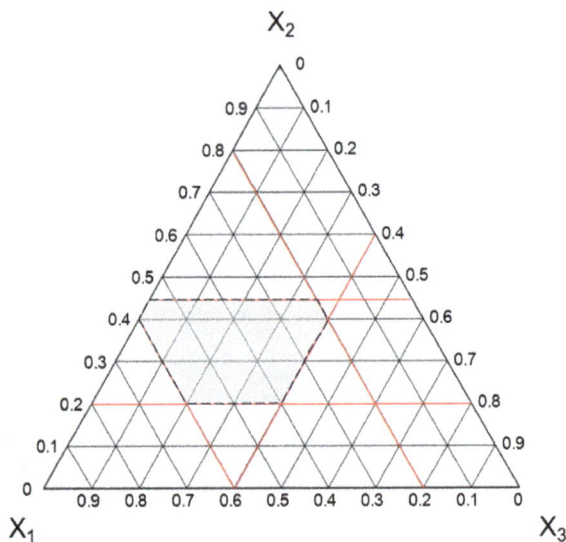

Figure 4.28. Graphical representation of the experimental domain after applying constraints. Independent unilateral constraints (red lines): $x_3 \leq 0.4$; bilateral constraints: $0.2 \leq x_1 \leq 0.6$; $0.2 \leq x_2 \leq 0.45$.

Table 4.13. Experimental matrix of a three-component simplex centroid design in pseudocomponents.

#	x_1	x_2	x_3	pseudocomp$_1$	pseudocomp$_2$	pseudocomp$_3$
1	0.8	0.1	0.1	1	0	0
2	0.6	0.3	0.1	0	1	0
3	0.6	0.1	0.3	0	0	1
4	0.7	0.2	0.1	0.5	0.5	0
5	0.7	0.1	0.2	0.5	0	0.5
6	0.6	0.2	0.2	0	0.5	0.5
7	0.6666667	0.1666667	0.1666667	0.3333	0.3333	0.3333

converting from pseudocomponents to actual values of the components involves multiplying each coordinate of the point of interest by the composition of its corresponding pseudocomponent. After obtaining the results, the sum of the coordinates provides the actual values of the components.

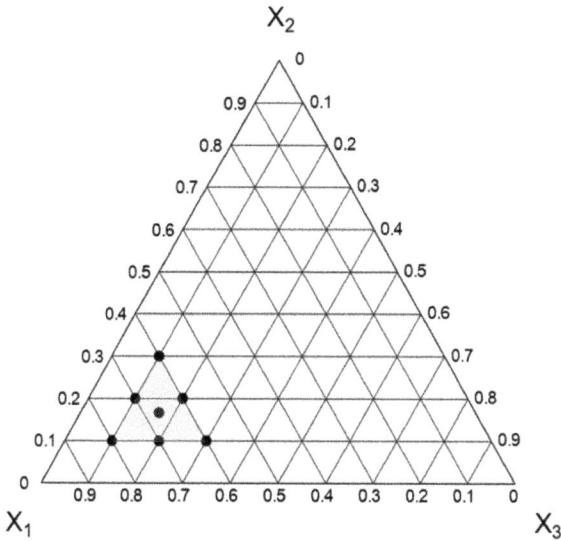

Figure 4.29. Graphical representation of a three-component simplex centroid design inside the global domain. This subdomain can be treated in pseudocomponents.

Relational constraints

Relational constraints involve relationships between two components. In the case of a ternary mixture, the drawn line would be oblique, representing a precise ratio between the two components. In the case shown in Figure 4.30 by the blue oblique line, x_1 must always be greater than or equal to x_2. For quaternary mixtures, relational constraints involve the sum of the components.

In cases with multiple constraints, the experimental domain can become irregular, making it impossible to conduct experiments at the vertices of the triangle. Therefore, to select the right experiments, an optimal design approach is necessary (Section 4.9). Taking the example shown in Figure 4.29 of a three-component mixture with unilateral constraints ($x_3 \leq 0.4$) and bilateral constraints ($0.2 \leq x_1 \leq 0.6$ and $0.2 \leq x_2 \leq 0.45$), Figure 4.31 shows the output along with the steps to accomplish the experimental matrix with CAT, from the grid creation of the candidate points to D-optimal design and model computation. Following these steps, one can calculate and generate a leverage surface plot, providing a visual representation of

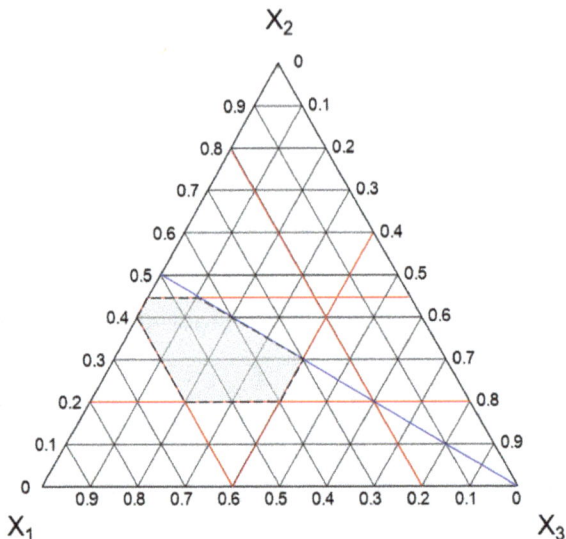

Figure 4.30. Graphical representation of the experimental domain after applying independent constraints (red lines) and the relational constraint $x_1 \geq x_2$ (the blue line). The relational constraint also narrowed the original range for x_1 (now between 0.3 and 0.6).

the experimental points selected by the algorithm. If the experimental variability is estimated with an adequate number of degrees of freedom, it becomes possible to calculate the semiamplitude of the confidence interval and generate the corresponding plot. The user should prepare the matrix of the candidate points through the "MLR-DOE" menu and "Mixture constraints setting" sub-menu.

Concerning mathematical modelling, the SCD with seven experiments allows for the estimation of the reduced cubic model (Scheffé polynomial model; the full cubic model would include the terms b_{112}, b_{113}, etc.):

$$y = b_1 x_1 + b_2 x_2 + b_3 x_3 + b_{12} x_1 x_2 + b_{13} x_1 x_3 + b_{23} x_2 x_3$$
$$+ b_{123} x_1 x_2 x_3 + \varepsilon. \tag{4.18}$$

When designs are used to cope with independent variables, the constant term (b_0) corresponds to the predicted response when all variables are equal to zero. In mixture-design models, this term is

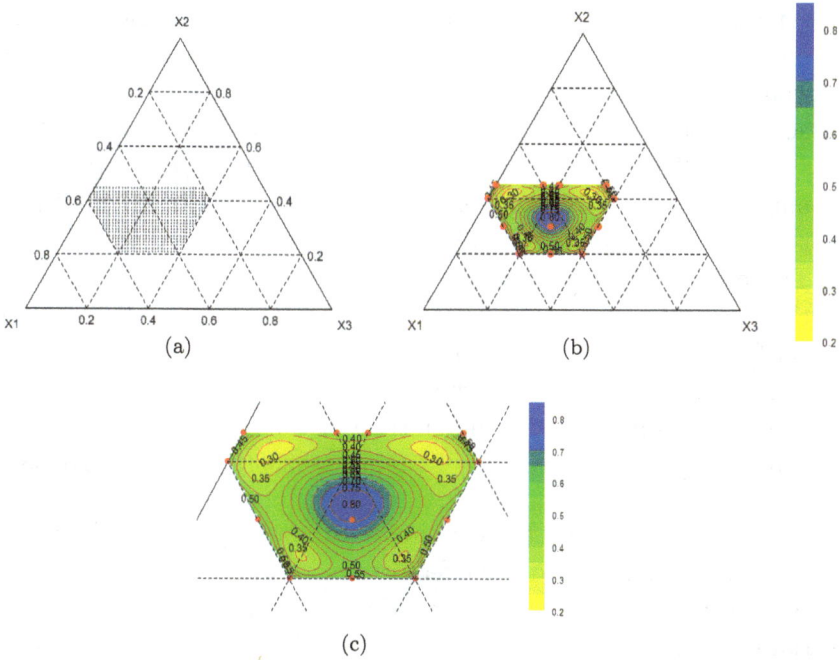

Figure 4.31. The output along with the steps to accomplish the experimental matrix with CAT: (a) define the list of the candidate points according to the predefined constraints within the desired step ($x_3 \leq 0.4$; $0.2 \leq x_1 \leq 0.6$; $0.2 \leq x_2 \leq 0.45$) and run the D-optimal design to select the number of experiments within the candidate points; (b) after the model computation on the selected points, display the experimental points together with the leverage (and, if experimental variability is present, also the semiamplitude of the confidence interval). (c) A zoom-in of the ternary plot is also provided.

missing in the equation, as it does not represent any feasible condition within the experimental domain.

Concerning the model interpretation:

- The linear terms are estimated from the experiments conducted at the vertices of a triangle, which correspond to the pure components; however, they do not match the effects of those components.
- The quadratic terms are estimated from the experiments conducted at the centres of the sides, representing binary mixtures [0.5 0.5], to highlight any potential quadratic effects or the

synergic effect of the two components. If no synergic effect is present, the response of the mixture made by 0.5 x_1 and 0.5 x_2 would be the average of those of the pure components. If the actual response differs from the average, a positive or a negative synergic effect is present. This can be found in the term $x_1 x_2$. The magnitude of the synergic effects of the two-component mixtures is given by the coefficients divided by 4 ($b_{12} \times 0.5 \times 0.5$).

- The cubic term is estimated from the experiment conducted at the centroid of the domain, located at [0.3333 0.3333 0.3333], to verify a possible curvature at the centre of the domain. The same concept applies to the coefficient of the three-term interaction, which corresponds to the magnitude of the synergic effect of the three components divided by 27 ($b_{123} \times 0.3333 \times 0.3333 \times 0.3333$).

In mixture designs, there is no direct correspondence between the coefficient values and the effect of the variable. The latter is the difference in the response between the two extremes of its range of variation (from one extreme to the other). In this case, it compares the response when the component is absent (0) to that of when it is present as a pure component (1). As stated, the coefficients of the linear terms are associated with the responses obtained using pure components, but their magnitude does not directly represent the effects of those components. To understand the effects of the components, the isoresponse plot must be interpreted. In it, it is essential to observe how the response changes when moving from a composition without that specific component to one containing only that pure component, all while holding the relative amounts of the other components constant. On the plot, the observation begins at the midpoint of the edge opposite the component under study (e.g., a mixture formed by the two remaining components, each at a 50% concentration) and continues towards the vertex representing the pure component. Defining and calculating the effect is straightforward when there are no constraints or when working with pseudocomponents because we are dealing within a regular figure. However, when constraints are introduced, thus defining a non-regular domain, the interpretation becomes more intricate.

Furthermore, as the number of components increases, this task becomes more complex, but it remains feasible when dealing with four components. In this scenario, the domain forms a tetrahedron, and "slices" can be obtained by cutting it with planes parallel to one of its faces while keeping one component constant, thereby obtaining an isoresponse triangle to interpret. However, the complexity increases significantly for more than four components. Despite this, in the literature, there are studies that explore the use of mixtures with five components [15].

Case study: Formulation of pharmaceutical tablets

This case is about optimising a tablet formulation. The original mixture consists of seven components: the active pharmaceutical ingredient (4%), a disintegrant (0.2%), two types of lubricants (4% and 1%), and three ligands: monohydrated alpha-lactose (MAL), anhydrous beta-lactose (ABL), and modified rice starch (MRS). The first four minor components will be kept constant, resulting in a ternary system where the sum is 90.8%, which is considered 100% for the current study. No constraints are imposed on the three excipients. The objective is to maximise two responses: breaking load (the force a tablet can withstand without disintegrating) and dissolution rate (the speed at which a tablet dissolves to release the API in the gastric environment). Table 4.14 shows the experimental matrix along with both responses.

Table 4.14. Experimental matrix with the two responses.

#	MAL	ABL	MRS	Breaking load	Dissolution rate
1	1	0	0	30	45
2	0	1	0	99	28
3	0	0	1	39	52
4	0.5	0.5	0	42	44
5	0.5	0	0.5	40	66
6	0	0.5	0.5	69	32
7	0.3333	0.3333	0.3333	52	57

Contour Plot Breaking load   **Contour Plot Dissolution rate**

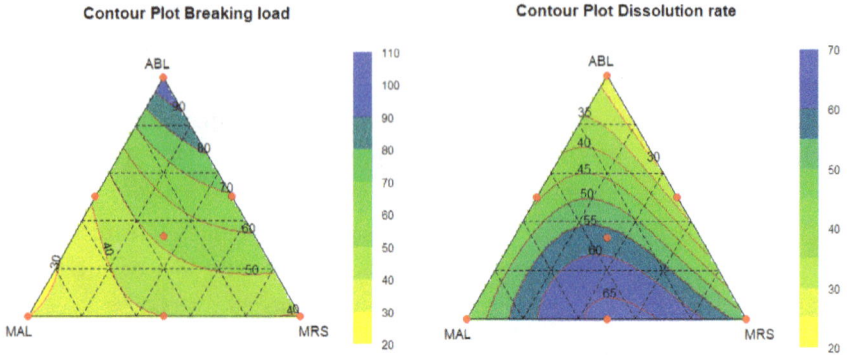

Figure 4.32. Response surface of both responses.

Following MLR, the equation relating the factors to the response are obtained:

$$\text{Br. Load} = 99x_{\text{MAL}} + 30x_{\text{ABL}} + 39x_{\text{MRS}} - 90x_{\text{MAL}}x_{\text{ABL}}$$
$$+ 0x_{\text{MAL}}x_{\text{MRS}} + 22x_{\text{ABL}}x_{\text{MRS}} + 96x_{\text{MAL}}x_{\text{ABL}}x_{\text{MRS}},$$

$$(4.19)$$

$$\text{Diss. Rate} = 28x_{\text{MAL}} + 45x_{\text{ABL}} + 52x_{\text{MRS}} - 30x_{\text{MAL}}x_{\text{ABL}}$$
$$- 32x_{\text{MAL}}x_{\text{MRS}} + 70x_{\text{ABL}}x_{\text{MRS}} + 210x_{\text{MAL}}x_{\text{ABL}}x_{\text{MRS}}.$$

$$(4.20)$$

From the analysis of the response surfaces in Figure 4.32, the ABL content has the most significant impact on both responses, displaying contrasting behaviours. Therefore, it is essential to find a compromise between the two responses using a Pareto front approach.

4.8.1. *Mixture–process designs*

In many cases, it is necessary to study mixture variables (i.e., ratios among components) in combination with process variables (e.g., time, temperature, and concentration). This strategy involves incorporating process variables into the mixture design. In such cases, it becomes crucial to investigate the blending properties of the mixture variables at different levels of all process variables. However, including process variables would significantly increase the experimental effort; therefore, it is advisable to limit the number of process variables, but still the choice of the experiments to be performed can be achieved through optimal designs. It is essential to consider process variables and those components which, despite being very important in the mixture, are characterised by a significantly smaller range of variability compared to the other main ones in the mix. Therefore, these components must not be taken into account as mixture variables.

The combined experimental design can be visualised as either setting up a design at each point of a configuration in the mixture components or conducting mixture experiments at each point of a configuration in the process variables. Figure 4.33 illustrates an example concerning two process variables studied at three levels, combined with a $\{3, 2\}$ simplex lattice design. In this case, the total number of experimental points would be $63 (7 \times 9)$, thus encompassing all possible combinations. It can be represented in two different ways: mixture in process or process in mixture.

The second-order (cubic) Scheffé model is applied along with the second-order (full quadratic) model for the process variables:

$$y = b_1 x_1 + b_2 x_2 + b_3 x_3 + b_{13} x_1 x_3 + b_{23} x_2 x_3 + b_{123} x_1 x_2 x_3 + b_1 z_1$$
$$+ b_2 z_2 + b_{12} z_1 z_2 + b_{11} z_1^2 + b_{22} z_2^2 + \varepsilon, \tag{4.21}$$

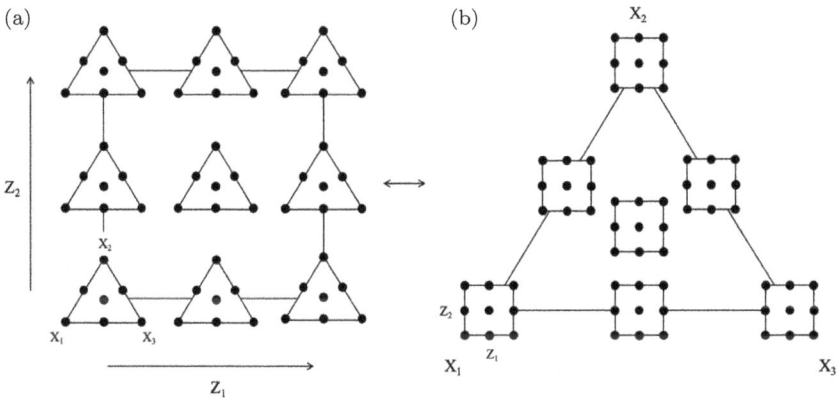

Figure 4.33. (a) Mixture in process and (b) process in mixture, each for the same 63 experiments.

where x denote the mixture factors and z denote the independent process factors.

The regression approach would also include interaction terms between mixture components and process factors (not reported in Equation (4.21)). However, interpreting these interactions can be complex and challenging. The primary focus of the regression approach lies in estimating the equation and then exploiting contour plots to visualise the effects of the three components within various process settings. One can interpret the interactions by making "local" models under the different process (or mixture) settings and then comparing the contour plots. The contour plots are produced through polynomial fitting, and shaded regions can be employed to highlight the study's objectives by facilitating the interpretation of interactions between ingredient ratios and process settings.

4.9. Optimal Designs

Optimal designs offer flexibility for the development of customised solutions. They are sometimes the only way to rationalise intricate case studies of high complexity. They have the capability to study quantitative, quantitative discrete, and qualitative factors with more than two levels while also accommodating non-regular experimental domains that may result from incompatibilities between certain factor levels. When dealing with quantitative variables, the domain may not be entirely accessible due to inherent constraints on the variables. For example, the temperature can vary within the range of 45–80°C, while the pressure ranges between 0.8 and 1.2 atm. However, the highest temperature compatible with a pressure of 1.2 atm is 65°C, and correspondingly, the maximum pressure sustainable at 80°C is 1.0 atm. Similarly, when considering qualitative variables, not all levels can align seamlessly due to inherent disparities, leading to asymmetrical combinations (e.g., instrument type and different speed treatment settings), where not all instrument types can be operated within identical speed levels.

Furthermore, in some real cases, researchers may have *a priori* knowledge about the non-significance of certain coefficients, allowing them to postulate the most appropriate mathematical model.

Optimal designs overcome effectively the limitations of standard designs. The latter are characterised typically by (a) regular geometric domains, such as cubes, spheres, hexagons, triangles, or tetrahedra, and (b) being able to estimate only specific types of coefficients, such as linear terms (Plackett–Burman designs), linear terms and interactions (factorial designs), and linear terms, interactions, and quadratic terms (central composite designs).

The first step when employing optimal designs is to define a set of possible experiments in a coded format (i.e., candidate points through MLR-DOE: Grid creation, as outlined in Section 4.6.2; column transformation: coding of factors within the range between -1 and $+1$, as outlined in Section 4.4), which in complex cases can reach thousands of points. The challenge is to identify a subset of the candidate points whose number lies between the number of postulated model terms and the number of feasible experiments. It is crucial to define the postulated model beforehand, as the selection of experiments depends on the specific model under consideration.

This leads to one important question that must be addressed: *how many* experiments and *which* ones are sufficient to yield information of acceptable quality while optimising experimental effort?

Various optimality criteria can be employed. These criteria include A-optimality (minimising trace of the dispersion matrix), D-optimality (maximising the determinant of the information matrix), G-optimality (minimising the maximum prediction variance by minimising the maximum leverage), and I-optimality (minimising the average prediction variance over the design space). The most commonly used optimisation approach is the D-optimal design, which is described in detail in this section and available in CAT. These criteria are useful to compare different subsets of experiments having the same number of experiments. Of course, conducting more experiments provides greater information; however, they also require higher experimental effort. To ensure reliable outcomes, an *a priori* evaluation of the precision of each coefficient for each experimental matrix becomes essential, and a commonly used criterion to compare experimental matrices with the same number of experiments according to the D-optimality is the determinant of the information matrix, which must be maximised. However, to compare experimental matrices that have a different number of experiments, the normalised determinant (M) is employed:

$$M = \frac{\det(\boldsymbol{X}^T \boldsymbol{X})}{N^p}, \qquad (4.22)$$

where $\boldsymbol{X}^T \boldsymbol{X}$ is the information matrix, N is the number of experiments, and p is the number of parameters. The increase in the number of experiments directly impacts both the numerator and the denominator in this calculation. The value of M rises when the addition of an experiment enhances the quality of information more than the corresponding increase in experimental effort. For each number of experiments, the algorithm searches for the best subset according to the normalised determinant; furthermore, one must also double-check the VIFs of the different subsets, as illustrated in Figure 4.34. Regarding mixture designs, the information matrix differs significantly from that of a centred matrix, causing VIFs to give anomalous values. For this reason, these indicators cannot be considered and are therefore not shown. After that, the researcher will choose a subset of experiments, the number of which will depend

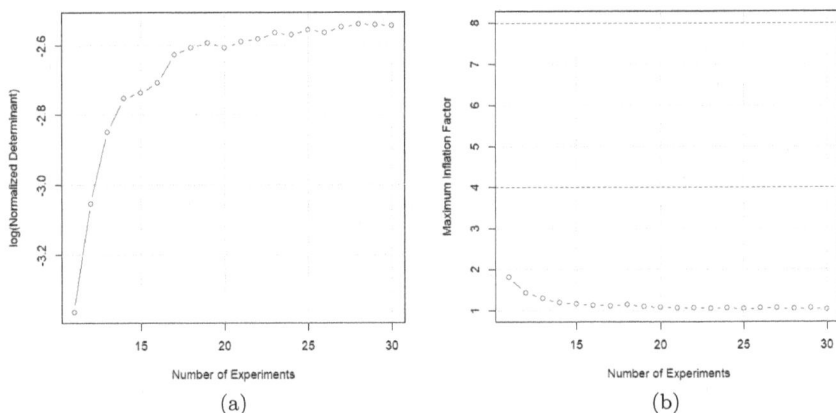

Figure 4.34. (a) Normalised determinant (logarithmic scale) plot and (b) VIFs plot with the maximum VIF for each solution; both plots are created using CAT. The normalised determinant "weights" the increase in information (connected with the increase in the number of experiments) by the experimental effort and therefore suggests the best compromise.

on the quality of information and the availability of resources. Nevertheless, it is important to note that requesting the exact number of experiments to be conducted prior to subset selection, as some commercial chemometric software may suggest, is not the correct approach.

A widely used algorithm, developed by Fedorov, employs a single-point exchange approach to achieve these optimal designs through leverage computation. Typically, the experiment selections are strategically made at the edges of the experimental domain to investigate linear terms and interactions, while adding central points aids in studying curvature.

D-optimal designs can also play a crucial role in "repairing" data matrices when experiments were previously conducted without following a suitable DoE approach. In such cases, the algorithm selects the most informative experiments from the pool of unperformed ones (i.e., candidate points). These selected experiments, when added to the subset of already performed ones, allow for a correct estimation of the model terms. This approach is known as "D-optimal design by addition". This method is also useful for selecting the right experiments to be replicated among the candidate points, serving the purpose of both estimating experimental variability and constructing a more robust model. Additionally, the D-optimal design by addition approach can be employed to select test points for model validation (Section 4.4.1).

It is clear that D-optimal designs maximise the information obtained by efficiently finding the best compromise between information quality and the required experimental effort, making them suitable for a wide range of goals, from screening to optimisation. By adhering to this approach, the researcher can carefully choose the number of experiments, replicates, and test points based on the available resources. However, to employ the D-optimal design effectively, the experimenter must specify a plausible model and be aware of the accessible experimental domain to provide candidate points, hence again necessitating a good deal of (chemical) knowledge of the system.

4.10.   Bridge between Experimental Design and Multivariate Analysis

When analysing multiple responses, before computing the MLR model, it is essential to carry out a thorough examination of the results. If one is working with two responses, a Cartesian plane is suitable to ascertain the correlation type and how samples lie within the bivariate space. However, when dealing with more than two responses, it becomes necessary to employ multivariate analysis techniques, such as PCA. PCA is central to understanding the relationships between responses, which often exhibit correlations. These correlations are visually represented in the loading plot and quantified by the variance explained in the significant components. When the first component (PC_1) accounts for a substantial portion of the total variance, it indicates a strong correlation among the responses, implying that they convey similar information. Understanding these correlations intuitively guides the interpretation of results and facilitates the subsequent selection and computation of the most suitable models. All these concepts are detailed in Chapters 6 and 7 of this book.

Case study: Organic synthesis

This case concerns the optimisation of an organic synthesis reaction, where the responses have conflicting objectives: maximising the yield of the main product while minimising various impurities. Ideally, yield and impurities would exhibit a negative correlation, where an increase in yield corresponds to a decrease in impurities. However, in real-world scenarios, the yield might be positively correlated with one impurity, while the others may be either partially correlated or entirely uncorrelated. In such cases, it becomes crucial to obtain a trade-off between maximising yield and minimising impurities, considering the relative significance of each response. One must weigh the trade-offs: is it preferable to increase the reaction yield, even if it means tolerating a higher level of impurity, or does a critical impurity pose a significant risk to the entire production, necessitating stringent control to ensure it remains below a specific threshold?

In the proposed case, a CCF design was applied to study three independent factors. Before proceeding with the regressions, PCA should be applied to the multiple responses, as shown in Table 4.15. As stated, PCA offers valuable insights into several fundamental aspects: the loading plot can provide hints on the correlation structure between responses, while the score plot can reveal the distribution of the experiments and, particularly, the experimental variability, both depicted in Figure 4.35. (These terms are studied thoroughly in Chapters 6 and 7.) The global variability arises from the variability induced into the system through the experimental design. Ideally, it is advantageous for replicates to form a tight cluster within this global domain, with its dispersion being significantly lower than the overall system variability. Otherwise, it will not be

Table 4.15. Experimental matrix, in the standard order, together with the responses.

#	x_1	x_2	x_3	Conversion	Product	Imp. A	Imp. B	Imp. C	Imp. D
1	−1	−1	−1	75.70	72.79	0.70	0.47	0.00	0.31
2	1	−1	−1	75.30	71.96	0.67	0.53	0.00	0.34
3	−1	1	−1	86.80	83.24	1.23	0.78	0.05	0.56
4	1	1	−1	89.70	85.89	1.26	1.00	0.00	0.69
5	−1	−1	1	96.70	89.06	1.15	3.54	0.00	2.39
6	1	−1	1	98.10	90.20	1.15	3.71	0.00	2.56
7	−1	1	1	99.80	84.78	1.22	6.52	0.27	4.78
8	1	1	1	99.70	82.54	1.12	7.52	0.34	5.74
9	−1	0	0	94.10	87.54	1.37	1.77	0.00	1.22
10	1	0	0	98.50	90.83	1.27	2.98	0.15	2.09
11	0	−1	0	92.60	88.29	1.07	1.73	0.00	1.10
12	0	1	0	99.40	90.71	1.41	3.26	0.11	2.38
13	0	0	−1	83.20	79.87	1.07	0.78	0.04	0.53
14	0	0	1	99.50	86.51	1.16	6.05	0.19	4.36
15	0	0	0	97.90	90.45	1.40	2.81	0.09	1.95
16	0	0	0	97.40	91.12	1.28	2.46	0.12	1.73
17	0	0	0	97.60	90.43	1.36	2.69	0.11	1.91

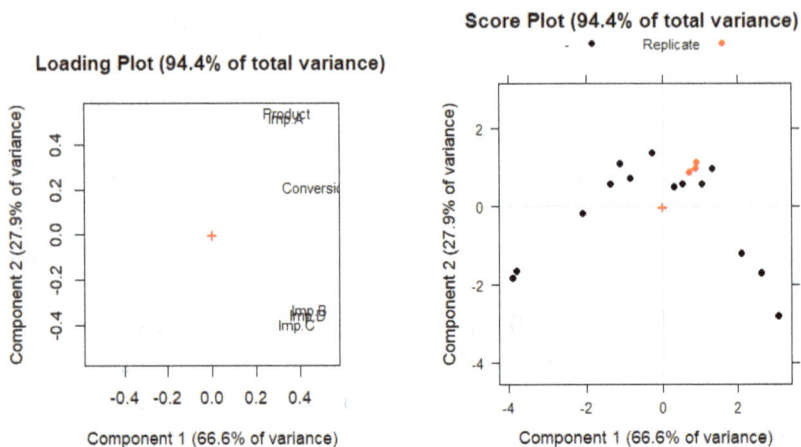

Figure 4.35. Loading plot and score plot obtained from a DoE study involving 14 experiments (organic syntheses) plus three replicates of the central point. In the loading plot, the main product yield (response named "product") is positively correlated with one of the impurities (response named "imp. A"). In the score plot, the samples projected into the plane PC_1–PC_2 define the variability of the experiments in the whole experimental domain, with the three replicates highlighted in red. It is evident that the experimental variability is much smaller than the induced variability.

possible to interpret and/or estimate any effect of the investigated DoE factors when the variability between replicates is of a similar magnitude to that of the natural variability of the system.

References

[1]   Leardi, R. (2013). Experimental design. In *Data Handling in Science and Technology*, Vol. 28. Elsevier Ltd, Chennai, pp. 9–53. doi: 10.1016/B978-0-444-59528-7.00002-8.

[2]   Leardi, R. (2009). Experimental design in chemistry: A tutorial. *Analytica Chimica Acta*, 652(1–2), 161–172. doi: 10.1016/j.aca. 2009.06.015.

[3]   Slutsky, B. (1998). Handbook of chemometrics and qualimetrics: Part A by D. L. Massart, B. G. M. Vandeginste, L. M. C. Buydens, S. De Jong, P. J. Lewi, and J. Smeyers-Verbeke. Data handling in science and technology volume 20A. Elsevier: Amsterdam. 1997. Xvii + 867 pp. ISBN 0-444-89724-0. $293.25. *Journal of Chemical Information and Computer Sciences*, 38(6), 1254–1254. doi: 10.1021/ci980427d.

[4]   Plackett, R. L. and Burman, J. P. (1946). The design of optimum multifactorial experiments. *Biometrika*, 33(4), 305. doi: 10.2307/2332195.

[5]   Parlier, G. (2011). *Transforming U.S. Army Supply Chains: Strategies for Management Innovation*. Business Expert Press, Hampton, New Jersey, United States. doi: 10.4128/9781606492369.

[6]   Burke, M. W., Leardi, R., Judge, R. A. and Pusey, M. L. (2001). Quantifying main trends in lysozyme nucleation: The effect of precipitant concentration, supersaturation, and impurities. *Crystal Growth and Design*, 1(4), 333–337. doi: 10.1021/cg0155088.

[7]   Reguera, C., Sánchez, M. S., Ortiz, M. C., and Sarabia, L. A. (2008). Pareto-optimal front as a tool to study the behaviour of experimental factors in multi-response analytical procedures. *Analytica Chimica Acta*, 624(2), 210–222. doi: 10.1016/j.aca.2008.07.006.

[8]   Starkweather, J. Categorical variables in regression: Implementation and interpretation. Available at: https://www.yumpu.com/en/doc ument/view/12353149/categorical-variables-in-regression-university-of-north-# (consulted March, 2025).

[9]   Mathieu, D., Nony, J., and Phan-Than-Lu, R. (2015). NEMRODW. Marseille, France: LPRAI.

[10] Drava, G., Leardi, R., Portesani, A., and Sales, E. (1996). Application of chemometrics to the production of friction materials: Analysis of previous data and search of new formulations. *Chemometrics and Intelligent Laboratory Systems*.

[11] Cornell, J. A. (1988). Analyzing data from mixture experiments containing process variables: A split-plot approach. *Journal of Quality Technology*, 20(1), 2–23. doi: 10.1080/00224065.1988.11979079.

[12] Næs, T., Faergestad, E. M., and Cornell, J. (1998). A comparison of methods for analyzing data from a three component mixture experiment in the presence of variation created by two process variables. *Chemometrics and Intelligent Laboratory Systems*, 41(2), 221–235.

[13] Næs, T., Bjerke, F., and Faergestad, E. M. (1998). A comparison of design and analysis techniques for mixtures. Available at: www.elsevier.com/locate/foodqual.

[14] Cornell, J. A. (1990). *Experiments with Mixtures*. New York: John Wiley and Sons.

[15] Coduri, M., Magnaghi, L. R., Fracchia, M., Biesuz, R., and Anselmi-Tamburini, U. (2024). Assessing phase stability in high-entropy materials by design of experiments: The case of the (Mg, Ni, Co, Cu, Zn)O system. *Chemistry of Materials*. doi: 10.1021/acs.chemmater.3c02120.

Chapter 5

Data Pre-Processing: The Importance of Properly Transforming Your Data Before Further Insights

Ricard Boqué

> ## Objectives and Scope
>
> This chapter describes the most common preprocessing techniques for multivariate data, including scaling, normalisation, baseline and/or scattering correction, smoothing, and derivatives. The chapter also provides tips and clues to identify typical deviations and artefacts in the data and to select the optimal preprocessing technique(s) for the problem(s) at hand. Finally, some guided examples using the CAT software are included to illustrate the ideas presented hereinafter.

5.1. Introduction

If we look at any dictionary, we will find that preprocessing is defined as any type of processing performed on raw data to prepare it for another data processing procedure. In chemometrics, data preprocessing refers to a series of techniques used to transform and improve raw data before applying multivariate statistical analysis. The main goal of data preprocessing is to reduce the effects of sources of variation that are irrelevant or unwanted for the analysis and to highlight the meaningful variation related to the analytical problem at hand. Data preprocessing in chemometrics typically involves a series of steps, including data cleaning, scaling, normalisation, and/or noise and baseline correction. Each step has its own purpose and can be tailored to address the specific needs of the data at hand and the analysis to be carried out.

If not applied properly, some preprocessing techniques can hide or even remove useful information; however, if correctly performed, data preprocessing can lead to more accurate and reliable statistical models as well as better insights into the underlying data structure. In the following, a simple and guided example is shown to illustrate this idea.

The well-known "Wines" dataset, consisting of 178 samples of three types of Italian wines (Barbera, Grignolino, and Barolo) described by 13 physicochemical attributes, will be used [1]. Each row corresponds to a sample, while the columns correspond to the variables used to describe the samples.

To load the dataset, please refer to the instructions given in Chapter 1 (on how to import Excel files from the "Data Handling" menu).

Then, choose the folder where the file to be imported is located is, in this case, "working". The screen depicted in Figure 5.A1 (Appendix) will appear. The name of the matrix will appear in the first row; if left unchanged, we will have to use that name to call the matrix throughout the work session.

It is possible to set which page of the Excel-type worksheet should contain the imported data, and in our case, we will leave it at page 1. The following two items ask us if we want to omit importing specific rows or columns, but in our case, that is not necessary. The next two items ask us if the names of the columns and rows are present. In this case, the first ones are present, but not the second ones. Finally, accept the orders by clicking "OK".

By writing the name of the matrix (i.e., "wines") in the command line of the R console, you can view it. The characteristics of the matrix can also be viewed through the sequential menus: Data Handling → Workspace management → Tell me. The matrix has 178 rows and 15 columns. The first column contains the names of the samples, while the second lists the type of wine.

To exemplify the effects of the different preprocessing options, we consider an unsupervised exploratory multivariate technique called principal component analysis (PCA). This technique will be discussed in detail in Chapter 7 (and briefly in Chapter 6); therefore, no further details will be given on it here. For now, just consider that PCA yields certain factors (called principal components, PC) that can be plotted to show the distribution of the samples and the variables. Further, consider it as merely an instrument to study how results can be affected by data processing. Additional comments will be provided as necessary, following the explanations. Also, please note that we will not discuss how to optimise the model or how to interpret it.

Building a PCA model on this dataset using CAT is straightforward. For that, go to the PCA menu, click on "Model computation", and then select the PCA option. A dialogue box opens again (Appendix, Figure 5.A2).

The first row contains the name of the data matrix. In the second row, you can choose which rows (samples) you want to include in the model. In our case, we consider all of them. The third row asks which variables you want to take into consideration in the model, and in our case the variables to be used span from column 3 to column 15 since the first two columns contain the names of the samples and the type of wine, respectively. The fourth line asks how many PCs you want to calculate; the default is 5.

Finally, there are two boxes to check. If you select the one on the left, the "Centred" item, the data will be column-centred with respect to the mean value (this is called *mean centring* and will be discussed later). Press "OK". Now, look at the model results provided in the command window. The percentages of explained variance

```
[1] Note: Data are saved in the PCA object, write PCA to see all
[1] Variance explained by 5 components: 100%
[1] % Variance explained by each component:
[1] 99.81   0.17   0.01   0.01   0.00
> |
```

Figure 5.1. PCA output of the mean-centred wines dataset.

(also called *information*) by each of the five PCs are reported. As will be explained in Chapters 6 and 7, the first PC explains the maximum variability, the second PC explains the second-highest variability not explained by PC_1, and so forth. You can observe in Figure 5.1 that PC_1 explains 99.81% of the variance; therefore, it accounts for almost all the variance (information) in the data. In such cases, we must suspect that something is wrong.

Aside from the numerical outputs, what makes PCA widely used in many scientific contexts is its graphical interpretation of the results. So, we interpret the model graphically. A way to do this is by selecting the "biplot" option; to obtain it from the PCA menu, go to "Plots" and select "Biplot". The plot shown in Figure 5.2 will appear.

The biplot is a joint plot of the scores and loadings of a PCA model. In a snapshot, the scores represent the projections of the samples onto the space of the PCs and allow us to identify groups of samples or trends and outliers. The loadings show the correlations between the variables used to describe the samples, as well as those of the PCs, and they indicate which variables are best able to distinguish the samples from one another. For more information, see Chapters 6 and 7. In short, the biplot allows observing and interpreting the relationships between samples and variables.

By looking at Figure 5.2, we can observe that the samples are basically scattered along PC_1, which explains most of the variance. But the more curious thing is that two variables, "Proline" and "Mg", appear at the extremes of PC_1 and PC_2, respectively. It may seem that these variables have a big influence on the model; however, remember that PC_2 explains only 0.2% of the total variance. So,

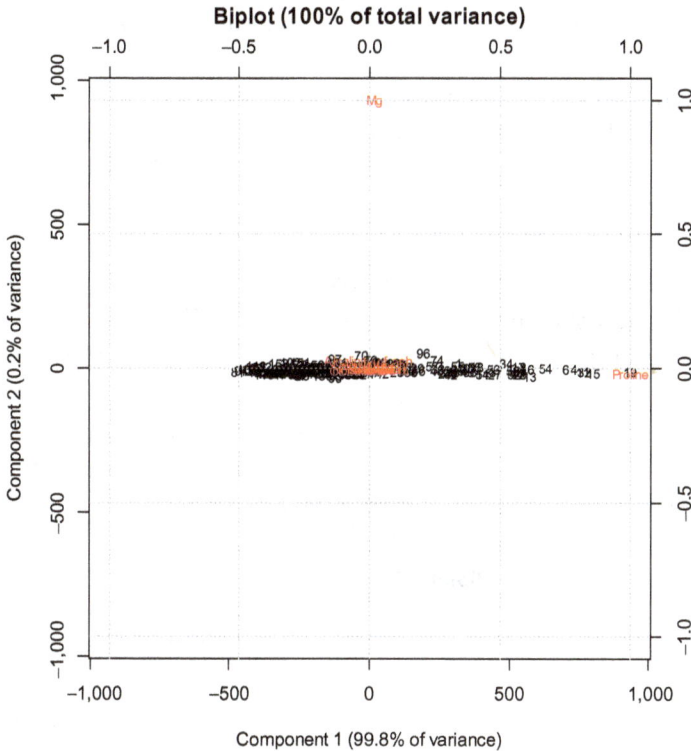

Figure 5.2. Biplot of the PCA model of the "Wines" dataset (mean-centred data).

does this mean that "Proline" is the most influential variable characterising the Italian wines? Curiously, the answer is no.

To see what is happening, go to the R console, write the name of the matrix in the command line, and take a look at the values of the 13 variables. The one with the largest values is "Proline" and the next is "Mg", which are coincidentally, the variables with the most weight in the PCA model we have built. This does not make sense, as a difference of 0.2 units in the content of "Malic acid" might be more important than a difference of 100 units in the content of "Proline" to differentiate the wine samples. This means that, clearly, we have a problem with the scaling of the data, and we need to correct for it.

To do so, return to the PCA menu, click on "Model computation", next in PCA, but now tick the "Scaled" box. Press OK.

Observe that, in the command line, the variance explained by the new PCs has changed substantially with respect to the previous situation. PC_1 and PC_2 explain together 55.4% of the total variance (36.2% and 19.2%, respectively), which is a more logical result for this type of data. Now, represent again the biplot of PC_1 versus PC_2 (remember, PCA → Plots → Biplot), and set it up as shown in Figure 5.A3 (Appendix).

In this case, we label the samples to identify the type of wine in the biplot and thus interpret the results best. For that, we define an external vector, wines [,1], which is the first column of the matrix. We obtain the plot as shown in Figure 5.3.

By studying Figure 5.3, we can observe that the information is far more interpretable than before. The samples are clearly grouped by the type of wine, and variables (in red) can be related to the samples

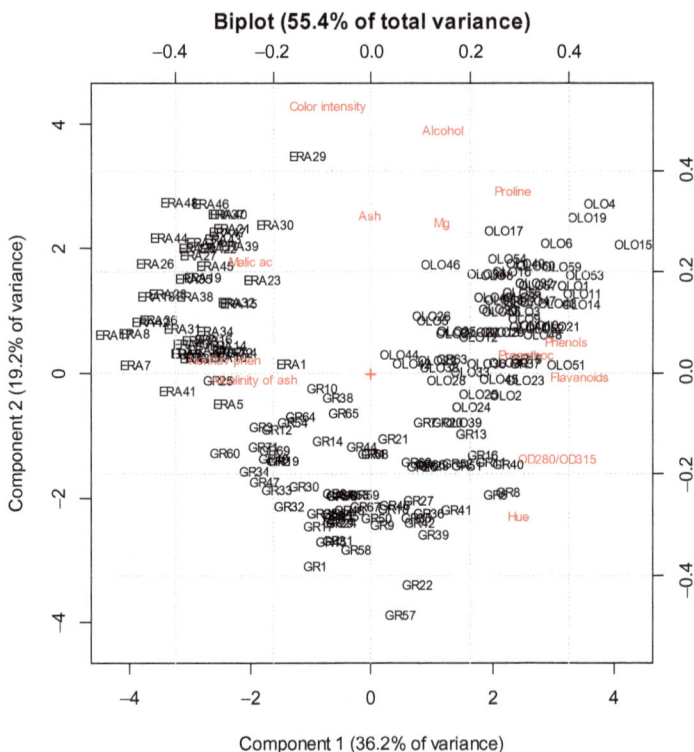

Figure 5.3. Biplot of the PCA model of the "Wines" dataset (mean-centred and scaled data).

in a more obvious way (for more details on the interpretation of scores and loadings plots, refer to Chapters 6 and 7).

5.2. Column Scaling

In the previous example, what we did using scaling was to divide the individual values of each column of the data matrix by the standard deviation of the values in that column. This preprocessing is called *standardisation*, and so the new (standardised) values are calculated as

$$x_{ij}^{s} = \frac{x_{ij}}{s_{j}},$$

where:

- x_{ij} is the original element of the matrix in the ith row and jth column,
- s_{j} is the standard deviation of column j,
- x_{ij}^{s} is the new transformed (standardised) data.

All columns (variables) in the new standardised matrix have unit variance; therefore, they possess, *a priori*, an equal mathematical weight. Standardisation is always performed when variables are expressed in different units of measure or have very different variance values.

5.2.1. *Column centring*

The goal of mean centring is to remove the mean value of each variable, or column, from the dataset. This helps to reduce the impact of the differences in the scales of the variables on the analysis. The new (centred) values are calculated as

$$x_{ij}^{c} = x_{ij} - \overline{x}_{j},$$

where:

- x_{ij} is the original element of the matrix in the ith row and jth column,
- $\overline{x}_{j}$ is the mean of column j,
- x_{ij}^{c} is the new transformed (centred) data.

All columns (variables) in the new centred matrix have a zero mean, which can help reduce the impact of the differences in the magnitude of the variables between samples on the analysis. To show this with a simple example, it is less important that three samples have original values of 100, 200, and 300 for a given variable, but rather -100, 0, and 100 for the centred values. That is, after mean centring, each row of the mean-centred data includes only how that row differs from the average sample in the original data matrix (or, in other words, how different this sample is from the "average" values in our dataset). It is usually said that mean-centring translates the axes of the original coordinate system to the centroid of the data. This is illustrated in Figure 5.4.

Without mean centring, the first PC for the data in Figure 5.4 sets at the origin or centre of the data. After mean centring, PC_1 focuses on the differences between the samples.

Mean centring is applied by default in many cases and is usually performed after applying other preprocessing techniques (if they are necessary). It is especially useful when using modelling methods based on latent variables, such as PCA or PLS, because the final

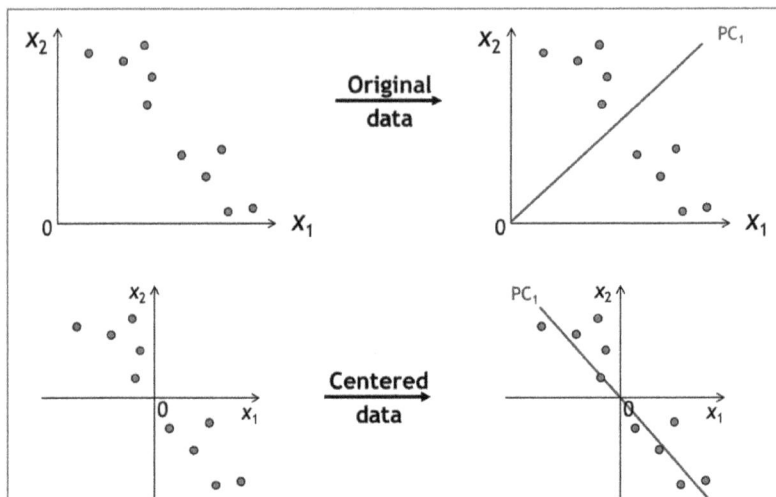

Figure 5.4. Calculation of the first PC for an original and a mean-centred dataset.

models become simpler than otherwise, that is, they require fewer latent variables. This is because the main source of variance, i.e., the differences around the mean, are removed. This can be graphically observed, for example, when performing a PCA on spectral data and inspecting the loadings of PC_1. If the spectra had not been mean-centred, the loadings of the first PC would take the shape of the average spectrum of the overall dataset.

5.2.2. *Autoscaling*

If you recall the initial example, data were mean centred and scaled (standardised). The simultaneous application of mean centring and standardisation is called *autoscaling*. The new (autoscaled) values are calculated as

$$x_{ij}^{a} = \frac{x_{ij} - \overline{x}_j}{s_j},$$

where:

- x_{ij} is the original element of the matrix in the ith row and jth column,
- $\overline{x}_j$ is the mean of column j,
- s_j is the standard deviation of column j,
- x_{ij}^{a} is the new transformed (autoscaled) data.

The goal of column autoscaling is to normalise the variables, or columns, of the dataset so that they have equal weight in the analysis and are not dominated by those with larger numerical values.

Column autoscaling is quite a common preprocessing technique in chemometrics, as it can improve the performance of statistical models and reduce the impact caused by differences in the scales of the variables on the analysis. However, it should be noted that column autoscaling does not necessarily improve the performance of all statistical models and may not be appropriate for all datasets. This is especially true for data containing noisy signals with low intensity, as autoscaling may artificially increase their importance and degrade model performance. For example, autoscaling may increase the mathematical relevance of spectral noise when very weak signals are measured.

Pareto scaling [2] is very similar to autoscaling. However, instead of the standard deviation, the square root of the standard deviation is used as the scaling factor.

5.2.3.   *Range scaling*

The basic idea of range scaling is to linearly transform each variable so that they have the same range of values, typically from 0 to 1. The new (range scaled) values are calculated as

$$x_{ij}^{\mathrm{rs}} = \frac{x_{ij} - \overline{x}_j}{x_{j,\mathrm{max}} - x_{j,\mathrm{min}}},$$

where:

- x_{ij} is the original element of the matrix in the ith row and jth column,
- $\overline{x}_j$ is the mean of column j,
- $x_{j,\mathrm{max}}$ and $x_{j,\mathrm{min}}$ are the maximum and minimum values of column j, respectively,
- x_{ij}^{rs} is the new transformed (range-scaled) data.

The goal of range scaling is to scale the variables of a dataset so that they have the same range or spread of values. This helps reduce the impact of differences in the scales on the analysis.

Range scaling is a common preprocessing technique in chemometrics, as it can improve the performance of statistical models and reduce the impact of differences in the scales on the analysis. However, as with autoscaling, care must be taken with data containing low-intensity and noisy signals, as noise can be artificially magnified and thus degrade the performance of subsequent models. Range scaling is also very sensitive to outliers.

5.3.   Row Scaling

In the previous section, we saw that the purpose of column scaling is to reduce the impact of differences in the scales of the variables on the analysis. In this section, we study row scaling, which is mainly used in chemometrics to remove systematic trends or baseline variations from spectral data and to improve the signal-to-noise ratio. These (undesired) trends can arise from a variety of sources, including variations in instrument performance, light scattering effects, sample preparation, or environmental conditions.

5.3.1. *Row autoscaling*

Light scattering effects, temperature changes, and other experimental or instrumental variations during signal measurement (or along a measurement session) can cause changes in the spectral (or chromato-graphic, electrical signal, etc.) baseline, which may induce unwanted errors in subsequent exploratory, classification, or regression models. These baseline shift trends are generally linear, but they can also be curvilinear. For example, a characteristic of NIR diffuse reflectance spectra is the increasing level of reflectance values over the 1100–2500 nm range. In general, most solids show absorbance spectra which are usually more intense at longer wavelengths since the probability of combination bands increases as the wavelength region approaches the fundamental vibrations.

To remove this kind of unwanted variation, several algorithms have been proposed. One of them is the standard normal variate (SNV) algorithm, which is probably the most common method used to reduce (physical) variability between samples due to scattering. The SNV procedure involves row-wise centring and scaling of the data in a way that eliminates the influence of these unwanted variations and makes the data more suitable for analysis. For a given sample i, the SNV-corrected spectrum is calculated as

$$x_{ij}^{\text{SNV}} = \frac{x_{ij} - \overline{x}_i}{s_i},$$

where:

- x_{ij} is the original element of the matrix in the ith row and jth column.
- $\overline{x}_i$ is the mean of the values of row (spectrum) i,

- s_i is the standard deviation of row (spectrum) i,
- x_{ij}^{SNV} is the new transformed (SNV) data.

SNV can be applied independently to each spectrum, and so it does not depend on the calibration set. In this sense, it is simpler to apply than, for instance, multiplicative scatter correction (MSC), a common alternative. However, it should be noted that SNV assumes that the variations in the data are linear and additive, in the sense that baseline variations can be effectively modelled as a linear function and that any deviation from the baseline (whether due to instrument noise, baseline drift, or other factors) is considered an additive component that can be represented by a constant value added to each data point. This may not always be the case; in addition, the method may not be effective at removing all sources of spectral variability. Other preprocessing techniques, such as baseline correction or normalisation, may be necessary for more complex datasets. Finally, since SNV does not involve a least-squares fitting in the estimation of its parameters, it can be sensitive to noisy values in the spectrum.

To check the effect of the SNV transformation, we will use the dataset "Moisture", which consists of 54 samples of soy wheat [3]. Column 1 contains the moisture content, and columns 2–176 (1104–2496 nm, step: 8 nm) contain the NIR spectra of the samples.

To load the dataset, go to the Data Handling menu in the CAT software. Import it as an Excel file from the "working" screen. The name of the matrix appears in the first row. The page of the Excel sheet should be 1. The following two items ask if we wish to omit the import of rows or columns; set them to 0. Then, check the "Header" box, and finally click OK. To plot the spectra, go to the Univariate menu → Plots → Row profile. Set the dialogue box as shown in Figure 5.A4 (Appendix).

After plotting the raw spectra (Figure 5.5), you can observe the increasing level of the absorbance values over the 1100–2500 nm range due to additive and multiplicative scatter effects. Then, to preprocess the spectra using SNV, go to the Transformations menu → Row → Autoscaling (SNV), and set up the dialogue box as shown in Figure 5.A5 (Appendix).

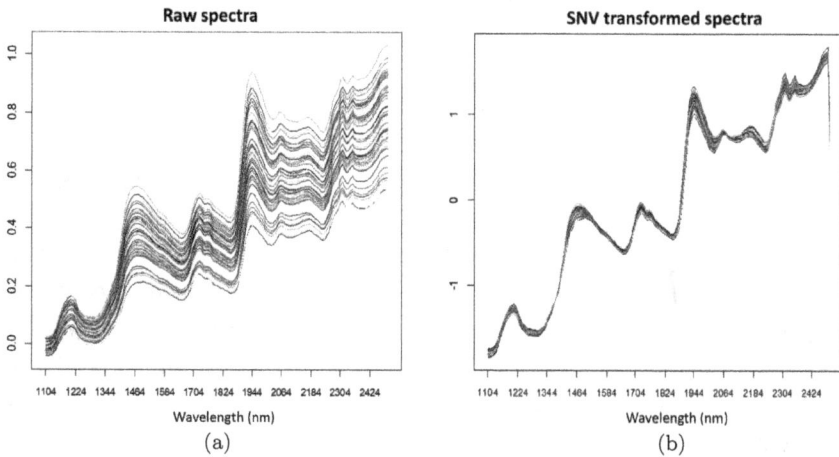

Figure 5.5. (a) Raw and (b) SNV-transformed spectra of the soy wheat samples (file "moisture").

The preprocessed spectra are saved in "moisture.snv". To plot them, follow the same steps as for plotting the raw spectra above (Figure 5.A4).

The preprocessed spectra will resemble those shown in Figure 5.5. You can observe that most of the constant and proportional offsets have been corrected for.

5.3.2. *Multiplicative scatter correction*

MSC is probably the most widely used preprocessing technique for NIR spectra, along with SNV and its derivatives. MSC was first introduced by Martens *et al.* [4]. The idea behind MSC is that nonlinearities in the data, caused by scatter from particulates in the samples, can be removed from the data matrix prior to data modelling. MSC is performed by regressing a measured spectrum for an individual sample against a reference spectrum and then correcting the measured spectrum using the slope (and possibly intercept) of this fit. MSC comprises two steps:

(a) estimating the correction coefficients (additive and multiplicative contributions):

$$x_{\text{raw}} = b_0 + b_{\text{ref},1} x_{\text{ref}} + e;$$

(b) correcting the measured spectrum:

$$x_{corr} = \frac{x_{raw} - b_0}{b_{ref,1}},$$

where the b's are the sample-dependent correction coefficients, e is the unmodelled part (noise), and x_{raw}, x_{ref}, and x_{corr} are the original, reference, and corrected spectra, respectively. Although the b-coefficients can be estimated from a smaller spectral range containing no chemical information, in practice, they are often estimated using the entire spectrum. This way, no subjective evaluation of the spectrum is required, and different users will obtain the same results using MSC. Another important aspect to consider when applying MSC is the definition of the reference spectrum. This can either be an *a priori*-defined reference spectrum or simply the average spectra of the calibration dataset. The latter option is the preferred, as it may be difficult to select an appropriate spectrum as the reference one.

MSC is particularly useful for dealing with spectral data that have been collected under different conditions or with different instruments, where differences in the scattering properties of the sample can affect the intensity of the spectra and make it difficult to compare different samples. MSC has the advantage of being relatively simple to apply and can be effective in removing the effects of scatter from spectral data. However, it should be noted that the method assumes the scatter properties of the sample are constant across all spectral bands, which may not always be the case. Finally, if the average spectrum of the calibration dataset is taken as the reference spectrum, it means this spectrum will have to be recalculated if the dataset changes, that is, if samples are added or removed from the original dataset.

5.3.3. *Detrending*

Detrending is a common preprocessing technique used in chemo-metrics to remove systematic trends or drifts from spectral data, especially in NIR and MIR spectroscopy. These trends can arise from a variety of sources, including variations in instrument performance, sample preparation, or environmental conditions (e.g., temperature variation).

There are several methods used for detrending in chemometrics, including polynomial fitting, moving average subtraction, and wavelet decomposition, with the first being the most common. It involves fitting a polynomial function of a given order to the spectral data and then subtracting it from the original data. This way, the visibility of analyte-specific information is enhanced, and a more accurate representation of the underlying spectral features is obtained. The order of the polynomial used for fitting can vary depending on the complexity of the baseline trend and the degree of noise in the data.

The detrending process can be described mathematically as follows:

1. Let $\mathbf{X}$ be an $I \times J$ matrix that represents the original spectral dataset, where I is the number of samples and J is the number of variables (i.e., spectral bands).
2. Fit a polynomial function of order k to each spectrum in the dataset. The function can be written as

$$y(\mathbf{x}) = c_0 + c_1\mathbf{x} + c_2\mathbf{x}^2 + \cdots + c_k\mathbf{x}^k,$$

where $\mathbf{x}$ is the raw spectrum vector and c_k's are the coefficients of the polynomial.
3. Subtract the polynomial function from each ith raw spectrum in the dataset to obtain the detrended spectra:

$$\mathbf{x}_i^{\text{dt}} = \mathbf{x}_i - y(\mathbf{x}_i).$$

The resulting matrix $\mathbf{X}^{\text{dt}}$ is the detrended version of the original spectral dataset, where each spectrum has been corrected for systematic trends or baseline variations.

Note that this algorithm does not fit only baseline points in the spectrum; instead, it fits the polynomial to all points, both baseline and signal. For this reason, the method is quite simple to apply; however, it tends to work only when the largest source of signal variation in each sample is background interference. In measurements

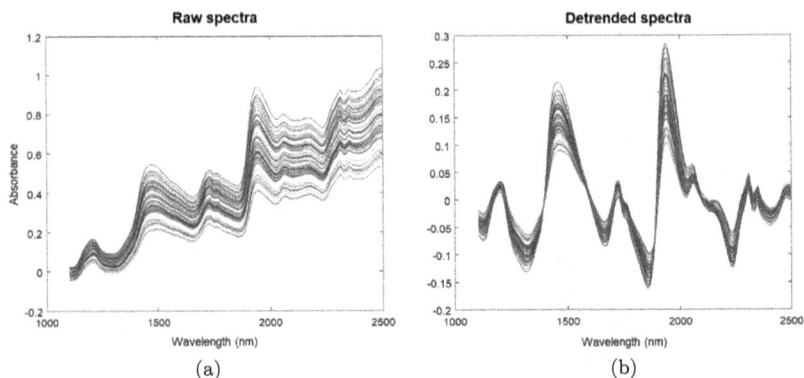

Figure 5.6. (a) Raw and (b) detrended spectra (polynomial of order 1) of the soy wheat samples (file "moisture").

where the variation of interest (i.e., the contribution of the analytes) is a reasonably significant portion of the variance, the detrending algorithm tends to remove variations which are useful for further modelling steps. In addition, the fact that an individual polynomial is fitted to each spectrum may increase the amount of interfering variance in a given dataset. Due to these reasons, the use of detrending is recommended only when the overall signal is dominated by backgrounds with similar shapes and is not "influenced" too much by chemical or physical variations of interest.

The detrending method has the advantage of being able to remove complex baseline variations from spectral data, and it can be effective in improving the accuracy of subsequent data analysis methods. However, care should be taken when choosing the order of the polynomial used for fitting, as overfitting can result in a serious loss of information and reduced accuracy. This procedure is not yet available in CAT, but it can be found in other software. Figure 5.6 shows the detrended spectra of the soy wheat samples (file "moisture"), where it is evident that the original trend in the raw spectra has been removed.

5.3.4. *Smoothing*

The noise of a continuous vector, such as a spectrum (amperogram, chromatogram, etc.), can be defined as the random variability that is not related to the properties of the sample. Usually, this variability is mainly due to different kinds of noise (including random fluctuation) of the measurement instrument. As it does not contain information of interest, its removal from the signal (e.g., a spectrum) will improve the signal-to-noise ratio of the data and make the chemometric algorithms work better. In chemometrics, the techniques used to remove or reduce noise are generally referred to as *denoising* or *smoothing* techniques.

Currently, smoothing works by averaging (or fitting a mathematical function) to the data points contained within a window of a certain size (i.e., a short series of sequential numerical values, e.g., from 450 to 440 nm, with one datum per nm; size = 11). There are different smoothing techniques that can be used in chemometrics, including moving average, Gaussian, and Savitzky–Golay (SG) smoothing. The choice of the smoothing technique will depend on the specific characteristics of the data and the goals of the analysis.

Moving average smoothing calculates the average of a set of data points over a specific window. The process involves selecting a window size, often represented by the number of data points to be included in the calculation. As new data points are added to the series, the unsmoothed data point within the window is dropped, and the average is recalculated. This results in a smoothed curve that follows the general direction of the original data.

Gaussian smoothing is a simple yet powerful method for smoothing spectral data. In this method, a Gaussian curve is generated first based on a specified standard deviation. Then, this curve is slid across the spectral data; at each point, it multiplies the spectral values by the corresponding weights from the Gaussian curve and sums them up. This process effectively averages the nearby spectral values according to the Gaussian distribution. High-frequency noise tends to be attenuated as neighbouring spectral values are averaged. This results in a smoother spectral curve, reducing noise while preserving underlying trends and important peaks.

In this chapter, we focus on SG smoothing since it is one of the most widely used preprocessing methods in chemometrics and has been already implemented in the CAT software. The algorithm, developed by Savitzky and Golay [5], can include a derivation step, which we will discuss in the following section. To find the smoothed value at a central point j in a window of size z, a polynomial of a certain degree is fitted to a symmetric window of the raw spectrum. Once the parameters of this polynomial have been calculated, the value of the central point (j) can then be estimated using the calculated equation. The original value is replaced by the new one, and the process continues. The procedure is illustrated in Figure 5.7.

This operation is applied sequentially to all points in the spectrum by moving the window. The number of points used to calculate the polynomial (window size) and the degree of the fitted polynomial are both decisions that need to be made by the analyst. The correct choice of the window size is crucial. A too-small window can introduce large artefacts in the corrected spectra and reduce the signal-to-noise ratio. However, if the size of the window is too large, small peaks that contain relevant information might get smoothed out. The optimisation of the window size by the analyst is usually based on experience and trial and error, as the optimal number of points to consider depends on the type of spectroscopy, the spectral resolution, or the sample characteristics (the shape of the spectrum). An example of the effect of the window size on the final spectra is shown in Figure 5.7.

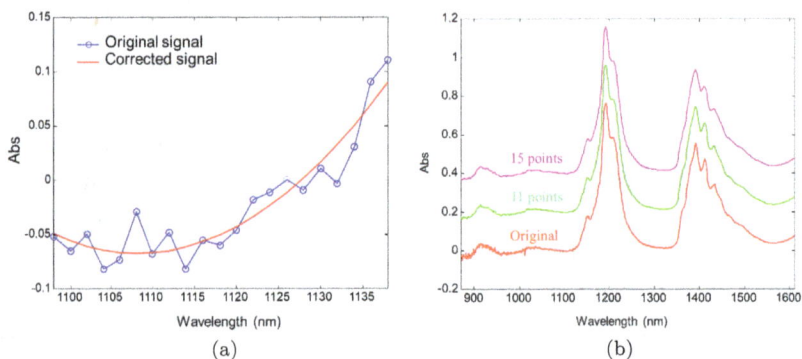

Figure 5.7. Savitzky–Golay smoothing: (a) original and SG-smoothed signals (window size: 11 points; polynomial degree: 2); (b) example of the effect of the window size on the NIR spectra of a gasoline sample.

Proposed Exercise

Perform SG smoothing on the NIR spectra of the "Moisture" dataset using the CAT software. You already know how to load the file. After that, go to Transformations menu → Row → Savitzky–Golay. Then, set up the parameters in the dialogue box (set "Derivative" to 0). The SG-preprocessed spectra will be saved in "moisture.sg". To plot them, go to Univariate menu → Plots → Row profile.

Try different window sizes and polynomial orders, and compare the plots obtained. Extract your own conclusions.

Note: if you take a look at the dimensions of the SG-transformed dataset, "moisture.sg", you will realise that the number of variables is lower than in the original dataset, "moisture". Try to guess why this happens. (*Hint*: to which point can the window be moved?)

5.3.5. *Derivatives*

Derivatives have the capability of removing both additive and multiplicative effects from spectra and have been used in analytical spectroscopy for many years. The first derivative removes only a constant baseline (also called offset); the second derivative removes both baseline offsets and linear trends. The most basic method for derivation is the differential or finite differences: the first derivative is estimated as the difference between two subsequent measurement points; the second-order derivative is then estimated by calculating the difference between two successive points of the first-order derivative spectrum:

$$x'_j = x_j - x_{j-1},$$

$$x''_j = x'_j - x'_{j-1} = x_{j-1} - 2 \cdot x_j + x_{j+1},$$

where x'_j denotes the first derivative and x''_j denotes the second derivative at point (wavelength) j.

The first derivative involves calculating the rate of change of the spectral intensity at each data point, which provides information about the shape of the spectral peaks. This can be particularly useful in resolving overlapping peaks in spectroscopic data, as the derivative can reveal subtle differences in peak shape and position that may be obscured in the raw data. However, it is worth noting that the first derivative can also amplify noise present in the data; hence, it is generally recommended to apply smoothing to the derivative data to reduce noise before further analysis.

Similar to the first derivative, the second derivative involves calculating the rate of change of the first derivative at each data point. This provides additional information about the shape of spectral peaks, such as the location of peak maxima and the inflection points of the curve. The second derivative can be particularly useful for resolving closely spaced peaks in spectroscopic data, as it can provide even greater resolution than the first derivative. However, it is also more sensitive to noise in the data than the first derivative; therefore, it is important to apply a preliminary smoothing step to reduce noise before proceeding with further analysis.

The most common technique and one that is implemented in most chemometric software to simultaneously apply smoothing and derivatives is the SG derivative method. This methodology follows the same process as SG smoothing, explained in the previous section, with an additional step. After fitting the polynomial to a window around a moving point, the derivative of any order of this function can easily be found analytically, and this value is subsequently used as the derivative estimate for this central point. The highest derivative that can be determined depends on the degree of the polynomial used during the fitting (i.e., a third-order polynomial can be used to estimate up to the third-order derivative).

To illustrate the effect of derivatives on the signal, we again use the "Moisture" dataset. Load the dataset (see Section 5.3.1), go to the Transformations menu → Row → Savitzky–Golay and set up the dialogue box as shown in Figure 5.A6 (Appendix).

Observe that the window size and polynomial degree are 5 and 2, respectively. This can be considered a moderate smoothing; however, as mentioned above, it will depend on the type of spectra, the number of data points, and the spectral resolution. The SG-preprocessed spectra have been saved in "moisture.sg". To plot them, go to Univariate menu → Plots → Row profile, and then set up the dialogue box as in Figure 5.A7 (Appendix).

A plot of the first derivative of the spectra will be displayed. Now, you have to repeat the procedure, but with the window size increased to 11. Do the same with the second derivative and try other polynomial orders. Figure 5.8 shows the first and second derivatives of the spectra of the "Moisture" dataset for different window sizes and a polynomial order of 2.

The "Moisture" dataset is not especially noisy, as you can observe by looking at Figure 5.5. However, a moderate smoothing may improve the performance of subsequent analysis models. On the contrary, a window size that is too large (in this case, 11) may remove useful information from the signal, as can be observed in Figure 5.8. In general, the appropriateness of a given preprocessing technique is assessed based on the performance parameters of the chemometric models (i.e., PCA or PLS) applied to the transformed data. We will return to this topic in the worked example at the end of the chapter.

Finally, a few words about retention time shifts, a source of variation typical in chromatographic analysis. Retention time shifts occur when the time at which a compound elutes from a chromatographic column varies between different runs or conditions. These shifts can result from changes in experimental parameters, column degradation, or variations in instrument conditions, impacting the accuracy and reproducibility of chromatographic analyses and making comparisons between chromatograms difficult.

Various techniques for aligning chromatographic data can be found in the literature. Among them, warping (or time warping) is one of the most popular [6]. In the context of chromatographic data alignment, time warping allows for the flexible adjustment of the time axis, accommodating variations in retention time across different chromatograms. Time warping allows for stretching or compressing of chromatographic profiles to align corresponding peaks and is particularly useful when dealing with nonlinear shifts in retention time.

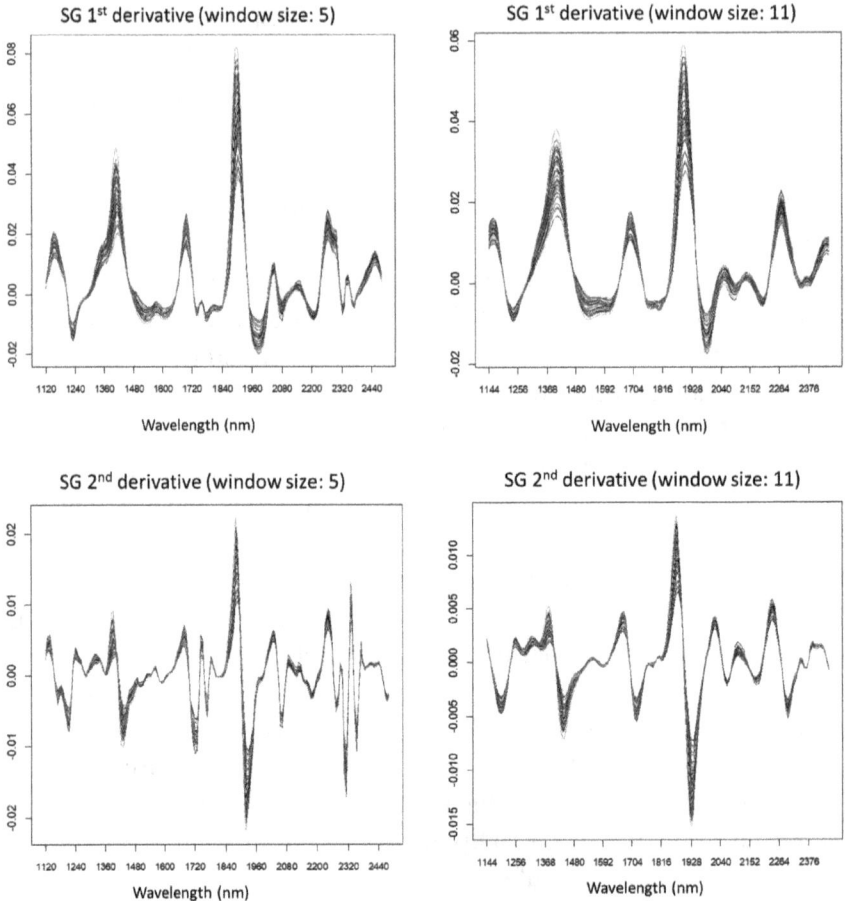

Figure 5.8. First and second derivatives of the spectra of the "Moisture" dataset for window sizes 5 and 11 and a polynomial order of 2.

5.4. The Impact of Signal Preprocessing

Data preprocessing is a crucial step in chemometrics and can significantly improve the accuracy and reliability of statistical models. By applying various preprocessing techniques such as smoothing, baseline correction, normalisation, or scaling, chemometricians can transform complex and almost useless raw data into a usable database and extract more accurate and meaningful information from the data, which can lead to more reliable and accurate predictions or classifications. In addition, preprocessing methods can help reduce the impact of external factors on the data, such as temperature or humidity, which can improve the reproducibility of the analysis.

The choice of the preprocessing technique depends on the characteristics of the data and the goals of the analysis; a careful consideration of each step is required for effective and meaningful analysis. Different preprocessing methods can have different effects on the data; it is crucial to choose an appropriate method that is suitable for the specific application. It is worth remembering that "a golden rule" does not exist.

Data preprocessing can play an important role in mitigating the risk of overfitting in chemometric analysis. Overfitting occurs when a model is too complex and is able to fit the training data too well, resulting in poor generalisation to new, unseen data. One of the main causes of overfitting is the presence of noise or irrelevant features in the data, which can cause the model to overemphasise these factors and lead to predictions that are overly complex and specific to the training data.

In addition, it is important to evaluate the impact of preprocessing on the results of the analysis. Preprocessing methods can introduce bias or distortions into the data, and it is important to ensure that the results of the analysis are robust and reproducible. It is therefore important to carefully evaluate the impact of preprocessing on the final models.

Furthermore, scaling and its artefacts can introduce challenges in the interpretation of chemometric models [7]. When scaling techniques are applied, such as normalisation or standardisation, they can impact the interpretability of the model in the following ways:

Loss of variable meaning: Scaling often transforms the original variables, making it challenging to interpret the impact of individual variables on the model's predictions. The scaled values might not have the same intuitive meaning as the original ones, leading to a potential loss of interpretability.

Non-linear transformations: Certain scaling methods, especially nonlinear transformations, can introduce complexities that make it difficult to understand the linear relationships between variables X and Y in prediction models.

Artefacts in variable importance: Scaling can influence the metrics used to assess the importance of variables. In some cases, features with large numeric values may be assigned higher importance solely due to their scale, even if they are not inherently more significant within the context of the problem.

Impact on visualisations: Interpretable visualisations, such as variable importance plots or X–Y relation plots, may be affected by scaling. Patterns in visualisations might be distorted, making it challenging to draw accurate conclusions about the relationships between features and predictions.

Addressing these challenges often involves a careful consideration of the scaling method used, documentation of the scaling process, and additional steps to ensure that interpretability is not compromised. It is essential to set a balance between the benefits of scaling for model performance and the need for transparent and interpretable models.

Unless a well-established rationale exists for choosing the appropriate preprocessing steps, trial-and-error approaches are often common practice for deciding which method should be applied to remove or reduce the influence of unwanted variation. The analyst must obtain some prior knowledge about the data or the measured signal in order to evaluate the preprocessed signal or the model performance.

Commercially available software such as PLS_Toolbox (Eigenvector Technologies, USA) can apply many pre-processing algorithms and combine them as selected by the analyst in an iterative way. Then, multivariate models can be applied to each of the new datasets thus generated. The most common way of judging whether a preprocessing method is beneficial for the analytical performance is to compute the prediction error for an independent test set, the root-mean-square error (RMSE), and afterwards, select the preprocessing method that gives the lowest RMSE. However, as datasets grow in size and the preprocessing and analysis algorithms become more complex, the computation time for iterating through all available options becomes significantly longer and impractical.

Methods that optimise the choice of the best preprocessing option have been proposed using various strategies, such as design of experiments [8], orthogonalisation (SPORT) [9], and spectral signal-to-error ratio [10]. However, they exceed the introductory level of this chapter, and readers are kindly referred to the references cited above.

5.5. Preprocessing "Fused Data"

Data fusion is a technique used to combine data from different sources to extract more accurate and comprehensive information [11]. Fusion of data from complementary techniques can provide more accurate knowledge about a sample and yield better inferences (classifications with less error rate and predictions with less uncertainty) than a single technique. The concept of data fusion is not new: humans combine multiple senses to achieve more accurate inferences about, for example, food suitability, thus improving their chances of survival. In fact, chemometricians combined data for a long time, for instance, single chemical parameters determined by classical or instrumental analysis to form a single matrix with the aim of improving food authentication results. Another typical example is the collection of the results of the very many analytical parameters that can be measured in a set of environmental samples to assess the natural quality of a given site. The challenge today is how to meaningfully combine not just single variables, as was done in the past, but blocks of them (e.g., near-infrared and mass spectra). Nowadays, analytical laboratories commonly have first-order (NIR, Raman, and HS-MS) and second-order (GC-MS, LC-DAD, and EEM fluorescence) instruments that can successively be used to analyse the same sample. The multivariate statistical analysis of fused data from these techniques can be a powerful tool for obtaining more reliable results. This requires developing new ideas for preprocessing data blocks, selecting variables, and validating models. Last, but not least, data to be fused must provide complementary information to be useful. This means that the chemical knowledge about the samples and the problem at hand is fundamental to selecting the suitable analytical techniques.

In chemometrics, the possible ways of fusing the data are generally grouped into three general strategies: low-level data fusion, when the individual raw matrices (of the instruments) are used as inputs; mid-level data fusion, when chemometrically extracted features (i.e., scores of the PCA on the individual matrices) of the data are used; and high-level data fusion, when the data are combined at the classification/prediction decision level.

Preprocessing in data fusion refers to the steps taken to prepare and combine data from multiple sources in a way that enhances the accuracy and reliability of the analysis. Since multivariate analysis is scale dependent, data from single techniques are usually preprocessed in order to properly scale the data, remove uninformative systematic variations, and reduce noise. Preprocessing of fused data is an important step in chemometric analysis that can help ensure that the data are comparable, unbiased, and informative. By carefully preprocessing the data, it is possible to improve the quality of the analysis and extract more accurate and meaningful information from the data.

Data from each source are treated specifically depending on their specific characteristics. For example, SNV, MSC, or derivatives are typically used to eliminate baseline shifts in infrared spectra; baseline corrections and derivatives are used with UV-vis spectra; and scaling/normalisation are usual for mass spectra. In addition, data fusion may require additional preprocessing aimed at compensating for the different measurement scales and variability of each technique to prevent any single block from becoming dominant. In this sense, each data block is weighted separately (block-scaling), usually using auto scaling, root-square scaling, and log scaling. Finally, after the data are merged, they are usually mean centred before building the model.

5.6. Worked Example

As indicated above, it is fundamental to carefully evaluate the impact of preprocessing on the final models. For that, we develop partial least-squares (PLS) regression models using the CAT software and the "Moisture" dataset and check the impact of data preprocessing on the final PLS prediction errors. PLS is treated in detail in Chapter 8; here, we apply it "blindly" without any comments or optimisations. Probably, you will not fully understand all the PLS terms, but do not worry; for now, our focus is on visualising the relevance of using different preprocessing techniques. You can (and should) return here after reading Chapter 8.

Load the "Moisture" dataset and go to Calibration menu → Model computation → PLS1 → single CV. As at this point we just want to compare the effects of preprocessing on the final models, it is not so critical to perform an extensive validation, so the option "single CV" (cross-validation) is suitable. Then, set up the dialogue menu as shown in Figure 5.A8 (Appendix).

The X variables, numbered from 2 to 176, are the NIR spectra of the samples, and the Y variable is the physicochemical variable to be predicted (moisture content in this case). The number of components (or factors) to be calculated is set at 10, although this value may be increased. This is not the optimal number of factors in the PLS regression model; it will be selected later, while the PLS algorithm will calculate the maximum number of factors. Finally, do not scale the data, and note that centring is performed by default, so in this case, we obtain the results of a mean-centred PLS model. The plots shown in Figure 5.9 should be obtained. For a full understanding of the terms, refer to Chapter 8.

A relevant plot for evaluating a PLS model is one that indicates how the average error (RMSECV) varies as a function of the number of factors included in the model. (See Chapter 8 to understand how the values for this plot are obtained.) Although a minimum can be observed for 10 components, a more parsimonious (non-overfitting) model with 5 components seems to be a better option. So, select 5 as the number of optimal components and look at the RGui console. The RMSECV value for this model is 0.8680, which, in relative terms to the average moisture values (calculated in the Excel file), represents an average of 7% error.

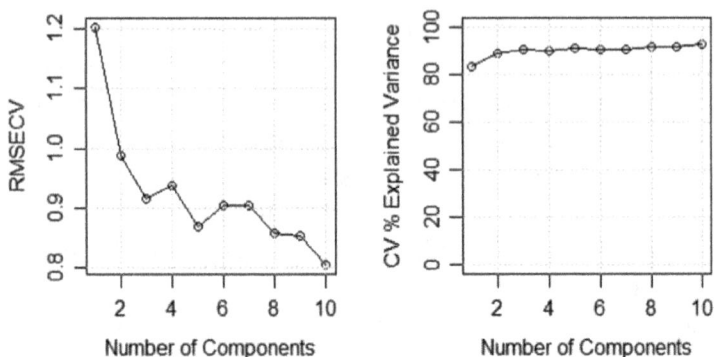

Figure 5.9. Root-mean-square error of cross-validation (RMSECV) and %CV explained variance for a mean-centred PLS model on the "Moisture" dataset. See details about validation in the text.

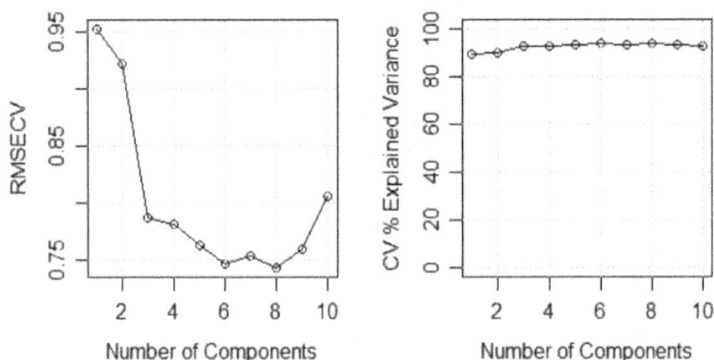

Figure 5.10. Root-mean-square error of cross-validation (RMSECV) and %CV explained variance for a SNV and mean-centred PLS model on the "Moisture" dataset. See details about validation in the text and in Chapter 8.

We now apply a simple preprocessing technique to the dataset: row autoscaling (SNV). So, go to the Transformations menu → Row → Autoscaling (SNV), and select First column: 2, and Last column: 176 of the file "Moisture". Then, go to Calibration menu → Model computation → PLS1 → single CV, and set the dialogue box as in Figure 5.A8 (Appendix).

Accordingly, in this case, we apply a PLS regression model with SNV and mean-centred data. The two plots in Figure 5.10 should be obtained.

The first local minimum occurs at 6 components; therefore, select this number as the optimal choice and observe the RMSECV values in the RGui console. The minimum RMSECV value is for 8 components; however, the difference is minimal, and a lower number of components is always preferred to avoid overfitting in the PLS model. The RMSECV value is 0.7466 (6% in relative terms), which is lower than that obtained for the PLS model with only mean centring. So, we have improved the overall prediction performance.

> **Proposed Exercise**
>
> Repeat the procedure above but with different preprocessing techniques, such as scaling, derivatives (with and without smoothing), and combinations of these methods (suggestion: apply SNV followed by a first derivative).
>
> Plot the different preprocessed data using the option Univariate menu → Plots → Row profile.
>
> Build a table that lists the various preprocessing techniques applied, along with the optimal number of factors and the corresponding RMSECV values obtained.
>
> Draw your own conclusions and determine the most suitable preprocessing option to achieve the best predictions.

Appendix

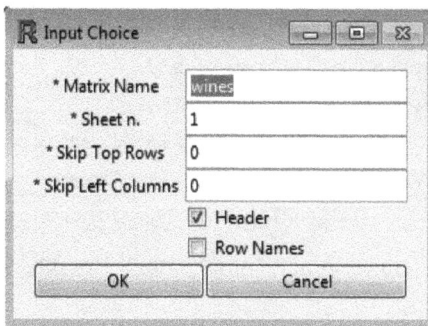

Figure 5.A1. Setup to import the "Wines" Excel-type file in CAT.

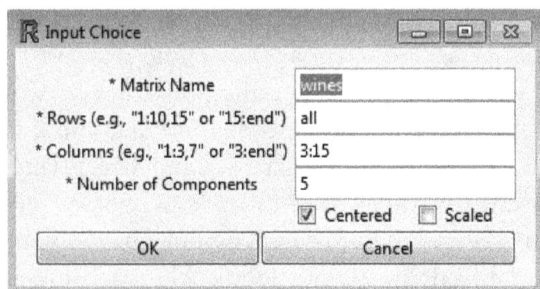

Figure 5.A2. Setup to start a PCA study.

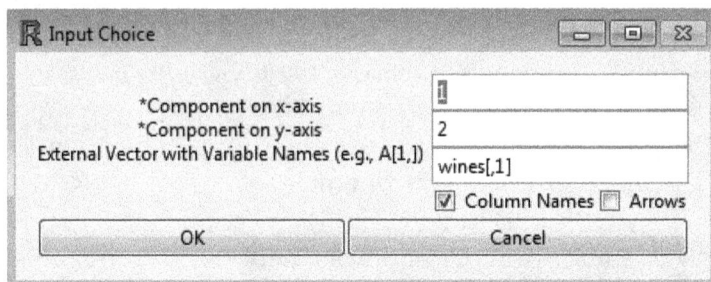

Figure 5.A3. Setup for the biplot using scaled data.

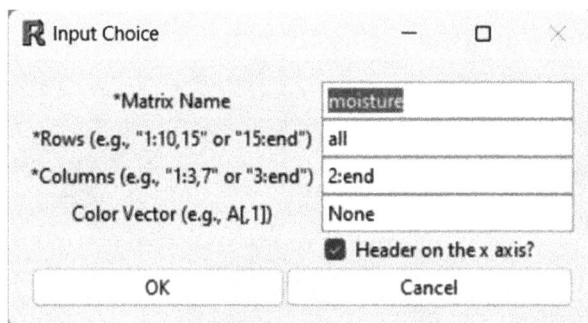

Figure 5.A4. Setup to plot the spectra of the "moisture" dataset.

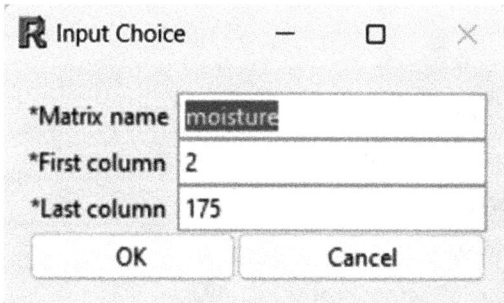

Figure 5.A5. Setup for SNV transformation.

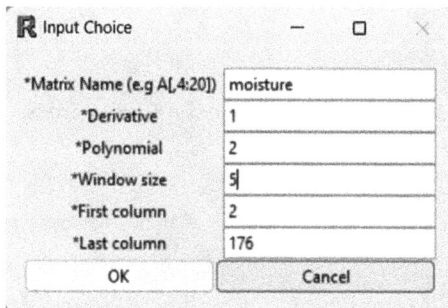

Figure 5.A6. Setup for a Savitzky–Golay first derivative.

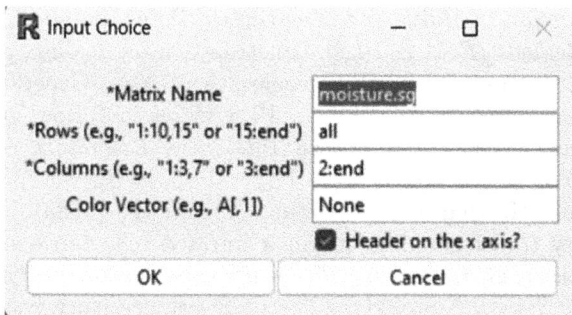

Figure 5.A7. Plot the first derivative of the moisture dataset.

Figure 5.A8.   Setup of a PLS analysis.

References

[1]   Forina, M., Armanino, C., Castino, M., and Ubigli, M. (1986). Multivariate data-analysis as a discriminating method of the origin of wines. *Vitis*, 25, 189–201.

[2]   Eriksson, L., Johansson, E., Kettaneh-Wold, N., and Wold, S. (1999). *Introduction to Multi- and Megavariate Data Analysis Using Projection Methods (PCA & PLS)*. Umetrics, Umeå; Scaling, pp. 213–225.

[3]   Leardi, R. and Lupiáñez González, A. (1998). Genetic algorithms applied to feature selection in PLS regression: How and when to use them. *Chemometrics and Intelligent Laboratory Systems*, 41, 195–207.

[4]   Martens, H., Jensen, S. A., and Geladi, P. (1983). Multivariate linearity transformations for near infrared reflectance spectroscopy. In Christie, O. H. J. (Ed.), *Proceedings of the Nordic Symposium on Applied Statistics*, Stokkland Forlag, Stavanger, Norway, pp. 205–234.

[5]   Savitzky, A. and Golay, M. J. E. (1964). Smoothing and differentiation of data by simplified least squares procedures. *Analytical Chemistry*, 36, 1627–1639.

[6]   Jellema, R. H. (2009). Variable shift and alignment. In Brown, S. D., Walczak, B., and Tauler, R. (Eds.), *Comprehensive Chemometrics*. Elsevier, Amsterdam, pp. 85–108.

[7] Oliveri, P., Malegori, C., Simonetti, R., and Casale, M. (2019). The impact of signal pre-processing on the final interpretation of analytical outcomes – A tutorial. *Analytica Chimica Acta*, 1058, 9–17.

[8] Gerretzen, J., Szymańska, E., Jansen, J. J., Bart, J., van Manen, H. J., van den Heuvel, E. R., and Buydens, L. M. C. (2015). Simple and effective way for data preprocessing selection based on design of experiments. *Analytical Chemistry*, 87, 12096–12103.

[9] Roger, J. M., Biancolillo, A., and Marini, F. (2020). Sequential preprocessing through ORThogonalization (SPORT) and its application to near infrared spectroscopy. *Chemometrics and Intelligent Laboratory Systems*, 199, 1103975.

[10] Skibsted, E. T. S., Boelens, H. F. M., Westerhuis, J. A., Witte, D. T., and Smilde, A. K. (2004). New indicator for optimal preprocessing and wavelength selection of near-infrared spectra. *Applied Spectroscopy*, 58, 264–271.

[11] Borràs, E., Ferré, J., Boqué, R., Mestres, M., Aceña, L., and Busto, O. (2015). Data fusion methodologies for food and beverage authentication and quality assessment – A review. *Analytica Chimica Acta*, 891, 1–14.

Further Reading

Engel, J., Gerretzen, J., Szymanska, E., Jansen, J. J., Downey, G., Blanchet, L., and Buydens, L. M. C. (2013). Breaking with trends in pre-processing? *TrAC Trends in Analytical Chemistry*, 50, 96–106.

Hibbert, D. B. (2016). Vocabulary of concepts and terms in chemometrics (IUPAC recommendations 2016). *Pure and Applied Chemistry*, 88, 407–443.

Oliveri, P., Malegori, C., Simonetti, R., and Casale, M. (2019). The impact of signal pre-processing on the final interpretation of analytical outcomes – A tutorial. *Analytica Chimica Acta*, 1058, 9–17.

Rinnan, Å., Nørgaard, L., van den Berg, F., Thygesen, J., Bro, R., and Engelsen, S. B. (2009). Data pre-processing. In Da-Wen, S. (Ed.), *Infrared Spectroscopy for Food Quality Analysis and Control*. Elsevier, Amsterdam.

Rinnan, A., van den Berg, F., and Engelsen, S. B. (2009). Review of the most common pre-processing techniques for near-infrared spectra. *TrAC Trends in Analytical Chemistry*, 28, 1201–1222.

Wise, B. M., Gallagher, N. B., Bro, R., Shaver, J. M., Windig, W., and Scott Koch, R. (2006). Chemometrics Tutorial for PLS_Toolbox and Solo. Eigenvector Research, Inc.

Chapter 6

Classification

Richard Brereton

Objectives and Scope

- To understand the difference between one- and two-class classifiers.
- To discuss the importance of preliminary exploratory analysis.
- To understand a simple two-class classifier, linear discriminant analysis (LDA).
- To show how an LDA model can be used to assign samples to different classes.
- To show how the model can differ according to the number of principal components.
- To understand a simple one-class classifier, quadratic discriminant analysis (QDA).
- To understand how the results of one-class and two-class classifiers are not comparable.
- To discuss the importance of validating class models.
- To describe how classification models can be used to determine which variables are most important for discriminating between groups.

General warning: Throughout the text, several exercises or calculations are provided for the student. These can be solved by taking into account the related descriptions given in the text and in Chapter 2. Calculations can be performed manually or using any common spreadsheet. Hints to verify the correctness of the solutions are given in the text.

6.1. Introduction

6.1.1. *Motivations*

Classification is often also called supervised pattern recognition (PR). In its original inception, PR had a wider definition, encompassing, for example, handwriting or facial recognition; however, in chemometrics and many other forms of applied statistical data analysis, the terminology is now primarily concerned with classification.

In its simplest form, classification involves determining whether an object, such as a spectrum or chromatogram, belongs to one or more predefined groups.

There are many different schools of thought or terminologies. In this chapter, we divide classification algorithms into one-class and two-class models. Some chemometricians like to use the term "discrimination" for one-class models and "classification" for two-class models, so be aware of the variety of differing terminologies used in the literature. Most one-class methods can also sometimes be called "soft models", while two-class methods can be called "hard models". Multiclass approaches, which involve more than two classes, are extensions of two-class models; however, for brevity, we restrict our discussion in this chapter to illustrating situations where there are only two classes in the data.

In traditional physical sciences, classification is usually quite a straightforward process. For example, we may be interested in whether a compound is a ketone or ester. There will be an unambiguous answer, usually obtained using a specific type of analytical measurement.

In chemometrics, the problem often becomes much harder. Unlike in traditional physical sciences, there is not always an unambiguous answer. Consider the example of using GCMS analysis on blood plasma extracts to determine whether a donor has a disease. Of course, the GCMS will carry some information based on the donor's disease status, but this will be mixed with signals due to the donor's diet, environment, genetics, etc. The outcome of the analysis will also depend on disease progression. There may also be a few misdiagnosed samples. Finally, we may not know how effective an analytical method is in diagnosing the disease. It may be effective, for example, in specific ethnic and age groups, but not in a wider population. In clinical sampling, obtaining a balanced and representative set of 100 samples is difficult. Although a chemist may be dealing with 1 mole, or 6×10^{23} molecules, which will exhibit a certain statistical distribution in their behaviour, because the population is so huge, the overall properties and behaviour can be predicted with great accuracy. A chemometrician might be dealing with a much smaller number of samples, which would have a considerable number of factors that influence variability. It is impossible to sample several million, or even billion, donors to obtain population statistics; therefore, we must base our predictions on only a small, and occasionally nonrepresentative, portion of the population.

> In traditional physical sciences, classification is usually quite a straightforward process. In chemometrics, the problem often becomes much harder. Unlike in traditional physical sciences, there is not always an unambiguous answer.

With such relatively small sample sizes, chemometrics comes to the rescue and proves helpful in practical situations. Classification methods play many practical roles, a few of which are given as follows:

- In the simplest form, they can involve analytical measurements to determine whether an unknown or new sample originates from one of two or more predefined groups. For example, can we use NIR of extracts to determine whether or not a sample of orange juice is adulterated?

- Another common enquiry is whether a model is valid, i.e., whether it is appropriate for a given situation. For example, can we use LCMS of extracts of blood plasma to determine whether or not an individual has a predefined disease? This would depend on sufficient availability of chemical information to distinguish the disease status, the appropriateness of the analytical procedure employed, or even the underlying metabolic processes.

- Classification methods can be used for outlier detection. A sample may not belong to any predefined groups, so its characteristics are far different from any of these groups. For example, if we define two characteristic subspecies of mice, one of our samples may have come from a new or unknown subspecies, and if the analytical and chemometric methods are adequate, it will not belong to either of the known classes.

- These methods are used for finding significant variables, which is crucial for most chemometrics applications. As an example, which biomarkers are responsible for detecting whether a donor is diseased or not? Often, extracts of bodily fluids are analysed through methods such as LCMS or GCMS, where hundreds of chromatographic peaks are detected, and multivariate approaches can be used to identify which best discriminate between two groups, thus aiding the investigator in determining the most likely biomarkers.

Classification methods have many practical roles. In the simplest form, they can determine whether an unknown sample originates from one of two or more groups. Another common enquiry is whether a model is valid. Classification can be used for outlier detection. It also plays a crucial role in finding significant variables.

We look at how multivariate classification can help in investigating these problems.

6.1.2.   *Case study*

> We aim to form a model that distinguishes between two groups,
> validate the model, look for outliers, and see which variables are
> most discriminatory.

In order to illustrate the methods discussed in this chapter, we
introduce a simulated case study, as presented in Table 6.1 (for those
with access to the electronic version of the text, the table is formatted
to allow for it to be directly imported into CAT simply through copy
and paste).

The data are arranged in a 40×10 matrix, representing:

- 40 samples, of which the first 20 (samples 1–20) are in class A,
  while the final 20 (samples 21–40) are in class B;
- 10 variables (A–J).

6.1.3.   *Preprocessing and exploratory data analysis*

The first step is to decide how to preprocess the data, that is, to
prepare it for analysis.

> It is always important to consider preprocessing of the data on
> a case-by-case basis prior to performing classification.

There are a large number of approaches, some of which were
detailed in Chapter 5, and almost all have their own pros and cons
according to the case study under consideration.

First, if necessary, we transform the rows of the matrix. Most
commonly, the sum of the variables in each row is transformed into a
constant total, e.g., 1. This is referred to as row scaling or, sometimes,
rather ambiguously, normalisation. This transformation is typically
performed if the samples are of different magnitudes; for example,
we may be measuring the elemental composition of a series of rocks,

Table 6.1. Values for the case study considered in this chapter (note that italized and bold characters differentiate the two classes of samples).

Samples	Class	A	B	C	D	E	F	G	H	I	J
						Variables					
1	*A*	*15.22*	*6.10*	*10.73*	*18.30*	*21.24*	*5.35*	*16.55*	*22.81*	*14.14*	*4.91*
2	*A*	*10.54*	*3.21*	*6.79*	*11.60*	*14.61*	*4.52*	*12.76*	*15.05*	*6.31*	*5.45*
3	*A*	*12.67*	*7.68*	*12.09*	*17.49*	*23.25*	*5.31*	*13.78*	*21.37*	*13.94*	*5.58*
4	*A*	*13.34*	*4.91*	*9.39*	*14.42*	*20.15*	*5.87*	*13.26*	*18.48*	*12.70*	*3.80*
5	*A*	*15.34*	*7.93*	*10.51*	*18.22*	*22.91*	*6.26*	*17.10*	*22.15*	*15.27*	*3.15*
6	*A*	*8.12*	*5.09*	*8.87*	*12.69*	*16.10*	*4.00*	*10.36*	*17.09*	*9.81*	*4.25*
7	*A*	*13.88*	*16.23*	*5.91*	*19.03*	*21.95*	*7.89*	*23.28*	*18.00*	*17.97*	*7.78*
8	*A*	*10.68*	*6.81*	*8.45*	*13.63*	*15.72*	*5.60*	*14.29*	*16.83*	*11.68*	*3.66*
9	*A*	*14.88*	*6.79*	*11.05*	*19.15*	*21.70*	*5.58*	*17.14*	*22.56*	*13.71*	*4.37*
10	*A*	*13.74*	*6.47*	*10.68*	*16.36*	*21.84*	*7.20*	*16.40*	*20.56*	*12.88*	*1.30*
11	*A*	*15.02*	*9.43*	*10.66*	*18.70*	*24.47*	*8.00*	*17.44*	*23.80*	*14.81*	*5.13*
12	*A*	*9.53*	*7.56*	*7.03*	*11.35*	*13.92*	*3.95*	*10.38*	*15.09*	*9.86*	*6.04*
13	*A*	*9.56*	*4.88*	*6.59*	*11.65*	*13.83*	*3.01*	*11.49*	*16.34*	*11.11*	*4.97*
14	*A*	*14.06*	*6.69*	*11.45*	*18.05*	*20.80*	*6.86*	*17.03*	*21.45*	*12.15*	*4.18*
15	*A*	*12.03*	*7.67*	*9.03*	*14.40*	*21.12*	*5.57*	*16.47*	*20.41*	*14.94*	*5.77*
16	*A*	*12.15*	*12.82*	*7.46*	*15.02*	*21.09*	*7.62*	*19.75*	*14.87*	*14.55*	*7.56*
17	*A*	*7.14*	*5.48*	*6.73*	*10.37*	*11.32*	*2.31*	*9.77*	*13.90*	*7.68*	*5.87*
18	*A*	*13.56*	*6.91*	*12.10*	*18.30*	*22.48*	*6.12*	*16.54*	*22.27*	*13.45*	*4.81*
19	*A*	*15.11*	*5.65*	*12.22*	*17.75*	*22.88*	*6.21*	*17.38*	*23.55*	*15.18*	*4.02*
20	*A*	*12.81*	*7.95*	*9.32*	*16.08*	*18.66*	*6.11*	*15.37*	*20.09*	*11.15*	*5.48*
21	**B**	**11.97**	**15.11**	**5.00**	**17.04**	**20.96**	**10.39**	**22.96**	**13.88**	**16.05**	**9.79**
22	**B**	**13.08**	**16.58**	**5.63**	**17.64**	**18.21**	**10.03**	**22.83**	**12.76**	**13.72**	**9.52**
23	**B**	**13.13**	**16.57**	**4.05**	**17.54**	**19.38**	**10.30**	**23.41**	**13.67**	**17.61**	**9.52**
24	**B**	**10.90**	**13.65**	**3.64**	**18.62**	**15.58**	**8.10**	**19.81**	**13.14**	**15.14**	**6.35**
25	**B**	**8.59**	**14.18**	**3.74**	**14.77**	**18.07**	**6.58**	**17.57**	**11.61**	**13.97**	**7.51**
26	**B**	**10.63**	**14.92**	**2.55**	**14.86**	**17.26**	**7.72**	**20.30**	**12.98**	**14.58**	**8.12**
27	**B**	**8.82**	**15.10**	**3.55**	**16.81**	**15.88**	**6.32**	**18.61**	**11.80**	**14.18**	**9.32**
28	**B**	**10.88**	**13.41**	**4.24**	**14.26**	**17.39**	**7.61**	**19.71**	**10.71**	**13.11**	**6.42**
29	**B**	**11.44**	**16.26**	**1.64**	**17.69**	**16.52**	**9.64**	**20.86**	**11.50**	**13.35**	**8.61**
30	**B**	**10.37**	**14.14**	**5.25**	**14.73**	**15.85**	**7.19**	**18.79**	**10.36**	**14.98**	**7.87**
31	**B**	**13.45**	**18.58**	**3.97**	**20.11**	**21.34**	**12.76**	**23.35**	**12.73**	**16.07**	**10.35**
32	**B**	**9.96**	**13.44**	**1.47**	**13.89**	**16.28**	**8.73**	**17.73**	**11.30**	**13.52**	**7.32**
33	**B**	**10.38**	**15.26**	**4.42**	**16.35**	**17.39**	**6.48**	**20.98**	**9.74**	**15.63**	**9.55**
34	**B**	**11.49**	**15.48**	**5.33**	**16.68**	**17.72**	**9.27**	**21.29**	**9.38**	**15.51**	**8.90**
35	**B**	**9.06**	**13.19**	**5.33**	**13.24**	**14.51**	**8.31**	**18.22**	**11.25**	**14.59**	**7.91**
36	**B**	**10.16**	**14.46**	**5.21**	**15.60**	**19.11**	**9.02**	**18.24**	**11.76**	**13.12**	**7.00**
37	**B**	**10.72**	**14.40**	**3.06**	**15.70**	**18.49**	**8.88**	**22.19**	**11.15**	**15.25**	**9.69**
38	**B**	**9.08**	**14.33**	**4.62**	**14.25**	**16.49**	**9.08**	**18.85**	**11.73**	**14.27**	**6.86**
39	**B**	**12.00**	**14.59**	**4.00**	**16.81**	**21.58**	**7.60**	**22.06**	**12.51**	**16.99**	**8.68**
40	**B**	**1.27**	**2.41**	**0.76**	**2.22**	**3.96**	**1.59**	**5.50**	**2.95**	**1.23**	**6.11**

where our interest lies in observing how the proportions change in each sample.

The second step is to transform the columns, which is most commonly carried out through mean centring, although standardisation (sometimes called autoscaling or, rather ambiguously, normalisation, not to be confused with row scaling) is often used when variables are on different scales. Column transformation must always be performed after row scaling.

It is not always necessary to transform the rows. In our case study, we do not need to do this, but every situation is different.

Equally, standardisation of the columns is also not quite necessary in our case, but mean centring can be an important transformation. Recalling Chapter 5, to do this, we calculate the means of each column as $\bar{x}_{ij} = (x_{ij} - \bar{x}_k)$, where:

- x_{ij} is the original element of the matrix in the ith row and jth column;
- $\bar{x}_j$ is the mean of column j;
- $\bar{x}_{ij}$ is the new transformed data.

In our example, if $i = 12$ and $j = 3$, then:

- $x_{ij} = 7.03$, as can be verified from Table 6.1;
- $\bar{x}_j = 6.61$, being the mean of column 3;
- $\bar{x}_{ij} = 7.03 - 6.61 = 0.42$, which is the new (transformed) data.

We will not discuss further transformations in this chapter, as these have already been described in Chapter 5; however, it is always important to consider preprocessing or preliminary transformation of the data on a case-by-case basis prior to performing classification.

Before doing pattern recognition, it is useful first to examine the data to ensure it makes sense. This procedure is often called exploratory data analysis (EDA), and it is most commonly done by performing PCA on the dataset.

> Before doing PR, it is useful first to examine the data using exploratory data analysis, most commonly through PCA.

We calculate the scores $\boldsymbol{T}$ and loadings $\boldsymbol{P}$ of this new transformed matrix (more details can be found in Chapters 2 and 7, the latter dedicated to PCA), which we denote as $\bar{\boldsymbol{x}} = \boldsymbol{TP} + \boldsymbol{R}$, where:

- $\boldsymbol{T}$ is a 40 (= I, or the number of samples) × A (the number of components retained in the model) scores matrix;

- P is an $A \times 10$ ($= J$, or the number of variables) loadings matrix;
- R is a 40×10 ($= I \times J$) residuals matrix.

Note that if the number of components equals 10 in this dataset, which is the lowest of the number of samples and variables, the residual matrix is $\mathbf{0}$. Note also that, in the unusual case where two variables are exactly correlated, this reduces the number of components by one, but it is not so in this case study.

To check your calculations, consider:

- the PC score for the eighth sample of the third PC, or $t_{8,3} = 0.333$ (note that some software will reverse the sign of all PCs, so you may get the answer as -0.333);
- and the PC loading for the second PC and third variable, or $p_{2,3} = 0.412$ (or -0.412, depending on your software).

Check also that the sum of the squares of the loadings for each PC equals 1. If you calculate a full model, the loadings matrix will have dimensions of 10×10, and in some software packages, this is transposed. This is a good way of ensuring that you have not accidentally transposed your loadings matrix.

At this point, it is a good idea to view the data. A common way is to plot the graphs of the scores and loadings of PC_2 versus PC_1, as shown in Figure 6.1.

> It is a good idea to view the data. A common way is to plot graphs of the scores and loadings of PC_2 versus PC_1.

We can see a number of features:

- On the whole, the two classes appear well separated but of different shapes.
- Samples 7 and 16, although originally classified as belonging to class A, appear to have some features of class B too.
- Sample 40 seems to belong to neither class and is called an outlier.
- We can compare the plot of the loadings with that of the scores. As an example, variables B and G are in the lower-right corner and appear to be mostly associated with class B. Variables H and C are

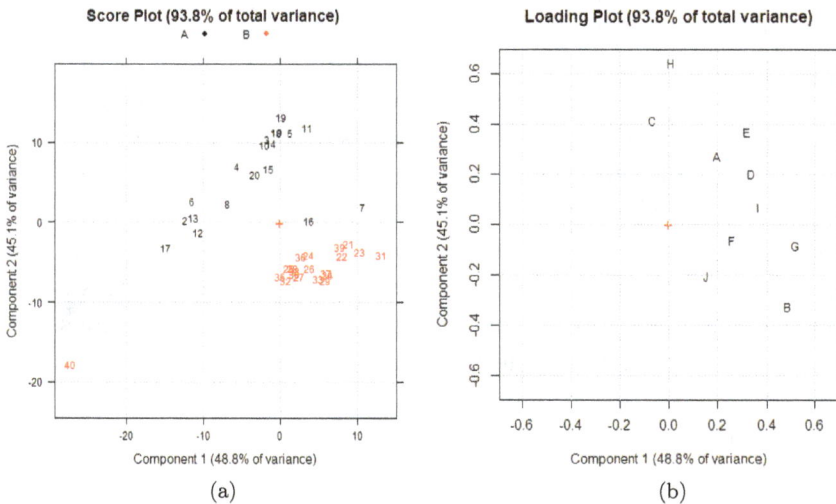

Figure 6.1. (a) Scores and (b) loadings of the first two PCs for the data in Table 6.1.

at the top and appear most closely associated with class A. These variables could be candidate markers for their respective classes; we examine them in greater detail later. It can also be seen that almost all the variables have positive loadings on PC_1.

Other deductions can also be made by examining these plots. For larger, more complicated datasets, they indeed provide a very valuable preview of the data in the initial stages. Sometimes, later components such as PC_3 or even further can be useful: every dataset is unique, but a preliminary exploration of all datasets is always recommended.

In our particular dataset, sample 40 appears to belong to no predefined class. In chemometrics, this occurs quite often. For example, we may be looking at a clinical sample of donors, some of whom have one disease while others may have another disease. But one of the donors may have been misdiagnosed, or they could have a third (unexpected) disease, or perhaps their metabolism is quite different due to factors such as age, genetics, or other underlying reasons that were not part of the general population already sampled.

There is no single correct way of deciding which samples, if any, to remove from a dataset; this depends very much on external decisions concerning the problem at hand. In hard sciences, such as mainstream chemistry, the decision is considerably simpler: if we have a class of ketones and another of amides, and if the molecule we are analysing is an ester, it may exhibit properties of neither of the other classes so that it can be quite unambiguously identified via its chemical structure. In most areas of chemometrics, this is not possible, as there can be considerable variation in populations. So, before performing PR using computational algorithms, it is always important first to work out why you are interested in the data and what sort of variation you can tolerate.

In our case, sample 40 has a significant influence on the analysis; therefore, a good strategy would be to remove it from the dataset, resulting in a new dataset:

- which has dimensions of 39×10,
- consisting of 20 samples from class A,
- and 19 from class B.

Before performing PR, first work out why you are interested in the data and what sort of variation you can tolerate.

All analyses in the remainder of this chapter will be on this reduced dataset.

We now need to preprocess the data again because the column means have changed since one sample has been removed. To check, if $i = 12$ and $j = 3$, then

- $x_{ij} = 7.03$, as can be verified from Table 6.1;
- $\bar{x}_j = 6.76$, being the mean of column 3, as there are now only 39 samples in this column, and removing sample 40 causes the column mean to vary;
- $\bar{x}_{ij} = 7.03 - 6.76 = 0.27$, which is the new (transformed) data.

Next, perform PCA again. For brevity, we only report the results for one PC calculation in the following. Note that you may get a result with sign reversal according to the package you use. To check, consider:

- the PC score for the eighth sample of the third PC, or $t_{8,3} = 0.589$ (note how this differs from the corresponding score when all 40 samples were included, which was 0.333, showing how influential the outlier was);
- and the PC loading for the second PC and third variable, or $p_{2,3} = 0.196$, which differs slightly from that for the full dataset.

The scores and loadings of the first two PCs are illustrated graphically in Figure 6.2.

- The scores now fill the entire space compared to Figure 6.1, as we are no longer distracted by sample 40. It can be noted that the simple removal of one sample reversed the sign of PC_1 when using the CAT software package; as previously said, this can happen and is not a problem and will depend on the package and algorithm used.
- PC_1 seems now to be the main source of variation between the two classes, suggesting that we are now looking primarily at the difference between classes A and B.

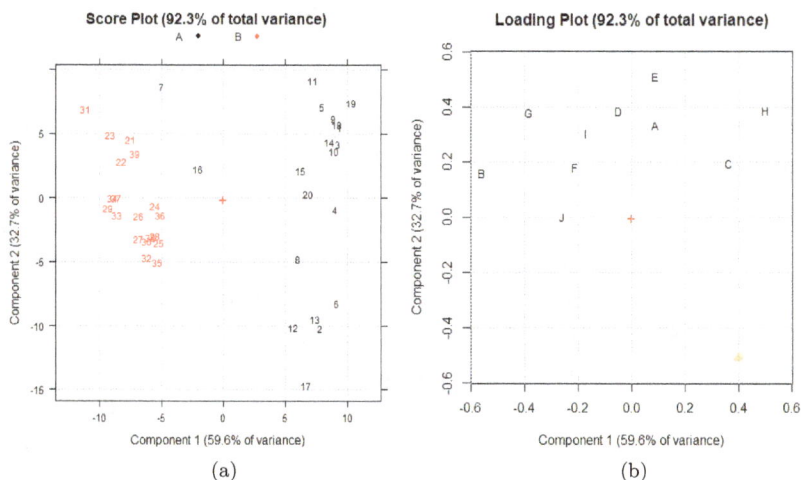

Figure 6.2. (a) Scores and (b) loadings of the first two PCs for the data of Table 6.1 after the removal of the outlier, sample 40.

- Apart from the reversal of PC_1, the relative positions of the loadings are fairly similar to those observed in Figure 6.1, except they are rotated in such a way that now they all have positive values on PC_2, with C and H, and B and G, at the right and left, rather than top and bottom of the plot, respectively, reflecting the change in positions of the two classes.

Of course, there is no strict rule regarding the retention or removal of samples prior to further analysis; however, a preliminary exploration of datasets, primarily using graphical tools such as PCA, can tell us if there are any unusual samples, or variables, that should be removed. They can also give us an idea about the presence of unusual or unexpected structure in the data and indicate whether any kind of pattern exists.

In this introductory text, space limitations prevent us from discussing all the tricks of the trade in depth, but it is worth emphasising the importance of always performing EDA prior to establishing classification models:

- First, preprocess the data.
- Perform PCA and examine the scores and loadings graphically.
- Remove any unusual samples, variables, or even unexpected groupings if necessary.
- Preprocess the new data again, and repeat PCA.
- Examine the scores and loadings again to see if the result is sensible.

> There is no strict rule regarding the retention or removal of samples prior to further analysis; however, a preliminary exploration can tell us if there are any unusual samples, or variables, that should be removed. They can also give us an idea about the presence of unusual or unexpected structure in the data and indicate whether any kind of pattern exists.

6.2. Two-Class Classifiers

6.2.1. *Hard boundaries*

Two-class classifiers aim to establish a boundary between two groups of samples. This is often called a hard boundary and is illustrated in Figure 6.3(a). Objects to the left of the boundary represent class A, illustrated by circle symbols, and to the right are objects representing class B, illustrated by diamond symbols. For a new sample, we can predict which class it belongs to based on its location relative to the boundary.

> Two-class classifiers produce a boundary between two groups of samples, which is often called a hard boundary.

In Figure 6.3(a), we assume that the boundary between different classes is linear and that they can be exactly separated. However, this

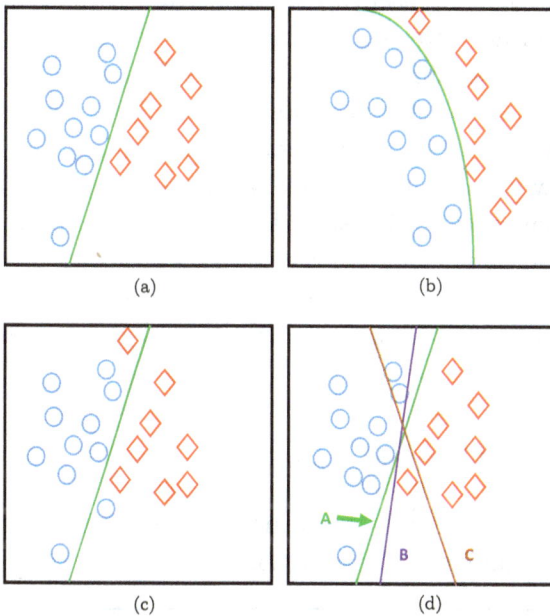

Figure 6.3. (a) Principle of hard boundary in a two-class classifier, (b) nonlinear boundary between two classes, (c) linearly inseparable classes, and (d) three different types of linear boundaries.

does not need to be the case. Two other situations are illustrated. In Figure 6.3(b), a nonlinear boundary is depicted. There is no requirement that classes be best separated by a linear model, although the majority of chemometric methods involve linear models. The situation in Figure 6.3(c) illustrates two classes that cannot be linearly separated, which frequently occurs.

Finally, there are numerous different mathematical models or algorithms. These all result in different types of boundaries between the classes. Figure 6.3(d) illustrates three different boundaries. Both boundaries A and B perfectly separate two classes using linear models, whereas boundary C does not succeed in doing so. Each boundary is formed using a different computational model or statistical criterion. There are hundreds of such models or approaches available, and experienced chemometric experts would carefully assess the options and choose the one they feel is most suitable for a particular problem. In this introductory text, we only describe one such model: linear discriminant analysis (LDA).

Of course, most datasets are obtained by recording several different variables or measurements. Hence, a dataset in which each sample is measured at 100 different spectral wavelengths can theoretically be represented by a 100-dimensional graph. We cannot, of course, visualise this on paper or a computer screen, but the boundary in this case will be a 99-dimensional hyperplane. A computer can certainly calculate on which side of this hyperplane each sample lies, thereby providing information about the provenance of the samples; however, to better understand the methods, we use simpler graphical approaches in this text to provide a preliminary understanding of the methods.

6.2.2. *Linear discriminant analysis*

LDA is one of the simplest and most well-established methods for determining boundaries between classes.

> LDA is one of the simplest methods for determining boundaries between classes.

If a dataset is:

- represented by a matrix X of dimensions $I \times J$,
- then if the mean of all samples for class A is given by $\bar{x}_A$, or in our case, a $1 \times J$ row vector,
- and the variance–covariance matrix of all samples represented by X is given by S,
- the Mahalanobis distance of sample i to the centre of class A is given by (refer also to Chapter 2) $d_{iA} = \sqrt{((x_i - \bar{x}_A)S^{-1}(x_j - \bar{x}_A)')}$.

Note that:

- in this chapter, we use the population variance–covariance, that is, we divide the numerator by I (the number of samples) rather than $I - 1$; this is because we are doing a scaling rather than statistical estimation;
- the variance–covariance matrix is for the entire dataset of I samples rather than just for each class: the latter, as described in the following section, is used for another method, namely quadratic discriminant analysis (QDA).

We illustrate this method graphically initially by using two variables: the scores t of the first two components of the column-centred data matrix of 39 samples, as calculated below.

> We illustrate this method graphically initially by using two variables: the scores of the first two components of a column-centred data matrix.

In this case, matrix X denotes the column-centred 39×2 matrix T.

The means of each class are $\bar{x}_A = [6.882\ 0.414]$ and $\bar{x}_B = [-7.244 - 0.436]$. Note that the means are not quite related as $\bar{x}_A \neq -\bar{x}_B$ since there are unequal numbers of samples in each class; however, you should verify that $(20/19)\bar{x}_A = -\bar{x}_B$ because the scores are centred so that the overall mean of each column is 0.

The variance–covariance matrix is $S = \begin{bmatrix} 58.34 & 0 \\ 0 & 32.04 \end{bmatrix}$.

- Note that the off-diagonal terms equal 0; this is because the PC scores are orthogonal – of course, when using raw data, there will usually be non-zero diagonal terms.
- While checking the calculations, remember to use the correct divisor for the variance, i.e., 39 rather than 38.

The Mahalanobis distance of sample 8 to the centre of class A is given by

$$d_{8A} = \sqrt{\left([(6.067 - 6.882) \quad (-4.809 - 0.414)] \begin{bmatrix} 0.0171 & 0 \\ 0 & 0.0312 \end{bmatrix} \begin{bmatrix} 6.067 & -6.882 \\ -4.809 & -0.414 \end{bmatrix} \right)}$$

$$= \sqrt{\left([(-0.815) \quad (-5.223)] \begin{bmatrix} 0.0171 & 0 \\ 0 & 0.0312 \end{bmatrix} \begin{bmatrix} (-0.815) \\ (-5.223) \end{bmatrix} \right)}$$

$$= \sqrt{0.862} = 0.929,$$

where:

- the inverse of the variance–covariance matrix $S^{-1} = \begin{bmatrix} 0.0171 & 0 \\ 0 & 0.0321 \end{bmatrix}$,
- the scores of the first two components of sample 8 is $t_8 = x_8 = [6.067 - 4.809]$,
- and the mean the scores of the first two components of class A is $\bar{t}_A = \bar{x}_A = [6.882 \ 0.414]$.

The distances of all 39 samples to each of the class centres can be calculated and are presented graphically in Figure 6.4. The equidistant line, where $d_A = d_B$, is also drawn:

- Samples above the line can be assigned to class A.
- Samples below the line can be assigned to class B.
- Samples 7 and 16 are, in this case, misassigned.

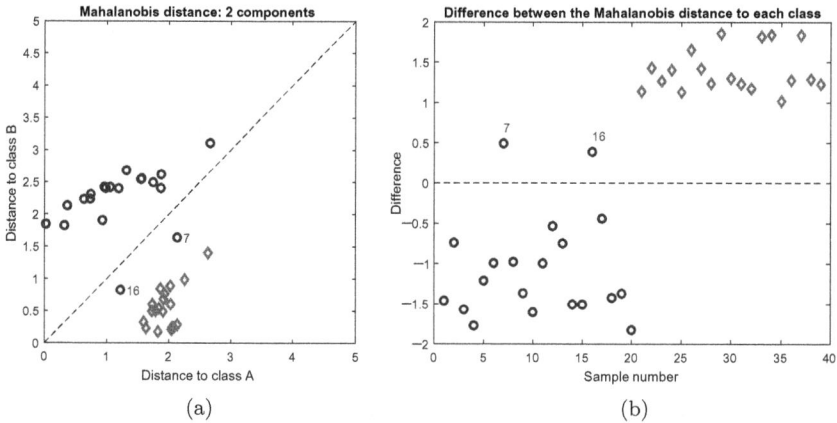

Figure 6.4. (a) Mahalanobis distance of the 39 samples to each of classes A and B, and (b) the difference between the Mahalanobis distances to classes A and B using 2 PC model.

The distances of all samples to the class centres can be calculated, and the equidistant line is also drawn. Samples above the line can be assigned to class A, while those below the line can be assigned to class B.

It is not really possible to come to definitive conclusions as to why samples 7 and 16 are misassigned. This may be a genuine mistake made by the original investigators; for example, a donor might have been misdiagnosed as having a disease. It may be a problem with the class model, and we will see that different models do provide different answers. It may also be possible that these are genuinely new sample types that do not belong to either of the original classes or even exhibit a mixture of properties.

The difference $d_A - d_B$ is presented in Figure 6.4(b); samples with a negative value are closer to the centroid of class A than that of class B. This difference could be considered a type of score or index; for example, for sample 23, $\Delta_{23} = d_{23A} - d_{23B} = 2.25 - 0.99 = 1.26$. As this is a positive number, it would be assigned to class B.

Samples close to the boundary (with a Δ score of less than about 0.5, in this case) might be ambiguous. It is not always necessary that a sample be ambiguously assigned to a single class; sometimes, they

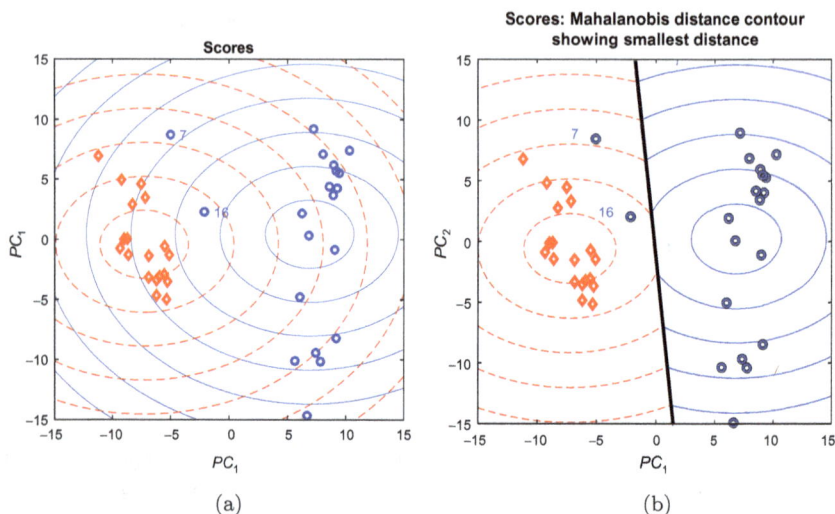

Figure 6.5. (a) Scores and Mahalanobis distance contours for classes A and B using LDA, and (b) Mahalanobis distances as in (a) but showing only the smallest distance to each centroid and the equidistant line between the classes.

may have mixed characteristics. This depends on the case study. For instance, a study to separate apples from oranges (perhaps using spectra of extracts) can only come to an unambiguous answer, whereas a study to distinguish diseased from healthy donors may not necessarily come to an unambiguous conclusion, as there may be varying degrees of the disease, and those at its very early stages may be hard to diagnose.

> It is not always necessary that a sample be ambiguously assigned to a single class. This depends on the case study.

We can also see how the samples are classified into groups using LDA and a scores plot. In Figure 6.5(a), we present the equidistant contours for the Mahalanobis distance from the centroids of each of the classes, using a two-component model.

In Figure 6.5(b), we illustrate the contours for the smallest of the two distances and the dividing line, where $d_A - d_B = 0$. This is another way of assessing which class a sample belongs to based on its Mahalanobis distance from the two centroids.

Of course, it is not necessary to use only 2 PCs in the model:

- Frequently, people use the raw data X rather than the PCs, T.
- In order to do this, the number of samples I must be more than the number of variables J in the dataset,
- Also, none of the variables must be collinear, that is, none should have a correlation of 1.
- However, if all the variables are used in the model, the calculated Mahalanobis distance is the same as when using a full PC model containing all non-zero components.
- In the case of typical spectroscopic or chromatographic data, there may be thousands of variables, far exceeding the number of samples, so it is inevitable that the data have to be reduced using PCA first.
- It is, of course, not necessary to use all variables or non-zero PCs in a model, and sometimes, just a few PCs or a selected number of variables can be used.

It is not necessary to use just 2 PCs in the model. Frequently, people use the raw data. In order to do this, the number of samples must be more than the number of variables, and none of the variables must be collinear.

A rather common, yet often unexpected, issue when many PCs or variables are employed in a model is that we rarely get a Mahalanobis distance to the centre of the data close to 0. The reasons are beyond the scope of this text, as they require understanding how the shape of the F distribution varies with an increase in the number of dimensions. Readers interested in exploring this further are referred to any text or webpage that illustrates the shape of the F distribution.

When many PCs or variables are employed in a model, we rarely get a Mahalanobis distance to the centre of the data close to 0.

However, we can calculate the Mahalanobis distance using all 10 variables for our dataset:

- First calculate the 10×10 variance–covariance matrix $\boldsymbol{S}$. Please ensure you use the population, not the sample formula, dividing by I and not $I - 1$, to reproduce numbers in this chapter.
- To check, $S_{33} = 9.79$ and $S_{62} = 7.69$. Note that $\boldsymbol{S}$ is a symmetric matrix so $S_{ij} = S_{ji}$.
- Calculate the inverse matrix $\boldsymbol{S}^{-1}$. Check that $S_{53}^{-1} = -0.266$ and that $\boldsymbol{S}\boldsymbol{S}^{-1} = \boldsymbol{I}$, or a unit matrix.
- Make sure that the 39×10 matrix $\boldsymbol{X}$ has been first centred down the columns to follow the calculations as given in the following.
- The mean for class A is $\bar{\boldsymbol{x}}_A = $ [0.79 −3.71 2.59 −0.25 0.83 −1.44 −2.48 3.68 − 1.05 1.70] and for class B is $\bar{\boldsymbol{x}}_B = $ [−0.83 3.91 −2.73 0.26 −0.88 1.52 2.6 −3.87 1.11 1.78]; check that you have correctly centred the overall data first – note that the variance–covariance matrix will be the same even if the data have not been first centred.
- Then, calculate $d_{iA} = \sqrt{((\boldsymbol{x}_i - \bar{\boldsymbol{x}}_A)\boldsymbol{S}^{-1}(\boldsymbol{x}_i - \bar{\boldsymbol{x}}_A)')}$ as usual.
- To check the value, the distance of sample 18 to the centre of class B $d_{18B} = 1.82$.

The plot of Mahalanobis distance to the centre of class B against the Mahalanobis distance to the centre of class A is given in Figure 6.6(a). Compare it with Figure 6.4(a).

- Samples 7 and 16 are now much closer to the equidistant boundary and thus assigned to class A.
- This does not mean it is appropriate to assign them to class A. Whether to have faith in the second PC or the full model depends on the aim of the analysis and can only be resolved through discussion. We also discuss validation later, which helps us decide how many PCs to retain in a model.
- None of the Mahalanobis distances are close to 0. As described above, this is a common consequence when there are many variables.

In Figure 6.6(b), the difference $d_A - d_B$ is plotted. In contrast to Figure 6.4(b), we find that the samples 7 and 16 are now assigned to class A. Whether this is meaningful can only be decided upon the basis of the actual application; nevertheless, this demonstrates

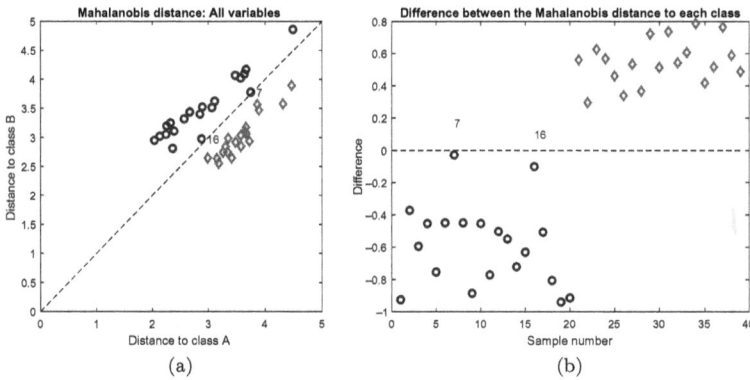

Figure 6.6. (a) Mahalanobis distance of the 39 samples to each of classes A and B using a full 10-variable model (a), and (b) as in (a) but with all 10 non-zero PCs.

that differences in how methods are applied can result in different outcomes.

Differences in how methods are applied can result in different outcomes.

If we use a full PC model (that is, all non-zero PCs), we would find that the Mahalanobis distance is the same as that when using raw data. That is, $d_{iA} = \sqrt{((\boldsymbol{x}_i - \bar{\boldsymbol{x}}_A)\boldsymbol{S}^{-1}(\boldsymbol{x}_i - \bar{\boldsymbol{x}}_A)')} = \sqrt{((\boldsymbol{t}_i - \bar{\boldsymbol{t}}_A)\boldsymbol{S}_t^{-1}(\boldsymbol{t}_i - \bar{\boldsymbol{t}}_A)')}$, where $\boldsymbol{S}$ is the variance–covariance matrix of the original data (column-centred in this case) and $\boldsymbol{S}_t$ is that of the scores. Note that $\boldsymbol{S}_t$ will be a diagonal matrix, whereas $\boldsymbol{S}$ will never be diagonal, and that this equality will only be valid if there are fewer variables than samples and none of the variables are correlated.

Check that you have done this correctly. The distance of sample 13 from class B should be 13.40 for both models.

6.3.   One-Class Classifiers

6.3.1.   *Soft boundaries*

One-class classifiers aim to produce a boundary around each group of samples. This is often called a soft boundary because the boundaries

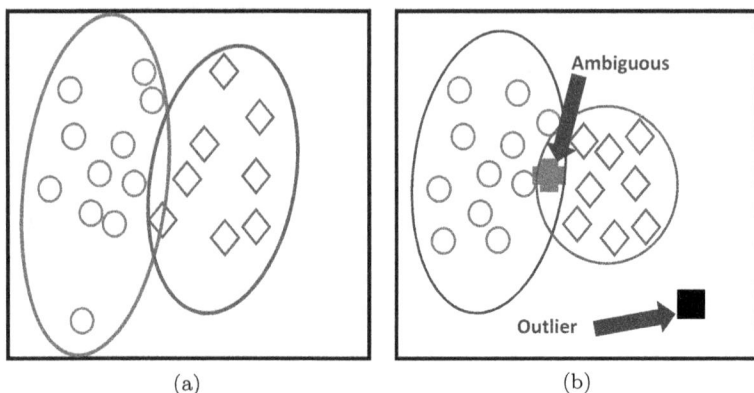

Figure 6.7. (a) Two one-class boundaries, and (b) different types of outcomes when using one-class classifiers.

between two classes can overlap. Figure 6.7(a) illustrates two one-class boundaries. In this case, there is an ambiguous area in which samples can belong to both classes, which is a realistic outcome in many situations. For example, an analytical technique may not be perfect and may be unable to unambiguously separate different groups. Alternatively, there may genuinely be groups with a mixture of characteristics. The interpretation of such situations depends on the specific case study of interest, and there are no hard and fast rules.

> One-class classifiers aim to produce a boundary around each group of samples, which is often called a soft boundary. There may be an ambiguous area in which samples can belong to more than one class.

Figure 6.7(b) illustrates a more complex case in which there are four different types of outcomes:

- 11 samples are unambiguously part of group A (circles).
- 8 samples are unambiguously part of group B (diamonds).
- 1 sample is ambiguous (cross).
- 1 sample is an outlier (square).

In a two-class model, as described in the previous section, we can only have two outcomes if there are two classes; hence, the type of verdict above provides more options for one-class models compared to two-class models.

> There are more options in the case of one-class models compared to two-class models.

Similar to two-class classifiers, there are also a number of common one-class models. Soft independent modelling of class analogy (SIMCA) is a well-known approach; however, in this text, we will restrict ourselves to QDA, which is actually one of the steps in SIMCA; an additional step, disjoint PCA, is not discussed here for brevity.

As above, we initially illustrate the method using a 2 PC model.

6.3.2. *Quadratic discriminant analysis*

QDA can be implemented either as a one-class or a two-class classifier. In this section, we focus only on describing its use as a one-class classifier.

Superficially, its description appears quite similar to LDA. If a dataset is:

- represented by a matrix $\boldsymbol{X}$ of dimensions $I \times J$,
- then if the mean of all samples in class A is given by $\bar{\boldsymbol{x}}_A$, or in our case, a $1 \times J$ row vector,
- and the variance–covariance matrix of class A $\boldsymbol{X}_A$ is given by $\boldsymbol{S}_A$,
- the Mahalanobis distance of sample i to the centre of class A is given by $d_{iA} = \sqrt{((\boldsymbol{x} - \bar{\boldsymbol{x}}_A)\boldsymbol{S}_A^{-1}(\boldsymbol{x}_i - \bar{\boldsymbol{x}}_A)')}$.

The difference from LDA is that the variance–covariance matrix $\boldsymbol{S}_A$ is calculated for each class separately.

> The difference between LDA and QDA is that the variance–covariance matrix for the latter method is calculated for each class separately.

In our case, $S_A = \begin{bmatrix} 13.82 & 1.98 \\ 1.98 & 51.33 \end{bmatrix}$ and $S_B = \begin{bmatrix} 2.86 & 4.07 \\ 4.07 & 11.36 \end{bmatrix}$, remembering to use the population variance and the scores of the first two PCs.

Note the following:

- The off-diagonal elements are no longer equal to 0, as the scores of the subset of samples are no longer orthogonal.
- The matrices remain diagonal.
- In both classes, for this case study, the variance of the second variable (t_{22}) is much greater than that of the first variable, as the samples are more spread out in PC_2 than in PC_1.
- PC_1 is primarily influenced by the variance between the classes, whereas PC_2 is primarily characterised by variance within each class.

We can now calculate the Mahalanobis distance to each class centroid, as we did in LDA, but this time using different variance–covariance matrices for each class separately. So, the Mahalanobis distance of sample 8 to the centre of class A is given by

$$d_{8A} = \sqrt{\left([(6.067 - 6.882)\ (-4.809 - 0.414)] \begin{bmatrix} 0.0728 & -0.0028 \\ -0.0028 & 0.0196 \end{bmatrix} \begin{bmatrix} 6.067 & -6.882 \\ -4.809 & -0.414 \end{bmatrix} \right)}$$

$$= \sqrt{\left([(-0.815)\ (-5.223)] \begin{bmatrix} 0.0728 & -0.0028 \\ -0.0028 & 0.0196 \end{bmatrix} \begin{bmatrix} (-0.815) \\ (-5.223) \end{bmatrix} \right)}$$

$$= \sqrt{0.6066} = 0.7789,$$

where:

- the inverse of the variance–covariance matrix $S_A^{-1} = \begin{bmatrix} 0.0728 & -0.0028 \\ -0.0028 & 0.0196 \end{bmatrix}$,
- the scores of the first two components of sample 8 is $t_8 = x_8 = [6.067 - 4.809]$,

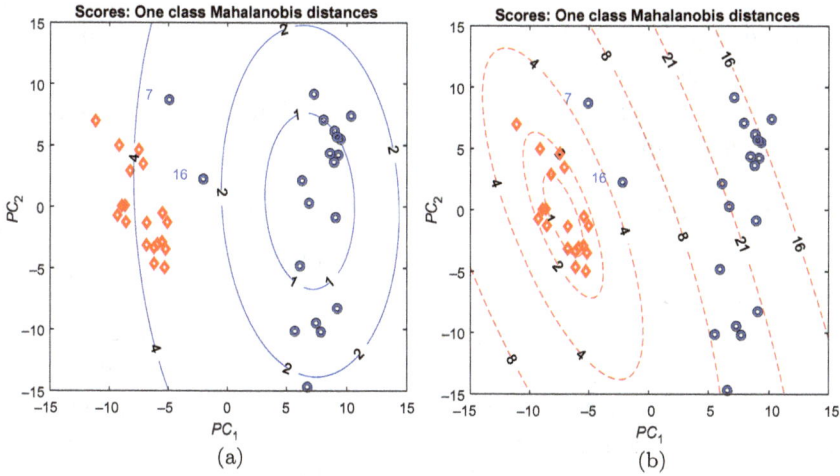

Figure 6.8. One-class Mahalanobis distance contours for class A (a) and class B (b), with distances indicated.

- and the mean of the scores of the first two components of class A is $\bar{t}_A = \bar{x}_A = [6.882 \quad 0.414]$.

We can now calculate the Mahalanobis distance of the samples to each class centroid using QDA. As these are one-class classifiers, they are best presented separately for each class, as shown in Figure 6.8. We can see several important features:

- The contours are in different directions and differently spaced for each class. This contrasts with Figure 6.5(a).
- For both classes, apart from samples 7 and 16, all but one of the samples are within 2 Mahalanobis distance units of their own centres.

To check your calculations, note that the sum of the squares of the Mahalanobis distances for each class equals $I_A J$, or the product of the number of samples and the number of variables. Hence, $\sum^{i \in A} d_{iA}^2 = 20 \times 2 = 40$ and $\sum^{i \in B} d_{iB}^2 = 19 \times 2 = 38$.

If you did not get this result, check your calculations again. For brevity, we will not prove this.

> The equidistant Mahalanobis distance contours for QDA are in different directions for each class, unlike those for LDA.
> The sum of the squares of the Mahalanobis distances for each class equals the product of the number of samples in each class and the number of variables.

The Mahalanobis distance can also be interpreted as a probability that a sample is a member of a class. Unlike hard models, soft or one-class boundaries do not assign samples uniquely to a class but can provide a probability or p-value indicating that a sample belongs to the group, usually (in our case and most situations) assuming a multi-normal distribution. Note that if the distribution differs strongly from multi-normality, this assumption will not be valid.

> Unlike hard models, soft or one-class boundaries do not assign samples uniquely to a class but can provide a probability or p-value that a sample belongs to the group.

- Samples with p-values less than 0.05 imply that, provided the samples are multi-normally distributed, they are predicted to lie at greater Mahalanobis distances (than those calculated for each of them) 5% of the time (or 1 in 20).
- This does not imply that the sample with a Mahalanobis distance whose p-value is less than 0.05 is not a member of the predefined group, but this occurrence is rare.
- If one analyses 20 samples, and one of them has a p-value of 0.01, the likelihood that it is a member of the predefined group is very low; therefore, it is probably an outlier or a member of another group.

A simple way of converting Mahalanobis distances to p-values is by using the χ^2 distribution, assuming that the samples are normally distributed. So long as there are a reasonable number of samples (10 or more), this is a good approximation. (If there are fewer samples, we use another distribution called Hotelling T^2, based on the F distribution, but we will not distinguish between these in this introductory text as there is no realistic distinction experimentally unless the sample size is very small.)

A simple way of converting Mahalanobis distances to p-values is by using the χ^2 distribution.

To do this, we use χ^2 distribution with as many degrees of freedom as there are variables. Consider an example:

- If we have two degrees of freedom, the critical value of χ^2 at $p = 0.05$ is 5.99.
- That means, if we calculate χ^2 for a series of multi-normally distributed variables with two degrees of freedom, only 1 in 20 will exceed 5.99.
- In our case, if we consider only the scores of the first two components, we have two degrees of freedom.
- To see what value of the Mahalanobis distance corresponds to a p-value of 0.05, take the square root of 5.99, which is 2.45. Any sample with more than 2.45 Mahalanobis distance units from the centre of the distribution has a p-value of less than 0.05.
- So, we expect only about 1 in 20 samples to have a distance greater than 2.45 if they are part of an underlying multi-normal distribution.
- The Mahalanobis distance for two degrees of freedom corresponding to $p = 0.5$ is 1.78, as you should be able to check, implying that roughly half of the samples will lie either side of this limit.

Hence:

- Sample 8 with a Mahalanobis distance of 0.78 to the centre of class A is within this critical limit.
- To check, its distance to the centre of class B is 10.02, which is well outside the $p = 0.05$ limit for class B.
- This sample is therefore clearly a member of class A but not class B.

We can alternatively use 1 − the cumulative χ^2 distribution function to determine the p-value of any sample based on its squared

Mahalanobis distance, which we call $1 - \text{cdf}(\chi^2)$. The cumulative distance function can be calculated in most software packages, including Excel, and depends on the number of degrees of freedom or variables. The shape of the χ^2 distribution depends on the number of degrees of freedom in the model.

Considering the example:

- For sample 8, the Mahalanobis distance to the centre of class A is 0.7789.
- Hence, $\chi^2 = 0.7789^2 = 0.6066$.
- $1 - \text{cdf}(0.6066, 2) = 0.738$, where 2 represents the number of degrees of freedom or variables.

This means we would expect to obtain a value of 0.7789 or more around three out of four occasions, with many samples being at this distance or further.

The contours for $p = 0.05$ and $p = 0.5$ are illustrated in Figure 6.9:

- In both cases, the vast majority of samples are within the $p = 0.05$ limits.

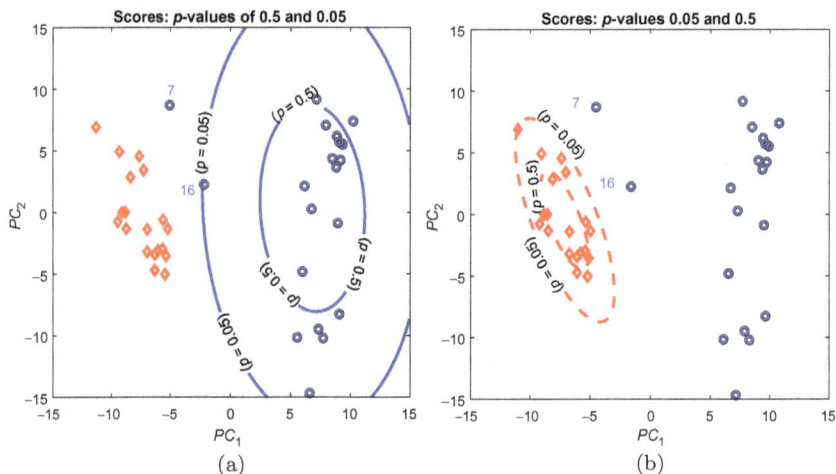

Figure 6.9. One-class Mahalanobis distance contours at two p-values for class A and class B. (a) Class A model, (b) class B model.

- Sample 7, however, is clearly outside the limit for class A, and as it is not within the $p = 0.05$ bounds of class A or B, it is likely to be an outlier.
- Sample 16 is just on the boundary of class A.
- Roughly half of the samples are within the $p = 0.5$ limits for both classes.
- There appear to be no ambiguous samples in this dataset.

If you calculate all the values of the Mahalanobis distance to each centre using a 2 PC model, you should find that:

- for class A, 19 are within the $p = 0.05$ limit (95% of samples),
- and 11 are within the $p = 0.5$ limit (55% of samples),
- whereas for class B, 19 are also within the $p = 0.05$ limit (100% of samples),
- but 7 are within the $p = 0.5$ limit (36% of samples).

Of course, the distributions are not perfectly normal; however, one can see graphically that there is a good approximation.

We can indeed choose any p-value to define the critical limits for our class. As there are two variables:

- If $p = 0.1$, the limiting Mahalanobis distance is $\sqrt{X_{0.1}^2(2)} = 2.146$.
- If $p = 0.05$, the limiting Mahalanobis distance is $\sqrt{X_{0.05}^2(2)} = 2.448$.
- If $p = 0.01$, the limiting Mahalanobis distance is $\sqrt{X_{0.01}^2(2)} = 3.035$, as can be verified using most statistical and spreadsheet packages.

> We can choose any p-value to define the critical limits for a class.

In Figures 6.10(a)–6.10(c), we can view the critical limits for three different p-values. It can be seen that if $p = 0.1$, both samples 7 and 16 lie outside the critical limits for both classes, as does one sample from class B. For $p = 0.05$, sample 7 is still outside the boundary of class A, as is one sample from class B (but is very close to the boundary). If $p = 0.01$, sample 7 still remains outside the limits for both classes and obviously cannot be assigned to either class, and

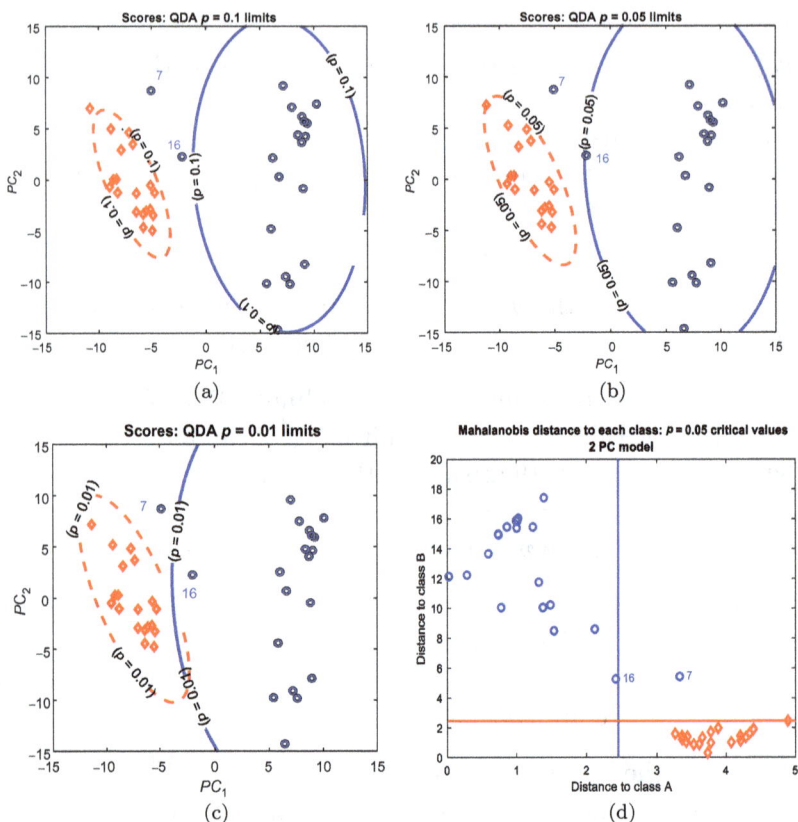

Figure 6.10. QDA limits for (a) $p = 0.1$, (b) $p = 0.05$, and (c) $p = 0.01$, and (d) class distance plot for 2 PC model.

there is a region where both class models overlap, although in this dataset there are no ambiguous samples.

We can also plot a class distance plot, as presented in Figure 6.10(d), using the Mahalanobis distance to each class centroid along the two axes and $p = 0.05$ as critical limits. Of course, different limits could be used according to the chosen p-value. The graph can be divided into four regions:

- Top left: class A,
- Bottom right: class B,
- Bottom left: ambiguous samples (none in this case),
- Top right: outliers (sample 7 in this case).

We can plot a class distance plot using the Mahalanobis distance to each class centroid along the two axes.

Moreover, just as in the case of LDA, there is no reason to restrict ourselves to a 2 PC model, and we can use all non-zero PCs or, alternatively, all variables if the number of variables is less than the number of samples in each class. Note that, as the number of samples in each class is less than the total number of samples in the overall dataset, in some cases this may restrict the number of variables that can be used; however, PC models overcome this limitation.

We can no longer visualise this as a 2D contour plot, as the scores are in 10 dimensions. For brevity, we only provide the results of some calculations, which the reader can use to reproduce these as appropriate:

- The Mahalanobis distance of sample 8 to class A is now $d_{8A} = 3.1276$.
- For the 2 PC model, this was 0.7789.
- However, the critical limits will be quite different from those of a 2 PC model, as there are more degrees of freedom. When there are 10 degrees of freedom, and $p = 0.05$ $\sqrt{X_{0.05}^2(10)} = 4.2787$.
- So, this sample is still within the $p = 0.05$ critical limit of class A.

The class distance plot is presented in Figure 6.11. Superficially, this looks much clearer and easier to interpret than the one plotted for the 2 PC model. However, as we include more PCs, the data can become prone to what is often called overfitting, that is, a solution is forced upon the data. This can be a serious problem in chemometrics when there are often many variables and may result in an overoptimistic solution. We discuss the problem of overfitting in the section on validation.

Superficially, a class distance plot using all PCs looks much clearer and easier to interpret than that plotted for the 2 PC model; however, as we include more PCs, the data can become prone to overfitting.

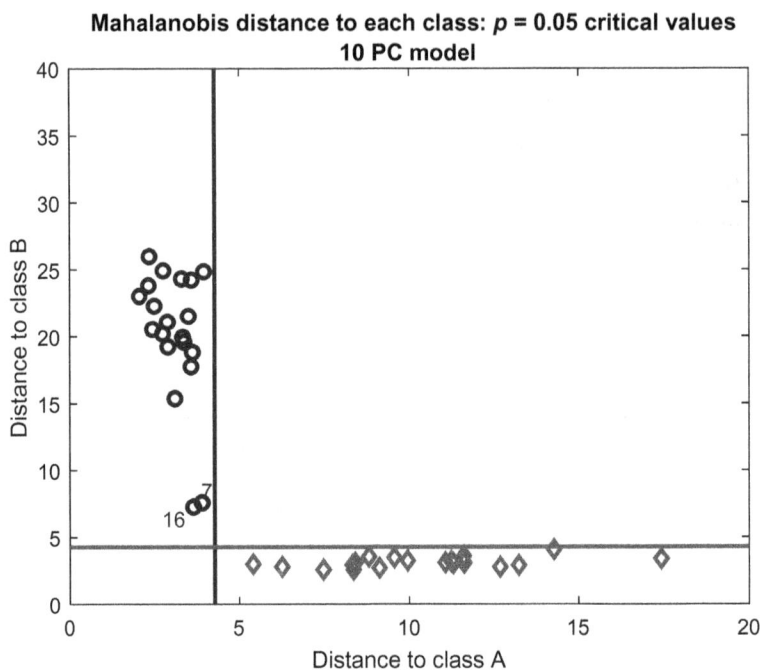

Figure 6.11.   Class distance plot for 10 PC model.

6.4. Validation

Validation plays an essential role in chemometrics. In many problems, there are many more variables (e.g., spectroscopic or chromatographic measurements) than samples. This means there are many possibilities of overfitting models. Hence, it is possible to obtain a result that is overoptimistic.

> In many problems, there are many more variables than samples. This means there are many possibilities of overfitting models.

Consider the following example:

- We wish to determine whether a coin is biased.
- We do this by tossing the coin 10 times and seeing how often it comes up heads (H) or tails (T).

- This experiment is repeated 1,000 times.
- Occasionally, we will obtain 8 or 9 Hs out of 10 tosses. This will happen only rarely if the coin is unbiased.
- However, out of the 1,000 experiments, there will be a few cases, even for an unbiased coin.
- It is possible to obtain a biased model of the coin by focusing more on the cases where there are many more Hs than Ts. This is equivalent to measuring 1,000 variables, choosing those variables that are discriminatory between two groups, and using only these variables in the model.
- Hence, it is possible to incorrectly decide that the coin is biased, or that there is a difference between two groups of samples, if the number of variables far exceeds the number of samples, even if there is no underlying difference or bias
- Thus, the more the variables in a model, the greater the chance of overfitting and coming to an incorrect conclusion unless one is careful.

To avoid erroneous conclusions, chemometrics experts usually validate their model.

A common way of doing this is to divide the samples into two groups:

- A training set is used to develop the mathematical model.
- A test set is used to determine how well the model performs on an independent group of samples.

> To avoid erroneous conclusions, chemometricians usually validate their models by dividing the samples into training and test sets.

Usually, between two-thirds and three-quarters of samples are placed in the training set, and the remainder is placed in the test set.

We exemplify this procedure by validating the QDA model using 10 components in our case study consisting of 39 samples, which the reader can reproduce:

- We establish a training set consisting of the first 15 samples (numbering 1–15) in class A and the first 15 samples in class B (numbering 21–35).
- We call this matrix $X_{(\textbf{train})}$.
- This 30×10 matrix must be column centred so that $\bar{x}_{(\text{train})ij} = (x_{ij} - \bar{x}_{(\text{train})j})$.

As an example, if $i = 12$ and $j = 3$, then:

- $x_{ij} = 7.03$, as can be verified from Table 6.1;
- $\bar{x}(\text{train})_j = 6.64$, being the mean of column 3;
- $\bar{x}(\text{train})_{ij} = 7.03 - 6.64 = 0.39$, which is the new (transformed) data;
- note that the mean of the columns of the training set is slightly different from that of the overall dataset because the samples are slightly different – for the full 39 samples, this value was 0.27, as calculated above;
- if you are performing these calculations, check that you have correctly selected and column centred the dataset.

Any model can be validated using any algorithm; however, for brevity, we initially check only one model, the one-class QDA model, using 10 PCs. Figure 6.11 suggested that if all samples are modelled together (often called auto-prediction), they appear to be assigned perfectly to their respective classes. This could be overoptimistic, but if all PCs are included in the model, this often forces the model to provide an unrealistically good answer. A model with 100% perfect prediction is not always the most appropriate.

> Any model can (should) be validated. A model with 100% perfect prediction is not always the most appropriate.

The 10 variable and 10 PC models provide identical answers in this case. For brevity, we only consider using a 10-variable model. Note that if the sample size is reduced and you are using PCs, it is necessary to recentre and calculate the PCs again over the training set, whereas if you are using variables, this is not necessary.

We now calculate the new variance–covariance matrices for the training set for each class separately. To check this, as an example, the (population) covariance between variables 5 and 3 (or E and C) and class A is given by

$$s(\text{train})_{53A} = 4.668,$$

which is the covariance between variables 5 and 3 for class A in the training set using samples 1–15.

The mean of variable 8 (or H) and class A is

$$\bar{x}(\text{train})_{8A} = 19.466.$$

To compare,

$$s_{53A} = 5.315$$

for the overall autopredictive model (consisting of all 20 samples as part of class A) and

$$\bar{x}_{8A} = 19.333$$

for the original dataset of 20 samples.

The test set consists of those samples left out, which are in this case samples 16–20 of class A and samples 36–39 of class B. Of course, it is not necessary to calculate the variance–covariance matrix of the test set samples, as we use the variance–covariance matrix and mean of the training set.

We can now calculate the Mahalanobis distances of the 30 training set samples to their respective class centres and present these graphically, as shown in Figure 6.12(a). To check your calculations,

$$d_{6B} = 26.80,$$

which is the Mahalanobis distance of the sixth sample (a member of class A) to the centroid of class B.

We can see several interesting features:

- Sample 7, although assigned to class A, is nevertheless the member that is closest to class B.

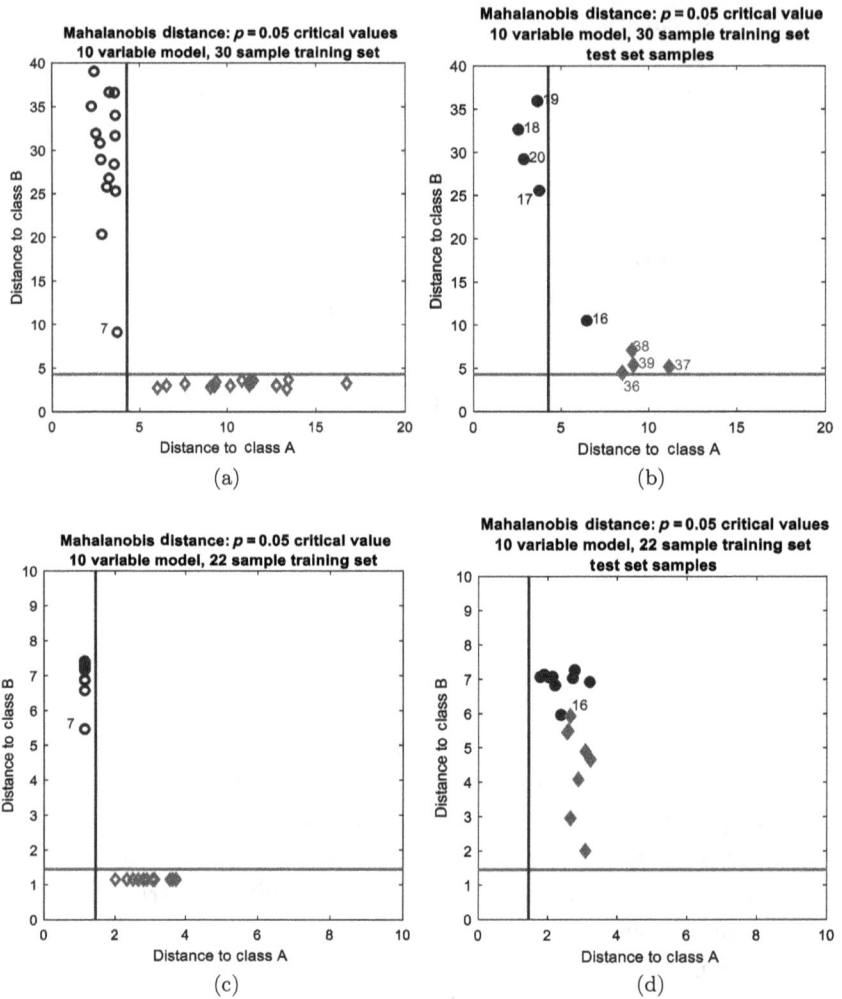

Figure 6.12. (a) Class distance plot for a 10 PC model using a training set of 15 samples each from class A and class B; (b) idem for 9 test set samples, as for (c) but training set of 11 samples for each class and as (d) but training set of 11 samples for each class.

- If you check the sum of the squares of all Mahalanobis distances of each class to itself, $\sum^{i \in A(\text{train})} d_{iA}^2 = 15 \times 10 = 150$, for both classes.
- All the distances of the training set samples appear to be within the $p = 0.05$ limit.

However, we can also calculate the distances for the nine test set samples (samples 16–20 of class A and 36–39 of class B). In this case, we do not recalculate the variance–covariance matrix or mean so that, for example,

$$d(\text{test})_{iA} = \sqrt{((x(\text{test})_i - \bar{x}(\text{train})_A)S(\text{train})_A^{-1}(x(\text{test})_i - \bar{x}(\text{train})_A)')}.$$

We leave it to the reader to perform these calculations in full.

To check, we present the values of d for the nine test samples in Table 6.2 and graphically in Figure 6.12(b):

- The first thing to note is that sample 16 is clearly a member of neither class, so it is an outlier.
- All the other four members of class A test set are within the critical value of this class.
- However, the four members of class B test set seem to be just slightly outside the $p = 0.05$ critical value, which, at first, is unexpected.

One issue is that, because 10 variables have been used to model 15 samples, each sample in the training set is very tightly modelled for

Table 6.2. Values of distance for nine test samples, using a 10-variable training set model of the remaining 30 samples from the case study. The critical value at $p = 0.05$ is 4.279, and samples outside these values are shaded.

	Distances	Class A	Class B
Class A	16	6.45	10.54
	17	3.75	25.57
	18	2.56	32.65
	19	3.65	35.93
	20	2.87	29.21
Class B	36	8.49	4.47
	37	11.16	5.15
	38	9.03	7.08
	39	9.11	5.36

class B, so any new sample is not likely to fit the model, and the test set samples are slightly outside the critical values.

This is not so for class A because sample 7 has a significant influence, being very far from the centre, and hence the training set model has to encompass this sample. The test set samples are much more similar to the remaining training set samples and thus appear as part of this rather widely modelled class. The number of training samples that are used to develop any model has a critical influence on the quality of the predictions that will arise from it.

> The number of samples in the training set has a crucial influence on the predictions.

The minimum number of samples, if we use all 10 variables, is 11 from each class. For brevity, we will not describe all the calculations in detail; however, the training set (samples 1–11 of class A and samples 20–31 of class B) is presented graphically in Figure 6.12(c), and the remaining test set samples are shown in Figure 6.12(d), and we leave these for the reader to verify.

We note the following:

- All the training set samples are at the same distance from their centres for both classes.
- This is given by $d = \sqrt{10} = 3.162$.
- This is because the sum of the squares of the distances of all 11 samples equals 110, so the average is 10.
- All the test set samples are outside the critical limits.
- This is because all training set samples are perfectly modelled, but none of the test set.
- Hence, in this case, the training set is overfitted, and the model is of no value; consequently, there are either too many variables or too few samples to obtain a meaningful answer in this case.

> When the number of samples in the training set is one more than the number of variables, all training set samples are perfectly modelled, but none of the test set.

In this case, we see that although it is important to validate models using independent training and test sets, the samples in the training set and the number of variables are crucial number of:

- Sample 7 has a considerable influence on the training set of class A.
- Too few samples in the training set leads to it being overfitted, so that the model is not valid for samples outside the training set.
- Too many samples, if these include outliers, can also lead to problems.
- Ideally, there must be several more samples than variables in the test set.

As this section is already long, we will not expand the results, and leave them as practice for the readers; however, an ideal approach might be as follows:

- Remove sample 7 from the training set of class A.
- Reduce the number of variables to 7 or 8 through PCA. When doing this, you must perform PCA only on the combined training sets, and estimate the scores of the test set samples from the training set model.

The main message is that it is usual to divide data into a training set, which is used to form a model, and a test set, which is used to determine how well the model performs. The training set normally classifies samples very well, while the test set often performs less well. If the classification performance of both training and test sets is very similar, it indicates that the model developed is a valid one. Readers who are interested may wish to expand on the types of models to find a suitable balance between training and test sets, as well as the number of variables.

An intermediate step between the classification ability (i.e., the capability of correctly classifying the samples of the training set, on which the model has been built) and the prediction ability (i.e., the capability of correctly classifying the samples of the test set, totally external to the samples on which the model has been built) is cross-validation (for more details about this, see Chapter 7 for PCA and Chapter 8 for PLS regression).

Since the results obtained in classification are often overoptimistic, CAT only reports the results from cross-validation and those on an external test set (as seen later for a real case study).

The number of variables and which samples are included in the training set can have a strong influence on the resultant predictions. Training and test sets can, of course, be developed for any type of model; we only illustrate one situation in this section, together with the methods for calculation. An experienced chemometrician may try to develop many different types of training set models.

Additionally, because the nature of the samples in the training set can influence the model, many now use iterative methods, in which training and test sets are repeatedly generated, using a different split of samples between training and test set each time. With modern computers, this can be done within minutes or even less. After typically 100 iterations, an average can be obtained to give an overview, which can be used to ensure that the influence, for example of outliers, is not excessive. This is a vast subject, and we have only provided a simple example in this section.

The training set normally classifies samples very well, but the test set often performs less well. Many now use iterative methods, in which training and test sets are repeatedly generated. With modern computers, this can be done within minutes or even less.

6.5. Discriminatory Variables

The final topic relates to discriminatory variables.

In chemometrics, it is often very important to determine which variables are most diagnostic of differences between classes. For example, in metabolomic profiling, we may wish to determine which metabolites are likely to be markers for a disease. We analyse two groups of donors, one with a disease and one without, and take extracts, for example from blood serum, which we then analyse using a chromatographic method, such as LCMS. In addition to determining whether we can discriminate between the two groups, for example using PCA or LDA, we may seek insights into which metabolites are responsible for this difference. Of course, this does not immediately solve a biological problem but allows us to propose which metabolites are likely to be significant, thereby providing clues for further research.

> It is often very important to determine which variables are most diagnostic of differences between classes.

Among the numerous approaches available, we will only have room to describe two in this introductory text. We illustrate them using autopredictive models for simplicity (that is, models applied to the entire dataset without splitting it into test or training sets), although, as described in the previous section, this is often alternatively done only with a training set. The issue of variable selection was discussed in Chapter 3, albeit in relation to regression purposes.

We use the first 39 samples of Table 6.1, i.e., 20 from class A and 19 from class B, to illustrate the calculations.

The first is a simple univariate approach using the t-statistic. The greater the t-statistic, the more significant the variable is, i.e., the more different it is between the two groups. If we define class A as positive, a positive value of t implies a variable is a marker for class A, and *vice versa*:

- Calculate the mean and standard deviation of the 10 candidate variables with the samples in each class separately, remembering to use the population standard deviation. These will result in four 1×10 row vectors: $\bar{\boldsymbol{x}}_A$, $\bar{\boldsymbol{x}}_B$, $\boldsymbol{s}_A$, and $\boldsymbol{s}_B$.
- To check your calculations, the third elements of these vectors (relating to variable C) should be $\bar{x}_{3A} = 9.35$, $\bar{x}_{3B} = 4.04$, $s_{3A} = 2.00$, and $s_{3B} = 1.17$.
- The pooled standard deviation across both groups is calculated for each variable j, $s_{j\text{pool}}$, which is defined as $s_{j\text{pool}} = \sqrt{\dfrac{(I_A-1)s_{jA}^2+(I_B-1)s_{jB}^2}{(I_A+I_B-2)}}$.

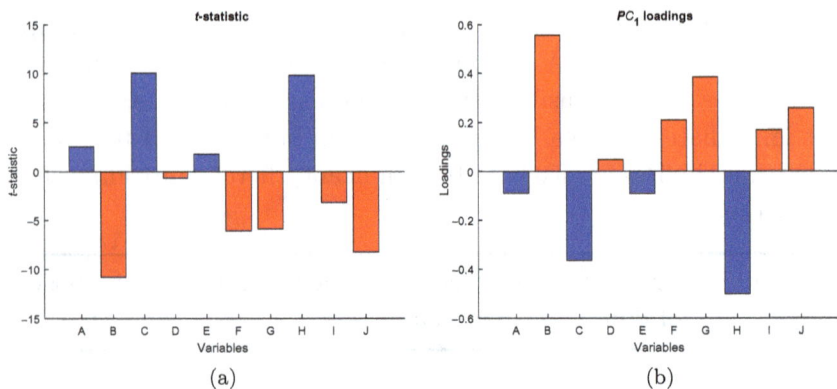

Figure 6.13. (a) The t-statistic of the 10 variables in the case study, and (b) the loadings for PC_1 of the 10 variables in the case study.

- In our case, if $j = 3$,

$$s_{3\text{pool}} = \sqrt{\frac{(30-1)2.0^2 + (29-1)1.17^2}{(30+29-2)}}$$

$$= \sqrt{\frac{29 \times 4.01 + 28 \times 1.38}{57}} = 1.649.$$

- The t-statistic can be calculated by $t_j = \dfrac{\bar{x}_{jA} - \bar{x}_{jB}}{s_{j\text{pool}}\sqrt{1/I_A + 1/I_B}}$.

- So, in our case, $t_3 = \dfrac{\bar{x}_{3A} - \bar{x}_{3B}}{s_{3\text{pool}}\sqrt{1/I_A + 1/I_B}} = \dfrac{9.35 - 4.04}{1.65 \times \sqrt{\left(\frac{1}{20} + \frac{1}{19}\right)}} =$

$\dfrac{5.32}{1.65 \times 0.32} = 10.06.$

We can calculate the t-statistic for each of the 10 variables, as presented in Figure 6.13(a).

- Variables with a positive value of t are diagnostic of class A and negative values of class B.
- However, some of the variables have relatively low values of t and so are unlikely to be strong markers.
- Visually, we see that variables C and H appear to be potentially strong markers for class A, whereas B and J appear to be the best for class B, followed possibly by F and G.

The t-statistic could be interpreted as a p-value indicating whether a variable is a significant marker, using a two-tailed t-test for the difference in means. Some chemometricians might not prefer to interpret these as probabilities because there may be interactions between variables; that is, some variables may be correlated. For example, two metabolites may originate from a single pathway, and so their concentrations could be partially or highly correlated. However, the t-statistic gives us an indication of which variables are most likely to differ between the two groups and is an important first step in a case study, and there are no hard and fast rules.

A simple univariate approach uses the t-statistic. The t-statistic could be interpreted as a p-value.

Alternative methods are based on multivariate indicators. In our case study, PC_1 appears to discriminate well between the two groups. This is not always the case, but if grouping is suggested using exploratory methods, such as a PCA scores plot, the scores can be compared with the loadings.

Alternative methods are based on multivariate indicators.

If we assume from Figure 6.2 that a positive score on PC_1 is most characteristic of class A and a negative score of class B, we should observe a similar trend in the loadings. If we consider PC_1 as the most diagnostic of the class difference, we can produce a bar chart of the loadings of PC_1, as shown in Figure 6.13(b).

As in Figure 6.13(a), class A is characterised by positive values of the PCs. We can see that variables C and H have strong positive loadings and are therefore likely to be most closely associated with class A, arriving at the same conclusion as the t-statistic. Variable B has the strongest negative loading and is most closely associated with class B, whereas variables G and J have the next strongest negative values, followed by F and I.

Our conclusions using both the univariate t-statistic and the multivariate loadings of PC_1 are fairly similar. We do not expect them to be identical, as they are based on different criteria, but they provide preliminary indications of which variables are the most likely markers for each group.

> Our conclusions provide preliminary indications of which variables are markers for each group.

There are, of course, many different ways of predicting which variables are the best discriminators or markers for each group, and in this introductory text, we only have room to illustrate two rather common approaches. It is, however, important to remember that this is only an exploratory investigation. Sometimes, variables may seem to act as markers merely by coincidence. A p-value, for example, of 0.05 means that the achieved value or greater (e.g., a t-statistic) is obtained by chance about once in 20 attempts, even if there were no underlying bias. Hence, given a dataset with 100 variables, which is common for example in metabolomics where a large number of metabolites can be detected, we may obtain on average a p-value of 0.05 or less in five cases. While this does not provide definite proof that the five metabolites are markers, it can be considered a preliminary stage for further investigation to narrow down the possibilities to a small number of candidate compounds.

6.6. A Real Case Study

In this section, we illustrate a worked example using the CAT software and introduce the concept of confusion matrices (which are sometimes called contingency tables, but this term has a wider meaning).

In the CAT folder on your computer, find the subfolder "working", wherein you have the dataset named "*winesclass*". It refers to data collected on three types of wines from Piedmont: Barbera (ERA), Grignolino (GR), and Barolo (OLO). On each sample, 13 chemico-physical variables have been measured. Within the winesclass file, the datasheet "train" contains the samples of the training set (30 samples per class), while the datasheet "test" contains the samples of the test set (29, 41, and 18 samples, respectively).

The goal of the work is to verify, by using the 13 variables, if it is possible to correctly predict the class of an unknown sample. Note that, from a purely mathematical point of view, the number of samples (30 per class) and the number of variables (13) make it possible to use both LDA and QDA.

Load the file as mentioned in previous chapters (Data *Handling* → *load* → *xlx* → *winesclass*), but in this case, *activate the two boxes* called "Header" and "Row Names".

Then, carry out a preliminary PCA (not shown here for brevity). Begin the PCA by following the menu options *PCA* → *Model computation* → *PCA*. In the dialogue box, there is an indication for the name of the matrix; simply enter "winesclass". Indicate the rows to be used (all), and for the columns, indicate "2:end" (because the first column contains labels).

Note that, by default CAT standardises the columns of the raw data (the boxes "centered" and "scaled" are flagged). In general, this is a convenient option, though you can try different alternatives when studying your own data. This can be done as indicated with more details in Chapters 5 and 7. For instance, it is common to use only mean centring when spectra are considered. A number of components are suggested; you can modify the values, but the default values are good enough for this example.

The following conclusions can be drawn:

- The samples of the three classes are quite different, though the class Grignolino (located in the middle) shows some overlap with the other two classes, which are obtained from entirely different geographical regions;
- The size and orientation of the three clouds are quite similar, indicating that the variance–covariance matrices are also quite similar.

To visualise the samples, navigate to *PCA* → *Plots* → *Scores* → *2D Scores plot*, where, in the dialogue window, you can change the number of components to visualise (for this example, set as 1 and 2), elect a vector for the labels (this is not required here), set colours to the three classes (type "winesclass[,1]"), and also click on "Row Names" and "Ellipses".

We restrict the discussion here to an LDA model. To obtain it, follow the next sequence in the menus (*Classification → Model Computation → LDA → Single CV*). In the dialogue window, all the rows should be selected. The X-Variables to be used in the model need to be indicated as "2:end" (as for PCA).

The following output is displayed on the main screen of the software:

Confusion matrix in cross-validation			
	ERA	GR	OLO
ERA	30	0	0
GR	1	29	0
OLO	0	0	30

The confusion matrix is the most efficient way to evaluate the classification/prediction results. It is a square matrix, having as many rows and columns as classes. The values presented there correspond to the counts of numbers of samples. In our case, the values placed on the rows correspond to true (correct) assignments to each class, whereas those on the columns correspond to the class to which they have been assigned (predicted). Some authors transpose the rows and columns, so when reading the literature or websites, please be aware of which convention has been employed.

In this case, we can see that all 30 Barbera samples and all the 30 Barolo samples have been correctly assigned, while 29 out of the 30 Grignolino samples have been correctly assigned, with the mispredicted sample assigned to the class Barbera.

It is very important to check to which class the misclassified samples have been assigned because, in many real cases, the cost of a classification error depends on the classes involved in the error. Let us suppose there are the following classes: A = healthy, B = moderately ill, and C = severely ill. It is obvious that misclassifying a sample that should belong to A as C is a much more serious error than misclassifying it as B.

In the main screen of CAT, you will find information on the wrong classifications. Looking at "Labels of Samples with wrong assignment", you will see "GR25". Further, a list of the misclassified samples is given, so that further investigation on them can be performed.

The percentage of correct cross-validation predictions for each class and the global predictive ability (computed as the average of the percentages) are also displayed on the main screen, as follows:

% Correct predictions in cross-validation		
ERA	GR	OLO
100	96.7	100

% Total correct predictions in cross-validation: 98.9

This value is much more "compact" than the confusion matrix (which is just one single number); however, it provides no information about the class to which the misclassified samples have been assigned or the predictive ability for each single class. Unfortunately, this is the only indicator that many people look at.

Along with the numerical output, three plots are also available. Obtain them by following the menus *Classification → Plots → Model (CV)*. Figure 6.14(a) shows the first one (Mahalanobis Distance), in which the distance of each sample from the category to which it has been assigned is reported. For instance, one can see that sample GR11, although correctly predicted, is quite far from the model of any category, while sample OLO6 is very close to the model of its corresponding category.

The second plot (Mahalanobis Distance (category)) reports the Mahalanobis distance of each sample from a specified category. Figure 6.14(b) shows the distance from the category Barolo; it can be seen that the samples belonging to that class have the smallest distances, while those of category Barbera have the largest. The third plot (Mahalanobis Distance (object)) shows the Mahalanobis distances for a predefined sample from all the classes. From

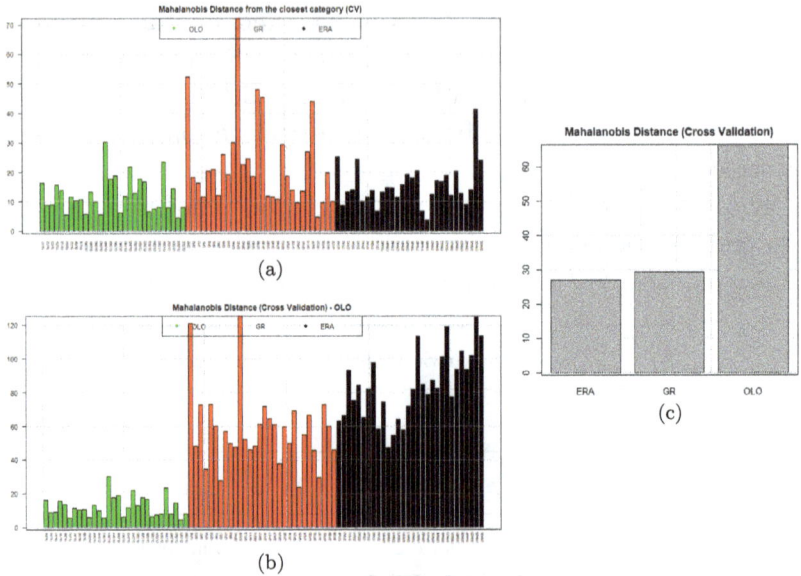

(a)

(c)

(b)

Figure 6.14. The Mahalanobis distance of each sample from (a) the closest category, and (b) the Barolo category and (c) an example of the distance from the three classes for sample GR25.

Figure 6.14(c), for sample GR25, one can see that the misclassified sample is very close to the delimiter between the categories Barbera and Grignolino since the difference between the two Mahalanobis distances is quite small.

When using the model for predictions on a test set (*Classification → Prediction → LDA*), the following output is obtained:

Confusion matrix in prediction

	ERA	GR	OLO
ERA	18	0	0
GR	1	39	1
OLO	0	0	29

Labels of samples with wrong assignment	
GR38	GR63

% Correct predictions		
ERA	GR	OLO
100	95.1	100

% Total correct predictions: 98.4

The performance of prediction is similar to that found in cross-validation. As for training, it is of utmost importance to study the samples of the test set using the same three plots described above (Figure 6.14), following the menus *Classification* → *Plots* → *Prediction*.

6.7. Conclusion

Classification, or PR, is a vast area of chemometrics. It has become especially prominent in recent years owing to the creation and availability of large datasets in application areas such as metabolomics and heritage science.

There are a huge number of classification approaches; in this chapter, we have only focused on LDA in the form of a two-class classifier and QDA as a one-class classifier. The methods can be extended to multi-class problems, in which there are more than two groups. Some common and classical approaches, such as SIMCA, a one-class classifier, and partial least squares discriminant analysis (PLS-DA), a two- or multi-class classifier, are well established within the chemometrics community and form part of most commonly used software packages.

To understand classification and the related algorithms, it is important to understand validation and also determine which variables are most likely to be the best discriminators or markers.

Working through this chapter will provide an understanding of the main principles, which can then be widened, as necessary, to more elaborate algorithms or approaches that guide the investigator in developing a case-by-case strategy for each study. There is, of course, no universal approach that can be used in all situations.

> There is no universal approach that can be used in all situations.

Bibliography

Brereton, R. G. (2009). *Chemometrics for Pattern Recognition*. Chichester: Wiley.

Brereton, R. G. (2011). One-class classifiers. *Journal of Chemometrics*, 25, 225–246.

De Maesschalck, R., Jouan-Rimbaud, D., and Massart, D. L. (2000). The Mahalanobis distance. *Chemometrics and Intelligent Laboratory Systems*, 50, 1–18.

Dixon, S. J. and Brereton, R. G. (2009). Comparison of performance of five common classifiers represented as boundary methods: Euclidean Distance to Centroids, Linear Discriminant Analysis, Quadratic Discriminant Analysis, Learning Vector Quantization and Support Vector Machines, as dependent on data structure. *Chemometrics and Intelligent Laboratory Systems*, 95, 1–17.

Fisher, R. A. (1936). The use of multiple measurements in taxonomic problems. *Annals of Eugenics*, 7, 179–188.

Johnson, R. A. and Wishern, D. W. (1988). *Applied Multivariate Statistical Analysis*. London: Prentice Hall.

Linear discriminant analysis. https://en.wikipedia.org/wiki/Linear_discriminant_analysis.

Quadratic classifier. https://en.wikipedia.org/wiki/Quadratic_classifier#Quadratic_discriminant_analysis.

Chapter 7

Exploring Multivariate Data: Models Based on Latent Variables

María Sagrario Sánchez, María de la Cruz Ortiz,
and Luis Antonio Sarabia

Objectives and Scope

This chapter presents the two most commonly used basic techniques for analysing chemical data: principal component analysis (PCA) for data tables and parallel factor analysis (PARAFAC) for data cubes. Starting from the variables recorded experimentally in a collection of chemical samples, PCA and PARAFAC share the strategy of constructing new variables, which are linear combinations of the original ones, in order to extract the underlying information in the chemical samples. These new variables are called the latent variables and constitute the structure of the data, which must be interpreted in chemical terms. In other words, PCA and PARAFAC are analogous to X-ray radiography; they allow us to see "inside" a data table or a data cube.

7.1.   Two-Way Models: PCA

In 1901, Karl Pearson [1] presented the procedure for obtaining the straight line or plane that best fits a cloud of points. The criterion used for fitting was to minimise the sum of squares of the distances of each point to the intended line or plane, a method currently known as the least-squares criterion. He found a unique mathematical solution whose computation (by hand at that time) was achievable up to four or five variables (that is, a cloud of points in a space with four or five dimensions). Remember that a line or a plane in an n-dimensional space is a linear combination of the n variables, $X_1, X_2, \ldots, X_n$, that define the space.

Hotelling [2], in 1933, using the concept of factor analysis, worked on finding a subset of factors that contains the least number of "fundamental independent variables ... that determine the values of the raw variables". Hotelling chose his "components" in such a way that each sequentially maximised its contribution to the total sum of the variance of the raw variables. He called the components of the variables thus determined "principal components" (PCs). In contrast to Pearson's proposal (which is the current definition of PC), Hotelling proposed the original variables $X_1, X_2, \ldots, X_n$ as a linear combination of the components. In fact, there remains a tradition of considering the PCs as a particular case of factor analysis, which is incorrect and leads to difficulties with mathematical formalisation.

The definitive impetus for the technique arose from Anderson's work [3] in 1963. The books by Jolliffe [4] and Jackson [5] are of interest for the technical details and interpretation of principal component analyses (PCAs). References [6] and [7] are more focused on applications of PCA in chemometrics and also contain some interesting historical notes.

7.1.1.   *Fundamentals of principal component analysis*

Suppose there are V $(X_1, X_2, \ldots, X_V)$ variables measured on N objects or chemical samples $(O_1, O_2, \ldots, O_N)$. Therefore, the data matrix $\mathbf{X} = (x_{ij})$ is an $N \times V$ matrix (with N rows or objects and V columns or variables). Sometimes, it will be denoted as $\mathbf{X}_{NV}$ to clearly specify its size. Each element x_{ij} $(i = 1, \ldots, N, j = 1, \ldots, V)$ is the current value that variable X_j assumes for object O_i. For example, if UV-visible spectra at V wavelengths are recorded for the objects, the ith row in $\mathbf{X}$ is the spectrum of O_i, i.e., the V absorbances $(x_{i1}, x_{i2}, \ldots, x_{iV})$ recorded for O_i at each wavelength.

As stated in Section 2.1.1 of Chapter 2, the N objects "depict" a cloud of points in the V-dimensional space defined by the variables X_j $(j = 1, \ldots, V)$, with each now acting as an individual axis in a Cartesian representation. To describe its structure, among all possible directions, the one following the maximum spread of the cloud of points is selected. This criterion of maximum spread responds to the reason why raw variables X_j are measured, which is to show the differences among objects O_i throughout the variability in the resulting measurements.

Note that the translation of the origin of the coordinate system to the centroid of the cloud of points does not modify the elongation, distance, or relative position of the objects. This translation is performed by subtracting the corresponding mean from the values of the jth variable, i.e., on a per-column basis. This is formally equivalent to working with data that are centred by columns, which is the situation assumed unless otherwise stated.

In that case, for a mean-centred $N \times V$ data matrix $\mathbf{X}$, the variance–covariance matrix, $\mathbf{S_X}$, is a $V \times V$ square and symmetric matrix equal to

$$\mathbf{S_X} = \left(\frac{1}{\sqrt{N-1}} \mathbf{X} \right)^T \left(\frac{1}{\sqrt{N-1}} \mathbf{X} \right). \tag{7.1}$$

With this notation, the first PC, PC_1, is defined as a linear combination of the X_j variables that accounts for the maximum possible variance, which is equivalent to finding the direction of maximum spread of the cloud of points. Formally, finding PC_1 corresponds to finding *loadings* p_{1j}, $j = 1, \ldots, V$, such that Equation (7.2) holds:

$$\max(\text{var}(PC_1)), \text{ where } PC_1 = p_{11}X_1 + p_{12}X_2 + \cdots + p_{1V}X_V$$

$$\text{constrained to } p_{11}^2 + p_{12}^2 + \cdots + p_{1V}^2 = 1. \tag{7.2}$$

The constraint in Equation (7.2) is necessary to avoid the trivial solution of maximising the variance simply by arbitrarily increasing the loadings.

For a given object $O_i = (x_{i1}, x_{i2}, \ldots, x_{iV})$ its *score* on PC_1, t_{i1}, is computed by substituting the actual values of x_{ij} in Equation (7.2), as shown in Equation (7.3):

$$t_{i1} = p_{11}x_{i1} + p_{12}x_{i2} + \cdots + p_{1V}x_{iV}, \quad i = 1, \ldots, N. \quad (7.3)$$

The variance of the scores (all N scores) is the variance of the first PC:

$$\text{var}(PC_1) = l_1. \quad (7.4)$$

Remember that, except for the constant $1/(N-1)$, $\text{var}(PC_1)$ is the sum of squares of the scores on PC_1, denoted as SS_{PC1}.

Table 7.1 shows a data matrix $\mathbf{X}$ with $N = 10$ rows consisting of objects O_i and $V = 2$ columns, which are the centred variables $\mathbf{x_1}$ and $\mathbf{x_2}$. The loadings of the first PC, PC_1, are in the row vector $\mathbf{p_1} = (0.87, -0.50)$ and $\mathbf{t_1}$ is the (column) vector with the scores of the 10 objects on PC_1, whose variance is 8.11 (penultimate row in Table 7.1).

Table 7.1. Dataset for the worked example.

	X		T		P		
	$\mathbf{x_1}$	$\mathbf{x_2}$	$\mathbf{t_1}$	$\mathbf{t_2}$	$\mathbf{p_1}$	0.87	−0.50
O_1	−3.74	2.34	−4.41	0.18	$\mathbf{p_2}$	0.50	0.87
O_2	−2.84	1.84	−3.38	0.19			
O_3	−2.13	0.49	−2.10	−0.63			
O_4	−1.15	1.04	−1.51	0.33			
O_5	−0.63	−0.41	−0.35	−0.67			
O_6	0.66	0.14	0.50	0.45			
O_7	1.27	−1.11	1.65	−0.34			
O_8	2.36	−0.76	2.42	0.51			
O_9	2.76	−1.16	2.97	0.36			
O_{10}	3.46	−2.41	4.20	−0.38			
Sum of squares	55.55	19.33	73.02	1.86			
Variance	6.17	2.15	8.11	0.21			
Total variance (%)	74.16	25.84	97.52	2.48			

The second PC, PC_2, is a new linear combination of variables X_j that maximise the variance that PC_1 did not explain. This idea is formalised with the constraint that PC_2 is uncorrelated with PC_1. Precisely, PC_2 is computed as in Equation (7.5):

$$\max(\text{var}(PC_2)), \text{ where } PC_2 = p_{21}X_1 + p_{22}X_2 + \cdots + p_{2V}X_V,$$

$$\text{with constraints } p_{21}^2 + p_{22}^2 + \cdots + p_{2V}^2 = 1,$$

$$\text{corr}(PC_1, PC_2) = 0. \tag{7.5}$$

Analogous to Equations (7.3) and (7.4), the scores of the ith object on PC_2 is computed as in Equation (7.6), and the variance of all the scores on PC_2 is the one in Equation (7.7):

$$t_{i2} = p_{21}x_{i1} + p_{22}x_{i2} + \cdots + p_{2V}x_{iV}, \quad i = 1, \ldots, N; \tag{7.6}$$

$$\text{var}(PC_2) = l_2. \tag{7.7}$$

Table 7.1 also contains loadings $\mathbf{p}_2$ (second row of $\mathbf{P}$) and scores $\mathbf{t}_2$ (second column of $\mathbf{T}$) for $\mathbf{X}$. This second component has a variance of 0.21, which is the 2.48% of the total variance of the scores.

The procedure explained continues until PC_V. However, if $N < V$ (fewer objects than variables), the cloud of points would be, at most, in an N-dimensional space, not in V dimensions. In that case, no more than N PCs make sense.

In what follows, it is assumed that $V < N$, so that V PCs are computable. This is the case of matrix $\mathbf{X}_{NV}$ of Table 7.1, which has $V = 2$, $N > 2$, so at most there are two PCs to be obtained.

In general, for $V < N$, calculating the last possible PC, PC_V, is about obtaining the vector of loadings $(p_{V1}, p_{V2}, \ldots, p_{VV})$ such that PC_V explains the maximum variance not explained by the previous PCs, as formally expressed in Equation (7.8):

$$\max(\text{var}(PC_V)), \text{ where } PC_V = p_{V1}X_1 + p_{V2}X_2 + \cdots + p_{VV}X_V,$$

$$\text{with constraints } p_{V1}^2 + p_{V2}^2 + \cdots + p_{VV}^2 = 1,$$

$$\text{corr}(PC_1, PC_V) = 0,$$

$$\mathrm{corr}(PC_2, PC_V) = 0,$$

$$\cdots$$

$$\mathrm{corr}(PC_{V-1}, PC_V) = 0. \tag{7.8}$$

As in the first two PCs, Equations (7.9) and (7.10) hold:

$$t_{iV} = p_{V1}x_{i1} + p_{V2}x_{i2} + \cdots + p_{VV}x_{iV}, \quad i = 1,\dots,N; \tag{7.9}$$

$$\mathrm{var}(PC_V) = l_V. \tag{7.10}$$

The principal components (PCs) are a set of new coordinates built as linear combinations of the original variables, orthogonal to one another, and accounting for decreasing amounts of variance of the data.

Taken together, $PC_1, PC_2, \dots, PC_V$ are a new set of orthogonal axes in the original space in which the cloud of points was represented with X_j. This can be seen in Figure 7.1(a) for $\mathbf{X}$ in Table 7.1, where the solid lines represent $\mathbf{x}_1$ and $\mathbf{x}_2$ and the dotted lines are PC_1 and PC_2.

The scores calculated with Equations (7.3), (7.6), and (7.9) are the coordinates of the objects on these new axes (Figure 7.1(b)). In matrix form, Equation (7.11) holds:

$$\mathbf{T}_{NV} = \mathbf{X}_{NV}\mathbf{P}_{VV}^T. \tag{7.11}$$

In Equation (7.11), the jth row $\mathbf{p}_j$ of $\mathbf{P}_{VV}$ contains the loadings of PC_j. By construction, $\mathbf{P}$ is an orthonormal matrix, that is, its rows have unitary length and $\mathbf{P}_{VV}^{-1} = \mathbf{P}_{VV}^T$ (Section 2.2.1 in Chapter 2). Therefore, by multiplying Equation (7.11) on the right with $\mathbf{P}_{VV}$, we obtain Equation (7.12), the essential equation of the PCA of matrix $\mathbf{X}$:

$$\mathbf{T}_{NV}\mathbf{P}_{VV} = \mathbf{X}_{NV}. \tag{7.12}$$

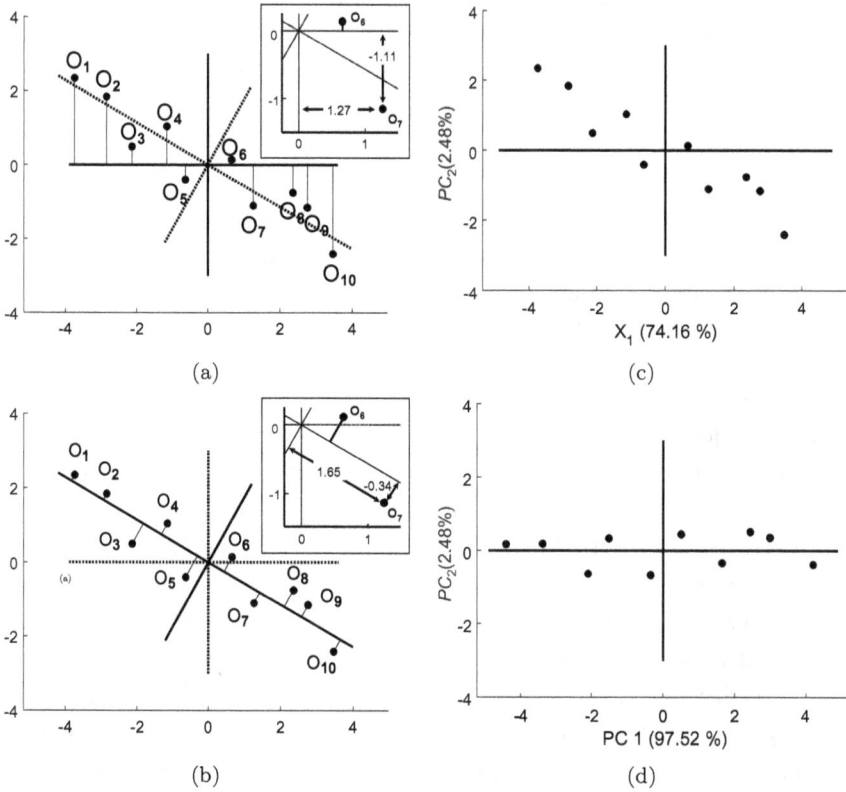

Figure 7.1. Representation of the samples in **X** from Table 7.1. (a) Projection of each object onto the axes in solid lines defined by the variables X_1 (on abscissas) and X_2 (on ordinates). The enlarged area corresponds to the values of object O_7. (b) Projection of the same objects onto the PCs, solid lines. The enlarged area corresponds to the scores of O_7. (c) Values of the original variables. (d) Scores on the PCs. In (c) and (d), the corresponding percentage of variance per axis is indicated.

The relations defined so far between the original dataset and loadings and scores of its PCA decomposition can be summarised as

$$\mathbf{X} = \mathbf{TP} \quad \text{or} \quad \mathbf{T} = \mathbf{XP}^T,$$

where $\mathbf{X}$ is a full rank $N \times V$ matrix with centred data $(V < N)$, $\mathbf{T}$ is the $N \times V$ matrix of scores, and $\mathbf{P}$ is the $V \times V$ matrix of loadings (arranged as rows).

Consequently, with a vector $\mathbf{x}_{\text{new}}$, with the values of X_j in a new object, different from the ones used to compute the PC decomposition in Equation (7.12), its scores are obtained by applying Equation (7.11) to obtain $\mathbf{t}_{\text{new}} = \mathbf{x}_{\text{new}}\mathbf{P}^T$.

Figure 7.1(a) depicts the 10 samples of $\mathbf{X}$ in Table 7.1 in terms of $\mathbf{x}_1$ and $\mathbf{x}_2$. The enlarged region illustrates the corresponding coordinates with one of the objects, the seventh one, O_7. The dotted lines mark the two PCs, whose loadings, i.e., the vectors that define the two directions, are the $\mathbf{p}_1$ and $\mathbf{p}_2$ rows of matrix $\mathbf{P}$ in Table 7.1.

Figure 7.1(b) depicts the same cloud of points but with the scores highlighted, that is, the coordinates of the objects on the PCs, which are now represented by solid lines, with the original variables in dotted lines. The scores of object O_7 are indicated in the enlarged area of the graph. It is evident that the PCs are an orthogonal rotation of the initial axes and that the position and relative distance between the points is maintained irrespective of whether they are represented with the initial values of the variables in Figure 7.1(c) or represented in the PCs, as in Figure 7.1(d).

The numerical information on the dispersion of the cloud of points is also shown in the last rows of Table 7.1. For example, the scores on the first component explain much more variance than those on the second (8.11 against 0.21), representing a reduction by a factor of 38.6, while with the values of the original variables this factor is only 2.87.

Moreover, this variance distribution provides additional information: the scores on the first component increase regularly from the

first object to the last. This also occurs if we observe the values of
the variable X_1 but with much less dispersion. This can be visualised
in Figures 7.1(c) and 7.1(d), which compare the representation of the
same points on the original variables and on the PCs, respectively,
along with an indication of the percentage of variance explained by
each axis. The same scale has been used on the axes of both graphs
to facilitate the comparison.

However, there are differences: the sign of the scores on the second
component separates objects 3, 5, 7, and 10 (negative sign) from the
six remaining points (positive sign), whereas when the sign of the
objects in the second variable X_2 is considered, X_2 separates objects
5, 7, 8, 9, and 10 (negative sign) from the remaining five (positive
sign).

This information will lead to a study of the four objects,
separated by the second PC, to determine whether they share a
property that the others do not exhibit. This latent structure is
not visible in the original values; however, undoubtedly, its interest
depends on each problem and on the information available to the
researcher when assigning a chemical meaning to (i.e., interpretating)
each component.

Note that Figure 7.1(d) is the usual representation of the scores,
although in a typical representation, the axes do not share a common
scale (except in the CAT software, where the same scale on the
axes is maintained when plotting scores and also in the graphical
representation of loadings). Therefore, great attention must be paid
to scaling the axes; and remember, it is the axes – not the points –
that change when performing PCA. Recall, here, what you have
learnt in Chapter 6.

7.1.2. *Decomposition of the sum of squares*

The variance of the jth variable, var(X_j), is the jth element on the
diagonal of the variance–covariance matrix $\mathbf{S_X}$ in Equation (7.1).
Since the mean of the variables is zero because we are working with
column-wise mean-centred data, the sum of squares (SS) of the values
of X_j is equal to $SS_{Xj} = (N - 1)\text{var}(X_j)$, which is, except for a

constant factor, the jth diagonal term of matrix $\mathbf{X}^T\mathbf{X}$. Consequently, the sum of squares of all the terms in $\mathbf{X}$ is the trace of $\mathbf{X}^T\mathbf{X}$, which is a square symmetric matrix (remember that the trace, Tr, is the sum of the elements on the main diagonal of a square matrix, as in Equation (2.10) in Section 2.2 of Chapter 2).

Because of the way the PCs are constructed, i.e., pairwise uncorrelated, the variance–covariance matrix of the scores, $\mathbf{S_T}$, is a diagonal matrix whose diagonal elements are $l_1, l_2, \ldots, l_V$ in Equations (7.4), (7.7), and (7.10) and zero in all the off-diagonal elements. Since the jth term of the diagonal of $\mathbf{T}^T\mathbf{T}$ is the sum of squares of the scores on PC_j, SS_{PCj}, $TrTr(\mathbf{T}^T\mathbf{T})$ is the sum of squares of all PCs.

Let row vectors $\mathbf{x}_i$ and $\mathbf{t}_i$ denote the ith rows of $\mathbf{X}$ and $\mathbf{T}$, respectively, identifying object O_i. Considering their relation according to Equation (7.11) and that $\mathbf{P}$ is an orthonormal matrix, the equalities in Equation (7.13) hold, and the conclusion is that the sum of squares of $\mathbf{X}$ is equal to the sum of squares of $\mathbf{T}$:

$$SS_{\mathbf{X}} = \sum_{i=1}^{N} \mathbf{x}_i \mathbf{x}_i^T = \sum_{i=1}^{N} \mathbf{x}_i \mathbf{P}_{VV}^{-1} \mathbf{P}_{VV} \mathbf{x}_i^T = \sum_{i=1}^{N} \mathbf{x}_i \mathbf{P}_{VV}^T (\mathbf{x}_i \mathbf{P}_{VV}^T)^T$$

$$= \sum_{i=1}^{N} \mathbf{t}_i \mathbf{t}_i^T = SS_{\mathbf{T}}. \tag{7.13}$$

Thus as a consequence, the PCs maintain the total sum of squares $SS = SS_{X_1} + SS_{X_2} + \cdots + SS_{X_V}$, which is distributed in different addends $SS = SS_{PC_1} + SS_{PC_2} + \cdots + SS_{PC_V}$. The invariance of the total sum of squares together with the sequential construction of the PCs (hence, of the sums of squares) show that the PCs minimise the unexplained SS. This is so because PC_1, by construction, has the largest variance and, therefore, the largest sum of squares; however, since the total sum of squares is constant, it turns out that the sum of squares that PC_1 does not explain is the smallest possible. The same occurs with the other components.

Table 7.1 shows the equality between the total sums of squares of the original variables $(55.55 + 19.33)$ and those of the scores $(73.02 + 1.86)$.

7.1.3. *Bilinear model*

Equation (7.12) is also referred to as a bilinear decomposition or a bilinear model of $\mathbf{X}$. To describe this concept, note that Equation (7.12) expresses the generic element x_{ij} of $\mathbf{X}$, as in Equation (7.14), i.e., as the dot product (Equation (2.3) in Section 2.1.2 of Chapter 2) of the ith row of $\mathbf{T}$, $\mathbf{t}_i = (t_{i1}, t_{i2}, \ldots, t_{iV})$, and the jth column of $\mathbf{P}$, $\mathbf{p}_j = (p_{1j}, p_{2j}, \ldots, p_{Vj})^T$:

$$x_{ij} = \sum_{h=1}^{V} t_{ih} p_{hj}. \qquad (7.14)$$

As a consequence, $\mathbf{X}$ is decomposed as the sum of V matrices, $\mathbf{X}_h$ ($h = 1, \ldots, V$), defined by the scores $\mathbf{t}_h = (t_{1h}, t_{2h}, \ldots, t_{Nh})^T$ – the hth column of $\mathbf{T}$ – and loadings $\mathbf{p}_h = (p_{h1}, p_{h2}, \ldots, p_{hV})$ – the hth row of $\mathbf{P}$ – of the hth PC. Each product $\mathbf{t}_h \mathbf{p}_h$, $h = 1, \ldots, V$, is one of the matrices of the bilinear decomposition of $\mathbf{X}$.

For the data matrix in Table 7.1, Table 7.2 shows the two matrices of the bilinear model, $\mathbf{x}_1$ and $\mathbf{x}_2$. The reader can verify that their sum is $\mathbf{X}$ (except maybe for some rounding errors). The reader is also encouraged to graph both matrices.

Table 7.2. Bilinear decomposition of $\mathbf{X}$ in Table 7.1.

	$\mathbf{x}_1 = \mathbf{t}_1\,\mathbf{p}_1$		$\mathbf{x}_2 = \mathbf{t}_2\,\mathbf{p}_2$	
O_1	−3.83	2.19	0.09	0.15
O_2	−2.94	1.68	0.09	0.16
O_3	−1.82	1.04	−0.31	−0.55
O_4	−1.31	0.75	0.16	0.29
O_5	−0.30	0.17	−0.33	−0.58
O_6	0.43	−0.25	0.22	0.39
O_7	1.43	−0.82	−0.17	−0.29
O_8	2.10	−1.20	0.25	0.44
O_9	2.58	−1.47	0.18	0.31
O_{10}	3.65	−2.08	−0.19	−0.33

7.1.4. *Algebraic solution*

With the notation already established, the diagonal variance-covariance matrix of the scores, $\mathbf{S_T}$, already mentioned is

$$\mathbf{S_T} = \left(\frac{1}{\sqrt{N-1}}\mathbf{T}\right)^T \left(\frac{1}{\sqrt{N-1}}\mathbf{T}\right). \qquad (7.15)$$

If $\mathbf{T}$ from Equation (7.11) is substituted into Equation (7.15), the resulting Equation (7.16) is the relation between the loadings of the PCs and the variance–covariance matrix of $\mathbf{X}$ in Equation (7.1):

$$\mathbf{S_T} = \frac{1}{N-1}(\mathbf{XP}^T)^T(\mathbf{XP}^T) = \frac{1}{N-1}\mathbf{P}(\mathbf{X}^T\mathbf{X})\mathbf{P}^T = \mathbf{PS_XP}^T. $$
$$(7.16)$$

This relation shows that PCA is equivalent to the eigenvalue-eigenvector problem for $\mathbf{S_X}$, introduced in Section 2.3 of Chapter 2, together with the singular value decomposition (SVD). Also, the definition of positive semi-definite matrices is given in Section 2.2.

Because $\mathbf{S_X}$ is a positive semi-definite square symmetric matrix, the eigenanalysis also refers to the SVD introduced by C. Jordan in 1874 (as stated in Ref. [8]). Matrix $\mathbf{S_X}$ verifies the conditions of Jordan's theorem, and its SVD decomposition gives $\mathbf{P}$ (orthogonal) and $\mathbf{S_T}$ (diagonal) with the relation in Equation (7.16). Then, the matrix of scores, $\mathbf{T}$, is obtained from $\mathbf{P}$ and $\mathbf{X}$ using Equation (7.11).

7.1.5. *Algorithmic implementation*

There are several computational procedures to obtain the $\mathbf{T}$ and $\mathbf{P}$ of PCA for a given data matrix $\mathbf{X}$. Their detailed analysis is outside the scope of this chapter, which is intended to be an introduction for the use of this multivariate technique to analyse data, but it can be interesting to grasp the basics of the two main types of current algorithms.

The first type comprises those that sequentially compute each component. Among them, the so-called "power method" and NIPALS stand out. The power method computes the first eigenvector by maximising its associated eigenvalue, then computes the residual matrix, and repeats the procedure with the consecutive resulting residual matrices. In chemometrics, NIPALS is more widespread. This method works by computing the first loading, $\mathbf{p}_1$, and score, $\mathbf{t}_1$, vectors (the ones with the largest size of $\mathbf{t}$). Then the contribution of this PC_1 is subtracted from $\mathbf{X}$, and as in the power method, the computation is repeated with the new matrix to find vectors $\mathbf{p}_2$ and $\mathbf{t}_2$, and so on. Both algorithms use multivariate linear regression by least squares. Furthermore, NIPALS can handle missing data in $\mathbf{X}$ because it does not need to compute covariance matrices. These procedures are advisable for large V (many variables) if a few PCs are of interest.

NIPALS is the algorithm implemented in CAT. Therefore, as previously stated, CAT can perform PCA also on data sets with missing data.

The second type of algorithms are those that compute the eigenvalue-eigenvector decomposition, equivalent to PCA (as shown in the previous section). Currently, they have become very efficient and accurate, once the problems associated with working with many similar small eigenvalues have been solved.

7.1.6. *Some comments about autoscaling in PCA*

Useful pretreatments in chemometrics, both general and those specific to data generated by a particular analytical technique, have been described in Chapter 5. In this section, we focus on the effect of different numerical expressions on PCA for the same experimental variables.

As a general rule, different magnitudes for expressing the variables X_j, $j = 1, \ldots, V$, should be avoided. Different magnitudes may occur because the variables are of different nature and, as such, are measured on different scales, such as pH and absorbance. In addition, the effect can be due to the characteristics of the sample, even when the variables have been obtained using the same technique, for example, chromatography. The peaks corresponding to minor compounds will have systematically lower values than those of the others.

Table 7.3. PCA of data matrix **X**.

	X		T		P		
	x_1	x_2	t_1	t_2	$\mathbf{p_1}$	−0.16	0.99
O_1	−3.74	23.40	23.70	−0.02	$\mathbf{p_2}$	0.99	0.16
O_2	−2.84	18.38	18.60	0.08			
O_3	−2.13	4.93	5.20	−1.33			
O_4	−1.15	10.36	10.42	0.50			
O_5	−0.63	−4.06	−3.91	−1.26			
O_6	0.66	1.41	1.29	0.87			
O_7	1.27	−11.09	−11.16	−0.49			
O_8	2.36	−7.60	−7.88	1.13			
O_9	2.76	−11.61	−11.89	0.90			
O_{10}	3.46	−24.12	−24.37	−0.37			
Sum of squares	55.52	1932.80	1981.82	6.85			
Variance	6.17	214.76	220.20	0.76			
Total variance (%)	2.79	97.21	99.66	0.34			

This effect is important because the magnitude of the values of variables $X_1, \ldots, X_V$ can modify the interpretation of PCA. For example, Table 7.3 shows the PCA of the data in Table 7.1 with the only difference being that the second column of **X**, corresponding to variable X_2, is now expressed in a unit 10 times smaller (e.g., millilitres instead of centilitres). Conceptually, this change does not affect the chemical meaning of the two components or sources of variability in the data, but it does affect the variance of X_2, which in the case of Table 7.3 is a hundred times greater than in the case of Table 7.1. As a consequence, the first PC explains 99.7% of the variance, compared to 97.5% with the data in Table 7.1; moreover, the second PC has become more irrelevant in terms of explained variance.

Nevertheless, the most important effect is that now the loadings in $\mathbf{p_1} = (-0.16, 0.99)$ indicate that PC_1 depends almost only on the second variable, X_2, and very little on the first, X_1. However, in the original data in Table 7.1, the loadings of the first PC were $\mathbf{p_1} = (0.87, -0.50)$, which indicate the contribution of both variables in the construction of PC_1, although the first has a weight 1.74 times greater than the second. In addition, the direction of the first PC is completely different: the slope of the line that represents it varies from $-0.57 = -0.50/0.87$ with the $\mathbf{X}$ values in Table 7.1 to $-6.19 = 0.99/(-0.16)$ with the $\mathbf{X}$ values in Table 7.3.

There are several scaling procedures to avoid the effect of the different magnitudes. The most common in chemometrics is the so-called *autoscaling* (sometimes also referred to as *standardisation*), which consists of subtracting the mean and dividing by the standard deviation on a per-column basis (recall here Chapter 5). This is, in fact, the default option when performing PCA with CAT (boxes "Centered" and "Scaled" flagged).

Whether autoscaling is applied to the data in Table 7.1 or Table 7.3, the same $\mathbf{X}_{\text{aut}}$ is obtained. The reader can check this as well as compute the sums of squares of the autoscaled variables, which are equal to each other (9 in this case), and thus the variances are both equal to 1. When performing PCA with the autoscaled data, the variances explained by the PCs are 96.73% and 3.27%, with loading vectors equal to $\mathbf{p}_1 = (0.71, -0.71)$ and $\mathbf{p}_2 = (0.71, 0.71)$. Thus, both autoscaled variables weigh the same (in absolute value) in the first PC (as well as in the second).

Actually, when autoscaling the data, matrix $\mathbf{S}_{\mathbf{X}\text{aut}}$ in Equation (7.1) is in fact the correlation matrix of the original variables, with its diagonal filled with ones, and thus its trace equal to V (in all the data matrices in the worked example, $V = 2$). An important result is that, in this situation, the loadings are the correlation coefficients between the variable and the PC. In the example, say $p_{12} = -0.71$ is the correlation coefficient between PC_1 and the autoscaled variable X_2.

Attention should be paid when using autoscaled data because the PCs yield coefficients for standardised variables and are therefore less easily interpreted directly. To interpret the PCs in terms of the original variables, each coefficient must be divided by the standard deviation of the corresponding variable [4].

It is left to the reader as an exercise to compute the translation of the data in Table 7.1 or Table 7.3 so that the columns do not have zero mean but some (m_1, m_2) and then perform a PCA with raw and autoscaled data. *Hint*: the translation is done by adding m_1 to every element of the first column corresponding to the values of X_1 and m_2 to the values of X_2. The goal of the exercise is to compare the three analyses to observe the changes in the structure of the different PCs.

In this regard, it should be remembered that PCA is a multivariate technique that operates on sums of squares (variances) and is therefore very sensitive to the scale and origin (of the coordinate system) of the measurements of each experimental variable. This is independent of the fact that other specific pretreatments are used to eliminate baseline or dispersion effects in analytical signals, such as spectroscopic ones.

7.1.7. *How many principal components?*

The previous sections showed the mathematical foundations that, together with the algorithmic implementation, allow the user to study the structure of a covariance or correlation matrix. This is related to the rank of $\mathbf{X}$, as introduced in Section 2.1.3 of Chapter 2. However, the experimental data describe a chemical problem, and as such, the structure of interest is that with chemical meaning or interpretation. Suppose that the data, along with the intended information, contain "noise", which hinders its interpretability and that this noise has little variability when compared to that among objects. In this circumstance, PCA acts as a filter when selecting the first A PCs that contain the relevant information. A separate issue is relating these A selected components to the chemical sources of data variability. In chemometrics, it is usual to refer to A as the chemical rank, distinguishing it from the mathematical rank.

Returning to the issue of separating information from noise in $\mathbf{X}$, it suffices to partition the matrices in Equation (7.12) into submatrices of interest. The vertical line in Equation (7.17) denotes the separation of $\mathbf{T}$ into two blocks: the first A columns constitute a block, and the remaining ones form the other block. Similarly for $\mathbf{P}$, the horizontal line separates the first A rows from the others. Then, matrix multiplication can be performed by using the submatrices as if they were single numbers.

$$\mathbf{X}_{NV} = (\mathbf{T}_{NA}|\mathbf{T}_{N(V-A)}) \left(\frac{\mathbf{P}_{AV}}{\mathbf{P}_{(V-A)V}} \right)$$

$$= \mathbf{T}_{NA}\mathbf{P}_{AV} + \mathbf{T}_{N(V-A)}\mathbf{P}_{(V-A)V} = \hat{\mathbf{X}}_{NV} + \mathbf{E}_{NV}. \quad (7.17)$$

Equation (7.17) also shows the operations with the submatrices, and indicates that any x_{ij} can be written as the sum of a summand attributable to the structure with chemical information, $\hat{x}_{ij}$, and another summand, e_{ij}, related to the noise. The matrix $\mathbf{E}_{NV} = (e_{ij})$ has the least possible sum of squares when A PCs are considered according to Equation (7.13). Due to this property, sometimes PCA is used for data compression.

In any case, the analysis of the internal structure (or latent structure) of a data matrix $\mathbf{X}$ involves determining the number of relevant components A. This is not trivial, and what follows is a summary of the most common criteria, which are calculated directly from the eigenvalues $(l_1, l_2, \ldots, l_V)$:

Mean eigenvalue: It consists of retaining those PCs whose eigenvalue l_i is greater than "the mean eigenvalue" $\bar{l} = (\sum_{j=1}^{V} l_j)/V$. For autoscaled data, $\bar{l} = 1$, and the variables have a variance of one; thus, it seems reasonable to discard those PCs whose variance is less than that of any initial variable. In practice, to take the sample variability into account, 0.7 has been proposed as a threshold value.

Percentage of cumulative variance: The first A PCs explain $100\left[\left(\sum_{j=1}^{A} l_j\right)/\left(\sum_{j=1}^{V} l_j\right)\right]$ percent of the variance. It is a descriptive index, and there is no justification for considering a threshold value as, for instance, 95%.

Broken stick: If a segment of length 100 is randomly partitioned into V parts, the expected length of the kth part is $b_k = 100\left(\sum_{j=k}^{V} \frac{1}{j}\right)/V$. Consequently, the criterion consists of retaining PC_k only if it explains a percentage of variance greater than b_k.

K-index: It is the application of the K-correlation index for a data matrix, based on Shannon's concept of entropy [9]. It has the peculiarity of providing a minimum and maximum value for the number of PCs.

Scree plot: It is the graphical representation of the values l_j as a function of j $(j = 1, \ldots, V)$. It usually has the appearance of a "mountainside" with a very steep slope for the first eigenvalues and a smooth slope for the last ones. The criterion is to retain components up to the first PC that is considered the starting point of the second group in the flatter area. If the eigenvalues are of very different magnitudes, their logarithms are used instead. Although this practice is common, it presents practical difficulties because, sometimes, both groups of eigenvalues are not discernible or because there are several slopes separated by plateaus in the graph.

Among the methods that use the residual sum of squares, the most widespread is *cross-validation* (CV), despite being a computationally expensive procedure. It is a sequential method. Initially, all objects (rows of $\mathbf{X}$) are partitioned into g disjoint groups, called cancellation groups. In each iteration, the objects in the corresponding cancellation group are separated from $\mathbf{X}$ in a submatrix $\mathbf{X}_{\text{val}}$, and the remaining objects form a training matrix, $\mathbf{X}_{\text{tr}}$. Next, $\mathbf{X}_{\text{tr}}$ is used to compute scores and loadings, $\mathbf{T}$ and $\mathbf{P}$, of the corresponding PCA with A PCs. Applying Equation (7.11) to the excluded objects in $\mathbf{X}_{\text{val}}$, their scores are obtained in the model with A components. Finally, with Equation (7.12), the values of the "original" variables (scaled according to the pretreatment applied to data) are estimated, $\hat{x}_{ij}(A)$.

By repeating this process with all cancellation sets, the sum of squares of the differences between the values x_{ij} and their estimates with A PCs $\hat{x}_{ij}(A)$ is obtained as:

$$PRESS(A) = \sum_{i=1}^{N} \sum_{j=1}^{V} [\hat{x}_{ij}(A) - x_{ij}]^2. \tag{7.18}$$

The notation "*PRESS*" stands for "PREdiction Sum of Squares". It is also usual to express *PRESS* as root-mean-square error in cross-validation (*RMSECV*), that is,

$$RMSECV(A) = \sqrt{\frac{PRESS(A)}{NV}}. \tag{7.19}$$

In any case, the criterion is to choose the value of A in which *RMSECV* -or PRESS has a minimum. The presence of a more or less clearly defined minimum depends on the quantity of noise supposedly modelled with the last PCs. However, the presence of several minima separated by plateaus can also occur; this is usually due to the presence of groups of objects that alter the structure of the PCs when they are not in $\mathbf{X}_{tr}$ but in the corresponding cancellation set $\mathbf{X}_{val}$.

Analogous to Equation (7.19), the root-mean-square error ($RMSE$) for a model with A PCs is defined as $RMSE(A) = \sqrt{\frac{\sum_{i=A+1}^{V} SS_{PCi}}{NV}}$, which is equivalent, via Equation (7.17), to $RMSE(A) = \sqrt{\frac{\sum_{i=1}^{N} \sum_{j=1}^{V} e_{ij}^2}{NV}}$.

It is clear that, due to the construction of the PCs, $RMSE$ decreases when A increases. Therefore, it is useful to jointly draw the evolution of $RMSECV$ and $RMSE$ as a function of A because the first value of A for which there is an increase in the "gap" between both curves indicates the number of components to be retained.

Another issue thoroughly discussed in the literature is related to the selection of the number of cancellation groups, g, and their composition. An extreme situation, a particular case of CV, is the so-called "leave one out" (LOO or LOO-CV), in which each cancellation group consists of a single object. LOO tends to point to more components than necessary because it underestimates the ability of the model with A components to reproduce the data. The recommended value of g is between 3 and 7 cancellation groups, although this depends on the number of objects in $\mathbf{X}$.

Figure 7.2 exemplifies a study to decide the number of PCs A, following the procedures explained in the previous paragraphs. In other words, according to the mathematical structure of PCA. The data correspond to the worked example in Section 7.1.10.1 of this chapter about food colourants. Data matrix $\mathbf{X}$ is 16×47, that is, $N = 16$ objects defined by $V = 47$ variables that have been autoscaled.

The eigenvalues (l_i) of the PCs vary between 36.7 and 2.7×10^{-5}, so a logarithmic transformation has been applied. Figure 7.2(a) depicts the scree plot that shows the decimal logarithm of the explained variance $\log_{10}(l_i)$ *versus* the number of components.

Compared to the original scree plot, the logarithmic transformation decreases the slope of the "mountainside" and simultaneously increases the slope of the plateau, but now it allows detecting the

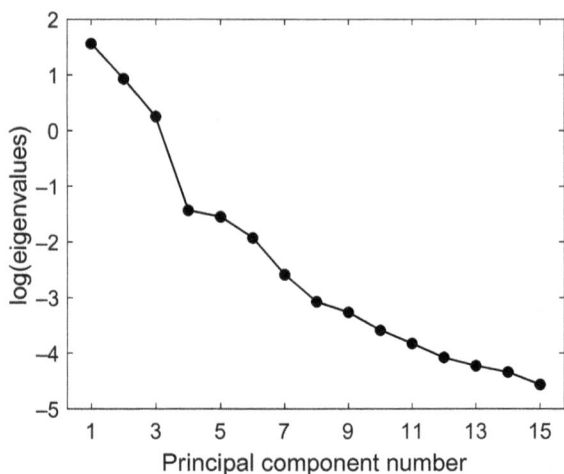

(a)

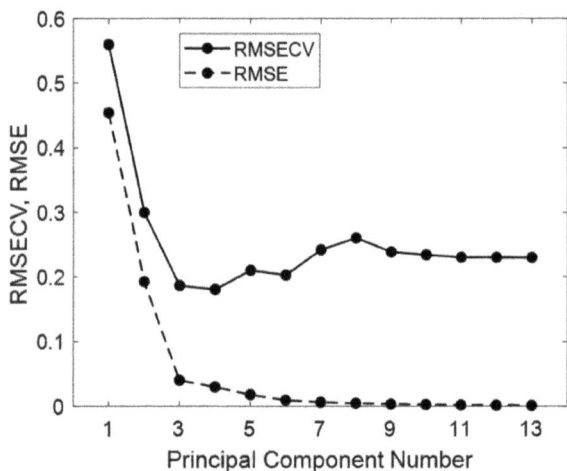

(b)

Figure 7.2. (a) Scree plot with logarithmic transformation of the eigenvalues and (b) root-mean-square errors for selecting the number of components.

position of the gap between both zones, occurring between 3 and 4 PCs. Therefore, the PCA model should have $A = 4$ PCs according to Figure 7.2(a).

Figure 7.2(b) shows the *RMSE* from CV with $g = 6$. The *RMSECV* is 0.1868 for $A = 3$ and 0.1807 for $A = 4$; hence, strictly,

4 PCs should be considered, although 3 would be admissible as well. This is also supported because both curves start separating more clearly for 3 PCs.

According to the mean eigenvalue criterion, $A = 3$ (Table 7.4 in Section 7.1.10). The percentage of variance explained by PC_3 is 3.79% while $b_3 = 12.12$. Thus, the broken stick criterion indicates that $A = 2$, for which $b_2 = 15.45 < 18.03$ that is the percentage explained by PC_2.

In summary, for these data, the usual criteria are not unanimous, though they point to A lying between 2 and 4, while the K-index indicates between 2 and 3 PCs.

Finally, there are also hypothesis tests to decide whether several eigenvalues are equal to one another or not, which serve to detect the absence of structure (sphericity) of the rejected components. Its use is not exempt from criticism and exceeds the scope of this chapter; Refs. [4] and [5] include detailed analyses of these tests.

7.1.8. *Indeterminacy and rotation in PCA*

Equation (7.12) shows the PCA decomposition as a product of loadings and scores. It is clear that for any k, $\mathbf{t}_k \mathbf{p}_k = (-\mathbf{t}_k)(-\mathbf{p}_k)$. That means that the loadings and scores of any PC exhibit an indeterminacy due to their sign, which should be considered when interpreting the results of a PCA.

However, rotational indeterminacy or ambiguity is more important. To explain its meaning, let $\mathbf{R}_{VV}$ be an orthonormal matrix, just like $\mathbf{P}_{VV}$ is. In particular, $\mathbf{R}_{VV}^T \mathbf{R}_{VV} = \mathbf{I}_V$, the identity matrix. Therefore, Equation (7.12) can also be written as

$$\mathbf{X}_{NV} = \mathbf{T}_{NV} \mathbf{P}_{VV} = \mathbf{T}_{NV} \mathbf{R}_{VV}^T \mathbf{R}_{VV} \mathbf{P}_{VV}$$

$$= (\mathbf{T}_{NV} \mathbf{R}_{VV}^T)(\mathbf{R}_{VV} \mathbf{P}_{VV}) = \mathbf{T}_{NV}^* \mathbf{P}_{VV}^*. \qquad (7.20)$$

Equation (7.20) shows new scores and loadings, $\mathbf{T}^*$ and $\mathbf{P}^*$, decomposing the same matrix $\mathbf{X}$. In practice, this rotational ambiguity implies that the loadings we get for a given $\mathbf{X}$ (the PCs) are not necessarily those related to the chemical information of interest. For example, if UV-visible spectra are recorded for mixtures of three compounds, a PCA can help in deciding that there are three sources of variability in the data matrix, but it is not expected that the three loading vectors will be similar to the spectra of the three compounds in the mixture.

Among the orthogonal rotations defined by matrix $\mathbf{R}$, varimax has gained some popularity in chemometrics. It consists of maximising the sum of the variance of the squared loadings [10] (i.e., the squared correlations between variables and PCs). Also, matrix $\mathbf{R}$ in Equation (7.20) can modify only some of the components, not necessarily all of them. Suppose that it affects PC_1 and PC_2, then their loadings and scores will be different, implying that the sum of squares of the scores on those rotated components SS_{PC1^*} and SS_{PC2^*} is different from the original, but their sum is the same because the total sum of squares is constant and the sum of squares of the non-rotated components has not changed. This property is useful for identifying chemical characteristics in the objects that, evidently, do not have to be linked to the distribution of the variance achieved by the mathematical model of PCA. Remember that this PCA model is the unique solution to the problem of finding lines or hyperplanes that best fit a cloud of points in the least-squares sense, i.e., the unique solution to the problem defined by Equations (7.2), (7.5), and (7.8), or the unique solution of the eigenvalue-eigenvector decomposition of the covariance matrix $\mathbf{S_X}$ in Equation (7.1).

On the other hand, PCA of chemical data, particularly from samples whose characteristics are known, must be consistent with that information. This means that the sources of variability present in the samples may be partially known, for example, the presence of at least three compounds in different quantities. However, in practice, specific PCs to detect and identify these sources of variability should not be expected. For example, if two vectors contain UV-visible spectra of two compounds and need to be uncorrelated (a condition that the PCs must fulfil), it is required that the positive coordinates of one vector (spectrum) correspond to the null coordinates of the other. That would mean both compounds absorb in different wavelength ranges. It is evident that, in this case, any multivariate exploratory analysis is superfluous because the signals are specific.

However, these notes of caution regarding the chemical interpretability of a PCA, particularly its loadings, do not reduce the
enormous usefulness of PCA to obtain a radiography of any data
matrix in order to, at least, facilitate the description of a structure
within it. This is shown in Section 7.1.10, which includes three
case studies addressing real problems. In particular, they illustrate
the need to apply with flexibility the rules that mathematically
define the PCs because, obviously, the components of chemical
interest do not have to coincide with the mathematical components,
nor are they expected to be ordered by the magnitude of the
eigenvalues.

7.1.9. *Projection space and its boundaries,*
Q and T^2 statistics

The PCA model (with A PCs) of any given $\mathbf{X}$ (a $N \times V$ column-
centred matrix) decomposes row vector $\mathbf{x}_i = (x_{i1}, x_{i2}, \ldots, x_{iV})$,
corresponding to the ith object, into two summands, as given in
Equation (7.17), $\hat{\mathbf{x}}_i$ defined from its projection onto the space of
the A PCs, i.e., the score $\mathbf{t}_i = (t_{i1}, t_{i2}, \ldots, t_{iA})$, and the residual
$\mathbf{e}_i = (e_{i1}, e_{i2}, \ldots, e_{iV})$.

Regarding $\mathbf{t}_i$, $T_i^2 = \sum_{j=1}^{A} (t_{ij}^2/l_j)$ is the Mahalanobis distance
(defined in Section 2.2.2 of Chapter 2) from $\mathbf{t}_i$ (the projection of $\mathbf{x}_i$)
to the centroid of the PCA space, which in general is the vector
containing the sample means – all zero in this case because the
data are column centred. It measures the position of the point
inside the projection space. But the metric used is not the usual
Euclidean one; rather, it is a metric that weights the distance with
the variance of each component, as seen in the definition of T_i^2. Under
the assumption of multivariate normality, a critical value T_c^2 can be
defined for a given confidence level c, for example, $c = 0.95$ (95%
confidence level).

In this way, any vector $\mathbf{x}$, not necessarily among those used to
build the PCA, whose distance $T_{\mathbf{x}}^2$ is greater than T_c^2 is considered to
be far from the cloud of points in $\mathbf{X}$. Geometrically, the set containing
all the scores whose value of T^2 is less than or equal to the critical

distance, $\{\mathbf{t}|T_\mathbf{t}^2 \leq T_c^2\}$, is a hyperellipsoid in the A-dimensional space, centred on the vector of means of $\mathbf{X}$, with the A PCs as principal axes and semi-axes equal to the corresponding eigenvalue l_i, $i = 1, \ldots, A$.

Regarding the residual $\boldsymbol{e}_i = (e_{i1}, e_{i2}, \ldots, e_{iV})$, its squared norm $Q_i = \sum_{j=1}^{V} e_{ij}^2$ (Equation (2.4) in Section 2.1.2 of Chapter 2) is the squared Euclidean distance between the point $\mathbf{x}_i$ and the PCA space with A PCs. Also, under normal distribution, a critical value Q_c can be defined for a given confidence level c, so that any point $\mathbf{x}$ whose distance Q to the space defined by the loadings $\mathbf{p}_i$, $i = 1, 2, \ldots, A$, is greater than Q_c is said to carry some variability which is not present in the data $\mathbf{X}$ used to build the PCA model with A PCs. In some texts, these points are called *leverage* points.

To emphasise the meaning, sometimes these statistics are referred to as Hotelling's T^2 and Q residuals. In any case, the two critical values T_c^2 and Q_c (for which usually the same confidence level is used) define a bounded region in the projection space related to the PCA model of $\mathbf{X}$ with A PCs. It is customary to present graphically the values of both statistics together with their critical values to identify the objects that surpass them, either in Q, T^2, or both (the so-called influence plot in CAT).

The region inside the projection space has become an essential tool for statistical process control (SPC) and quality assurance. In the former case, $\mathbf{X}$ would represent V process variables, used for producing N products when the process is in control, that is, the process is working properly. In general, these variables are correlated; they may even be redundant, but they obey a latent structure that, in a certain sense, presents an overall "picture" that considers the whole process. The goal of computing a PCA model with A PCs is to make this latent structure explicit and then use it to monitor the process to ensure that it remains under statistical control. This is done at a given confidence level c by checking that any new vector $\mathbf{x}_{\text{new}}$ of process variables verifies both $T_{\text{new}}^2 \leq T_c^2$ and $Q_{\text{new}} \leq Q_c$. This implies maintaining two control charts (plots of T_i^2 and Q_i versus i together with the threshold values T_c^2 and Q_c), which is very useful in practice, particularly when V is large.

Furthermore, and much more importantly, the distance T^2 gathers the covariance (correlation) between the original variables, an aspect that is ignored if the control is done with V individual control charts, one for each original variable. In addition, due to the properties of T^2, when $T^2_{\text{new}} > T^2_c$, it is possible to identify in the vector of scores $\mathbf{t}_{\text{new}}$ the individual coordinate(s) responsible for the out-of-control condition and, through the corresponding loading(s) $\mathbf{p}_i$, the process variable(s) that cause(s) this condition.

An out-of-control condition due to the residuals, $Q_{\text{new}} > Q_c$, in general, indicates that $\mathbf{x}_{\text{new}}$ represents an independent change of the underlying structure of the process. This makes it necessary to examine the vector $\mathbf{e}_{\text{new}}$ for the coordinate(s) that has (have) caused the out-of-control condition in order to inspect in which variables the change has occurred.

Regarding quality assurance, it is evident that the quality of a product is multivariate; it is difficult to think of a case in which quality is defined by a single characteristic. In this situation, each element x_{ij} of matrix $\mathbf{X}$ would be the value of the jth quality characteristic measured on the ith object. Then, the quality control of a new product $\mathbf{x}_{\text{new}}$ with the values of T^2 and Q statistics and the defined threshold values (T^2_c and Q_c) is the same as in the case of process control.

The bounded region inside the projection space formed by all the vectors with values of T^2 and Q statistics less than their threshold values, $\{\mathbf{x}|T^2_{\mathbf{x}} \leq T^2_c \text{ and } Q_{\mathbf{x}} \leq Q_c\}$, is also useful for detecting vectors $\mathbf{x}$ which are different from those in $\mathbf{X}$. There is a discrepancy in the nomenclature throughout the literature, so it is necessary to define the exact meaning of the terms used in each problem. For example, to state that $\mathbf{x}$ is an *outlier* if $T^2_{\mathbf{x}} > T^2_c$ and $Q_{\mathbf{x}} \leq Q_c$, and it is a *leverage* when $T^2_{\mathbf{x}} \leq T^2_c$ and $Q_{\mathbf{x}} > Q_c$. During the PCA construction process, it is usual to eliminate *outlier leverage* ($T^2_{\mathbf{x}} > T^2_c$ and $Q_{\mathbf{x}} > Q_c$) because such samples can have, although not always, a significant and disproportionate effect on the resulting PCA. They can even cause a component just to explain the presence of a single anomalous data. An example of application can be seen later on in Figure 7.4(b) in Section 7.1.10.1.

A projection subspace is bounded with the aid of Q and T^2 statistics. It helps detect outliers, whose presence (and sometimes the data pretreatment as well) can modify the PCA decomposition (i.e., the number of components, their direction, and/or their spread).

7.1.10. *Worked examples and case studies of PCA*

7.1.10.1. **A mixture of two food colourants**

Goal: To qualitatively describe the presence of two additives, tartrazine (E-102) and sunset yellow (E-110), contained in two products used as food colourants. UV-vis molecular absorption spectrophotometry is used to analyse the samples.

Training set: It consists of a 16×47 data matrix $\mathbf{X}$. The 16 objects are samples that contain mixtures of both colourants, prepared with four levels of concentration (0, 2, 4, and $6\,\mathrm{mg\,kg^{-1}}$) of E-102 and at every level increasing concentrations of E-110, also at four levels (0, 2, 4, and $6\,\mathrm{mg\,kg^{-1}}$). The 47 variables are the absorbances recorded by the spectrophotometer between 340 and 570 nm, every 5 nm.

Test set: There are two samples of different food colourants purchased in a local market, which are prepared in triplicate for measurement in the laboratory, recording their spectra in the same wavelength range as the 16 standard mixtures.

The dataset is in the file "colorants.xlsx", sheet PCA. The first column contains the name of the sample, and the remaining 47 columns the absorbances, labelled with the corresponding wavelength (nm). Be aware that the columns are sorted in decreasing order of wavelengths.

The first 16 samples are the training set, while samples 17–22 are the test set. The name of the samples of the training set encodes the levels of the two colourants: "M" followed by their levels (e.g., the sample M26 contains $2\,\mathrm{mg\,kg^{-1}}$ of E-102 and $6\,\mathrm{mg\,kg^{-1}}$ of E-110). Test samples, with unknown quantities of colourants, are also coded

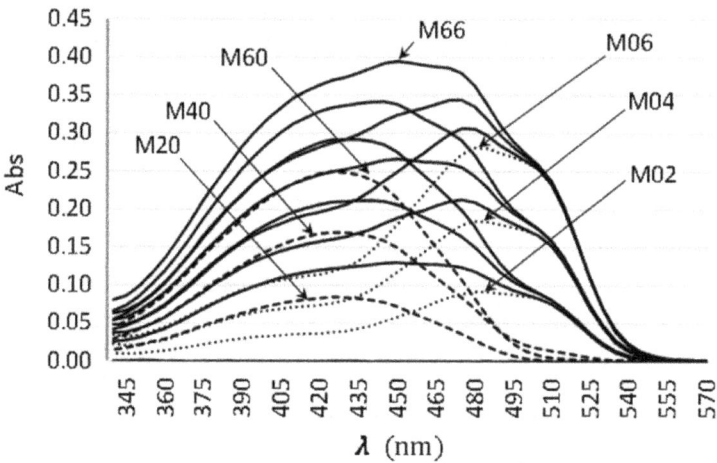

Figure 7.3. Spectra of 16 mixtures of colourants tartrazine and sunset yellow. The digits in the code refer to the quantities of E-102 and E-110, respectively, in mg kg^{-1}.

with M and two digits separated by a dot. The first digit is the sample, and the second is the replicate.

Figure 7.3 shows a clear overlap among the spectra of the 16 samples of the training set. The code to identify some of the spectra (with different line types) is the same as in the Excel file just explained. Looking at the figure, it can be seen that the absorbances at the lowest wavelengths seem to be related to E-102 (dashed lines for "pure" samples), while the absorbances at the highest wavelengths are related to E-110 (dotted lines), though the spectra of the mixture of the two compounds are not so clearly identified.

Guided solution: CAT directly imports the Excel file so that the reader can repeat all the tasks in the present worked example, where all tables and graphs correspond to CAT outputs.

PCA is computed after autoscaling the data in training set **X**. Although 16 PCs are theoretically possible, Table 7.4 shows only the details of the first four PCs, which already account for 99.90% of the total variance.

The analysis to decide the number of PCs to be retained, A, was already discussed in Section 7.1.7. The conclusion there was to use

Table 7.4. Eigenvalues and the explained and cumulative variances when adding principal components for the colourants dataset.

PC number	Eigenvalues	Explained variance (%)	Cumulative variance (%)
1	36.66	78.01	78.01
2	8.47	18.03	96.04
3	1.78	3.79	99.83
4	0.03	0.78	99.90

two or three PCs that jointly explain 96.04% or 99.83% of the total variance of the spectra.

The importance of incorporating chemical knowledge when deciding was also discussed. In this case, the cause of the variability among the signals observed in Figure 7.3 is that they are mixtures prepared in the laboratory containing two colourants. According to this, two PCs ($A = 2$) are selected, whose structure will be a consequence of the design of the mixtures.

> The theoretical methods should be used together with the chemical knowledge of the problem when deciding on the number of principal components.

The scores of the training samples on the first and second PCs are represented in Figure 7.4(a). The position of the scores, together with the chemical knowledge, leads to an interpretation of the components.

Figure 7.4(a) shows two characteristic patterns of the scores. The most extreme positions on the first component, PC_1, that explains 78.01% of the variance, are for the score of the sample with no colourant (M00) on the left, while on the right we have sample M66 (maximum concentration of both colourants). Between these limits, and following the direction of the component from left to right, we see that the scores correspond to samples with increasing total concentration of both analytes; for example, M02 and M20 have the same score in PC_1, approximately the same – greater – score for

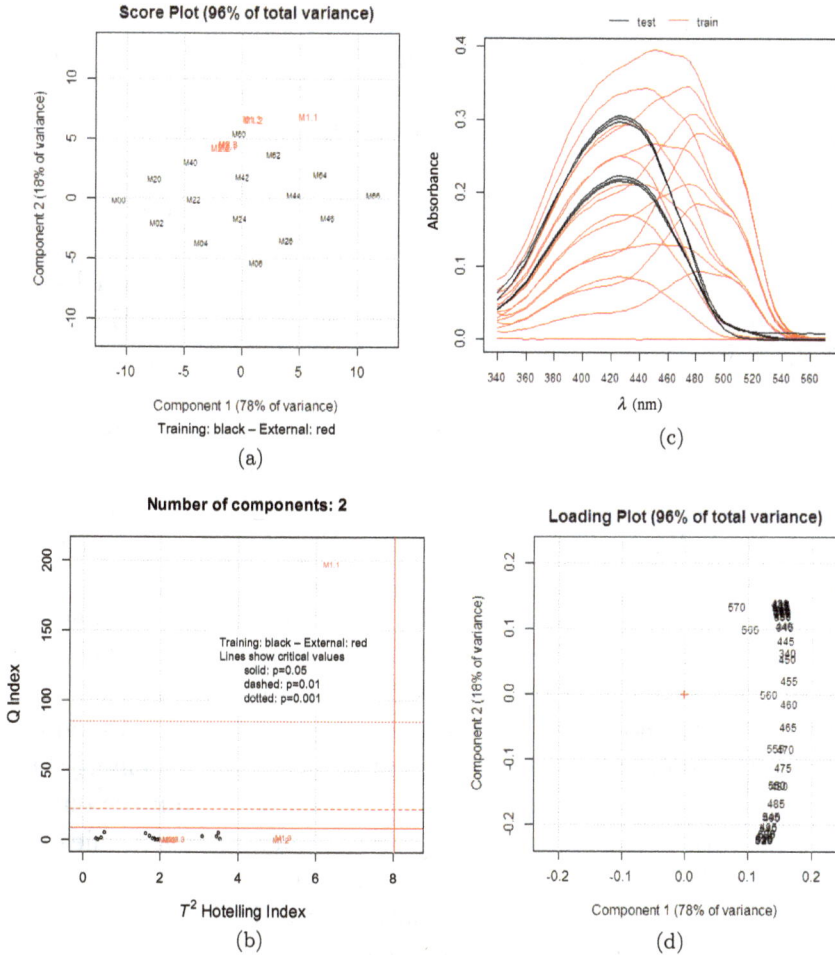

Figure 7.4. PCA for dataset colourants. (a) Scores on the first two PCs. The first digit is the concentration of tartrazine (E-102) and the second that of sunset yellow (E-110), both in mg kg^{-1}. Test samples are in red. (b) Values of T^2 and Q statistics; red lines show critical values (solid: $p = 0.05$; dashed: $p = 0.01$; dotted: $p = 0.001$). (c) Spectra, those of test samples in black. (d) Loadings for the two components, identified by the wavelength number.

samples M06, M24, M42, M60, and so forth. In general, the samples with the same "sum of the digits" have approximately the same score. As a consequence, we can infer that PC_1 represents the total content of colourants, related to the fact that the bigger the total amount of colourant substances, the higher the absorbance in the spectrum.

The scores on the second component PC_2, on the other hand, are approximately the same for samples whose difference between the digits is the same, for example, M04 and M26 or M40 and M62. In addition, the samples with more E-110 (the second digit in the codification) than E-102 have a negative score; the larger the score, the greater the difference between the colourants, whereas those samples with a higher amount of E-102 have a positive score, so that the higher the score is, the greater the difference with E-110. Consistently, the samples with the same concentration of E-110 and E-102 have scores of practically zero on PC_2. Therefore, we can say that PC_2 gathers the variability caused in the spectra by the different amounts of each colourant in the sample, or, in other words, PC_2 is related to the difference in the concentrations of both analytes.

The two components together in Figure 7.4(a) describe the design of the samples, although neither PC_1 nor PC_2 is independently related to a single colourant.

Finally, the red labels in Figure 7.4(a) identify the commercial samples (test samples) projected onto the plane of the first two components. There are three scores, M2.1, M2.2 and M2.3, graphically between M40 and M60 and close to each other, that correspond to the replicates of one of the commercial samples. The remaining three scores for the other commercial sample seen at the top are outside the calibration range prepared in the laboratory. Moreover, one of its replicates, M1.1, is clearly reflecting some anomaly since the scores of the three replicates should be similar, as with the other test sample. Furthermore, Figure 7.4(b) shows that the anomalous sample has a very large residual Q, well outside the 99.9% confidence limit of the Q statistic (horizontal dotted line).

By looking at the original spectra in Figure 7.4(c), it can be seen that one of the samples of the test set in black (indeed, it is sample M1.1) has strange absorbances from 520 nm onwards (actually, they are a non-null constant).

Since there are 47 variables (autoscaled absorbances), the two loading vectors $\mathbf{p}_1$ and $\mathbf{p}_2$ that made up the rows of matrix $\mathbf{P}$ have 47 coordinates. Figure 7.4(d) depicts the individual pairs (p_{1j}, p_{2j}), $j = 1, \ldots, 47$, of loadings of each variable, labelled with the corresponding wavelength. When the data are autoscaled, each loading is the correlation coefficient between the PC and the corresponding variable. That means that, for example, PC_1 has a similar correlation coefficient with all the variables, from 0.08 with the variable "570" to 0.165 with the variable "460", whereas PC_2 has rather different correlation coefficients with the different variables.

The analysis of Figure 7.4(d) allows for deciding the variables that contribute more (or weigh more) in each component and also which variables are correlated with each other and/or with the PCs. It has been pointed out that the 47 loadings, p_{1j}, on the first component are all positive and practically the same, indicating a size factor: all variables increase/decrease simultaneously in all samples. This is consistent with the meaning attributed to PC_1 in the analysis of the scores (Figure 7.4(a)); this component reproduces the sum of the concentrations, that is, the total absorbance of the mixtures. On the contrary, the 47 loadings, p_{2j}, of the second component are negative for the absorbances recorded between 460 and 560 nm and positive from 340 to 455 nm. Samples with a positive score in PC_2 will have a higher absorbance at wavelengths between 340 and 455 nm, while those with a negative score will have a higher value in the range of 460–560 nm. This is consistent with the range in which E-102 and E-110 have their corresponding maximum absorbance, 425 and 480 nm, respectively, and the comments regarding Figure 7.3. Finally, the loadings of wavelengths between 560 and 570 nm, somehow displaced in Figure 7.4(d), correspond to changes in the baseline in that region of the spectral range.

Regarding the correlation between two variables, j and k, it suffices to consider the angle between the lines that join the origin with points (p_{1j}, p_{2j}) and (p_{1k}, p_{2k}). If the angle is close to zero, there is a high positive correlation, while if it is close to 180° (π radians), the correlation is large and negative; angles of 90° ($\pi/2$ radians) indicate no correlation between variables.

Imagining the lines from the origin to the points in Figure 7.4(d), it is clear and coherent that all the wavelengths are positively correlated.

7.1.10.2. Fraud detection

Goal: The aims are the qualitative detection of tallow in adulterated lard and to study the feasibility of gas chromatography to detect fraud because when the amount of tallow in lard is less than 10%, the specific chromatographic peak of tallow may be masked. With these aims, analyses were performed using gas chromatography-flame ionisation detector (GC-FID) with three different chromatographic columns: column A (10% DEGS, WAW 80/100 chromosorb, 3 m, 1/8″), column B (20% DEGS, WAW 80/100 chromosorb, 2 m, 1/8″), and column C (20% DEGS, WAW 80/100 chromosorb, 3 m, 1/8″).

Training set: **X** is a 17 × 5 matrix. The 17 objects correspond to 7 samples of pure lard and 10 samples prepared in the laboratory by adulterating pure samples with increasing percentages of tallow, namely three samples with 5% tallow, another three with 10%, and four with 15%. Regarding the variables, although the percentage areas of the chromatographic peaks of 16 fatty acids were available [11], only five (corresponding to fatty acids C14:0, C16:0, C18:0, C18:1, and C18:2) will be used in the current study. They were chosen with a variable selection criterion that maintains the PCA structure as close as possible to the one with all the variables.

The training set is in the sheet labelled PCA of the "fraud.xlsx" file. Each column is labelled with the corresponding fatty acid. The first two columns contain labels: the chromatographic column used in the analysis in "Chr.Column" and the percentage of tallow added in the samples of lard in "Tallow".

Test set (not available in the Excel file): The matrix with data acquired using chromatograms obtained from "lard" samples of six meat producers that were suspected of adulterating the lard with ruminant tallow. This represents a major economic fraud because ruminant tallow is not used for human food, contrary to lard, which is a widely used product. The code for these manufacturers is 1, 2, 3, 4, 5, and 6.

Guided solution: Table 7.5 shows the decomposition of the PCA performed with autoscaled data. With the criteria explained in

Table 7.5. Eigenvalues and the explained and cumulative variance for dataset fraud.

PC number	Eigenvalues	Explained variance (%)	Cumulative variance (%)
1	4.0032	80.06	80.06
2	0.6199	12.40	92.46
3	0.3016	6.03	98.49
4	0.0567	1.13	99.62
5	0.0185	0.38	100.00

Section 7.1.7, the first three components are considered. They explain 98.5% of the variability in the predictor variables (fatty acids).

Figure 7.5(a) shows that the scores on the plane formed by the first and second PCs (coloured by the different quantities of adulteration) reflect the structure of the calibration samples prepared in the laboratory. These two components explain 80.06% and 12.40% of variance, respectively, as shown in Table 7.5.

Although none of the components is uniquely related to the percentage of adulteration, when considered jointly, we can see that the scores are grouped into four parallel directions related to the amount of adulteration, moving upwards and to the right from 0% tallow, in red, up to 15% of tallow added to the lard, in cyan.

According to the information available on the samples with which the experiment was carried out, the second component seems to be related to the characteristics of the lard being used for the experiments coming from pigs of different origins and breeds.

Figure 7.5(b) shows the scores on the PC_1–PC_3 plane, labelled according to the chromatographic column used for the experiment. The third component (6.03% variance explained, Table 7.5) reflects the variability due to the use of the three different columns. The samples measured with column A have negative scores on PC_3, while those measured with column B have a positive score. The samples measured with column C occupy intermediate positions on this component, which is coherent with its characteristics. Column C shares the same length as column A and possesses the same chemical

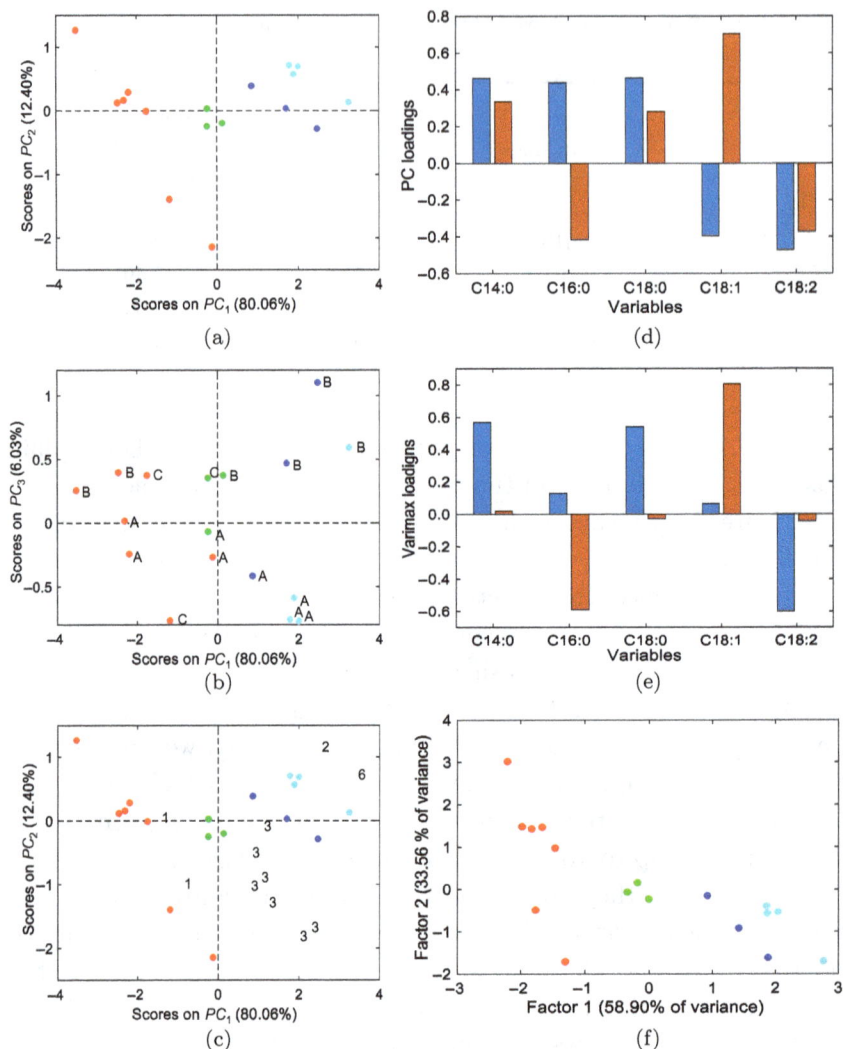

Figure 7.5. Dataset for fraud detection. The percentage of tallow is marked in colour: red for 0%, green for 5%, blue for 10%, and cyan for 15%. (a) Scores on the first two PCs. (b) Scores on PC_3 versus PC_1: A, B, and C are the different chromatographic columns. (c) Test samples projected onto the first PC plane. Numbers 1–6 identify the manufacturer. (d) PC loadings of PC_1 (blue) and PC_2 (orange). (e) Varimax loadings of Factor 1 (blue) and Factor 2 (orange). (f) Scores on the two factors after Varimax rotation.

composition as column B. Note that the variability in the data attributable to changes in the chromatographic method is 6.03%, whereas 92.46% of the variability is attributable to adulteration and type of lard.

Since the adulteration is described on the first PC plane, the samples from the six meat producers are projected onto this plane (Figure 7.5(c)). It can be concluded that samples from producer 1 are near the position occupied by those of unadulterated lard, which is marked in red. On the contrary, producer 3 would be adulterating between 5% and 10% (green and blue points, respectively), and producers 2 and 6 (which correspond to two different plants of the same meat company) would be adulterating with the same amount of ruminant tallow, possibly above 15% (in cyan). Finally, the samples from manufacturers 4 and 5 do not appear in the graph because they are outside the plane of the samples prepared in the laboratory.

To illustrate its use, a Varimax rotation of the first two components was performed. It has already been stated that these two PCs explain the variability attributable to the amount of tallow in the samples and the different origins of the lard (i.e., genetic variability among the pigs from which the lard was sourced for preparing the mixtures). In CAT, it is also possible to perform a Varimax rotation. The graphs of Varimax rotation were modified with respect to those obtained with CAT to maintain the established colour coding.

The goal of the Varimax rotation is to achieve loadings for the variables that are farthest away from zero in one factor and closest to zero in the other. In that way, the new factors (rotated components) account for the contributions of different variables, leading to better interpretability of the results.

The loadings of the two PCs are shown in Figure 7.5(d) in a bar plot in blue for PC_1 and in orange for PC_2. It is clear that all five variables significantly intervene in both PCs (regardless of the sign, the height of all bars is similar). Figure 7.5(e) on its part depicts the loadings of the two factors obtained after the Varimax rotation, again blue for the first factor and orange for the second. Now, in blue, variables C14:0 and C18:0, with positive loadings, and C18:2 with negative loading, contribute to define Factor 1, whereas only

C16:0 (negative loading) and C18:1 (positive loading), in orange, make up Factor 2.

The explained variance in this first plane is the same as that of the first two PCs, 92.46%. This is so because the two new factors are just new orthogonal axes that define the same plane as the one defined by the two PCs. Nevertheless, the distribution of the variance explained by each Varimax factor differs from that explained by each PC. Figure 7.5(f) shows the scores on the plane Factor 1–Factor 2 with an increase of the amount of tallow along with Factor 1. Factor 2, on the other hand, is related to the different genetics of the pigs.

As variables C14:0, C18:0, and C18:2 "make" Factor 1, they are the ones related to the amount of tallow in lard, whereas variables C16:0 and C18:1, defining Factor 2, are related to the origin of the lard.

7.1.10.3. Vinegars

Goal: To determine relevant variables in the ageing process of vinegars from two Spanish regions (Jerez and Rioja), including artisanal vinegars and unaged vinegars [12].

Training set: The dataset has 84 samples of vinegar, 66 from Rioja and 18 from Jerez. Among the Rioja vinegars, 8 are artisanal, 25 aged in barrel for five years, and 33 were bottled without ageing. Regarding the samples from Jerez with special ageing (*criadera-solera*): 6 aged for only one year and 12 aged for up to three years. The 20 measured variables were: acidity, acetic degree, dry extract, ash, dry extract/acetic degree ratio, dry extract/ash ratio, pH, Cl^-, Cu, K, Na, Fe, Ca, lactic acid, tartaric acid, succinic acid, citric acid, proline, acetoin, and glycerol. Therefore, the data matrix is 84×20.

Guided solution: Like in the previous cases, the data have been autoscaled to avoid the effect of different scales on the variables.

The autoscaled data are available in the "vinegars.xlsx" file, with identification of the variables and the type of vinegar (column "label"). The pretreatment applied to data does not hinder the

Table 7.6. Eigenvalues and the explained and cumulative variances for the vinegars dataset.

PC number	Eigenvalues	Explained variance (%)	Cumulative variance (%)
1	4.66	23.32	23.32
2	3.19	15.93	39.25
3	2.52	12.59	51.84
4	1.55	7.75	59.59

PCA structure but preserves the confidentiality of the raw data for proprietary reasons.

Table 7.6 shows the decomposition of the first four PCs (theoretically, up to 20 would be possible). These four PCs have eigenvalues greater than 1. This fact, together with the interpretation on them, leads to the selection of four PCs for the description of this dataset. Table 7.6 also shows that the variance explained by each of these components is distributed similarly among them, and the four together capture 59.6% of the total variance of the data.

Figure 7.6 displays the scores on the four PCs, and Table 7.7 shows the corresponding four vectors of loadings, $\mathbf{p}_i$, $i = 1, \ldots, 4$, as column vectors. The loadings with absolute values greater than or equal to 0.25 are shown in bold in Table 7.7.

The analysis of loadings and scores supports the interpretation of the PCs.

First component: The 33 bottled vinegars without ageing, depicted as red diamonds in Figure 7.6(a), have negative scores on PC_1, which clearly separate them from the other types of vinegars. To interpret the pattern, recall that their scores are negative and also take into account the sign of the coordinates of $\mathbf{p}_1^T$ in Table 7.7 (first column corresponding to the loadings on PC_1). Therefore, these vinegars stand out for containing more calcium than the others. On the other side, the remaining vinegars have higher contents of the variables whose loadings are positive (above all, acetic degree, dry extract, ash, K, and glycerol).

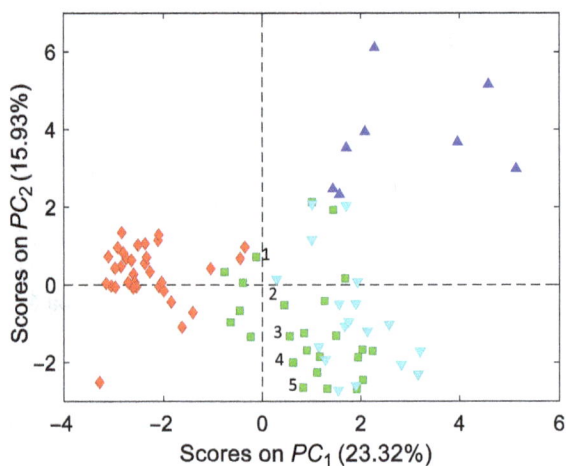

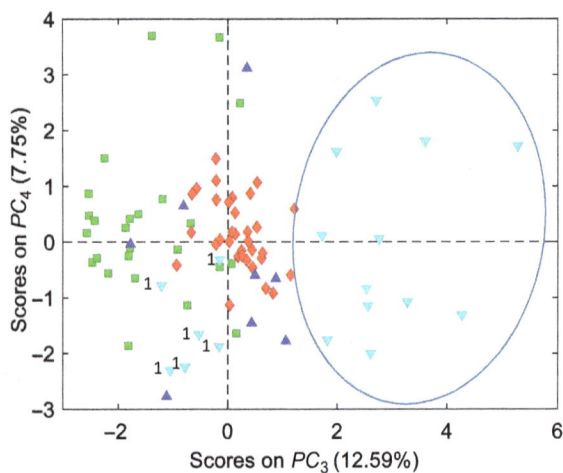

Figure 7.6. Dataset of vinegars: (a) PC_2 versus PC_1, and (b) PC_4 versus PC_3. Symbols: red diamond (Rioja bottles), blue triangle (Rioja artisanal), green square (aged Rioja), and cyan triangle (aged Jerez). The numbers denote the years of ageing.

Second component: The artisanal vinegars, blue triangles in Figure 7.6(a), have positive scores with values greater than 2 on PC_2, which separates them from the others, with scores less than 2. Observing the sign of the loadings on this second component

Table 7.7. Loadings for the first four PCs.

Variable	PC_1	PC_2	PC_3	PC_4
Acidity	0.25	**−0.25**	−0.01	**−0.26**
Acetic degree	**0.31**	0.21	0.15	−0.14
Dry extract	**0.43**	0.01	0.02	0.07
Ashes	**0.29**	**−0.34**	−0.10	−0.12
Dry extr/grade	0.08	−0.05	−0.03	**0.64**
Dry extr/ash	0.10	**0.43**	0.07	**0.26**
pH	0.23	−0.21	**−0.27**	−0.08
Cl⁻	0.11	−0.23	**0.46**	0.04
Cu	−0.04	0.17	0.26	**−0.30**
K	**0.37**	−0.17	0.16	−0.09
Na	0.00	**−0.26**	**0.50**	0.10
Fe	0.09	−0.19	−0.15	0.18
Ca	**−0.19**	−0.21	0.10	0.12
Lactic acid	0.19	−0.25	−0.07	0.32
Tartaric acid	0.19	0.14	**0.38**	0.11
Succinic acid	0.18	0.10	**−0.25**	0.09
Citric acid	0.17	0.18	−0.07	**0.30**
Proline	0.22	**0.30**	0.05	0.01
Acetoin	0.21	0.01	**−0.29**	−0.16
Glycerol	**0.28**	**0.27**	0.02	−0.15

(second column in Table 7.7), the conclusion is that these artisanal vinegars have a higher content of the variables with positive loading (dry extract/ash ratio, proline, and glycerol) than the other vinegars, which, on the contrary, have less acidity, ash, and Na content (the variables with negative loadings).

Another characteristic to be highlighted in Figure 7.6(a) is that PC_2 is mostly related to Rioja vinegars, in particular, those aged in barrels, depicted as green squares. The number next to the score is the number of years of ageing (only one series has been labelled, but this evolution is seen in the other aged vinegars as well). Following the years of ageing, moving downwards along PC_2, the evolution of ageing for these samples can be followed, and as such, the conclusion is that as vinegars age, they increase acidity, ash, and Na (negative loadings in Table 7.7).

Third component: Figure 7.6(b) is the projection of the samples in the plane made by the third and fourth PCs. Clearly, the vinegars from Jerez, aged for more than a year using the *criadera-solera* method (the cyan triangles surrounded by a circle), are separated from the others as a consequence of having the largest positive scores on PC_3. The variables that contribute the most to characterising these Jerez vinegars are the larger contents of Cl^-, Na, and tartaric acid, which is consistent with the proximity to the sea to the Jerez wineries.

Fourth component: The "youngest" Jerez vinegars that were aged for only one year (marked with a "1" next to the corresponding cyan triangle) are indistinguishable from other vinegars in the third component. However, when looking at the fourth component ("ignoring" the reds and blues already described by the first two PCs), it is seen that one-year-aged Jerez vinegars are separated from the other vinegars. The variables that most contribute to this difference (fourth column in Table 7.7) are dry extr/grade, Cu, and citric acid. For example, these vinegars have less content of citric acid (positive loading) and more content of Cu (negative loading) than the rest of the analysed vinegars.

7.2. Three-Way Models: PARAFAC

Several current analytical techniques provide not a vector but a data matrix per analysed sample. Some examples include excitation-emission molecular fluorescence, liquid chromatography with different detectors (HPLC-DAD, HPLC-FLD, and LC-MS), and gas chromatography with a mass spectrometry detector (GC-MS).

When analysing different samples with any of these analytical techniques, the resulting data form a data cube, that is, a three-dimensional array $\underline{\mathbf{X}} = (x_{ijk})$, instead of a data matrix. For example, if we record the fluorescent intensity at V excitation wavelengths and W emission wavelengths for N samples, the element x_{ijk} is the fluorescent intensity recorded at the jth excitation wavelength ($j = 1, \ldots, V$) and the kth emission wavelength ($k = 1, \ldots, W$) for the ith sample ($I = 1, \ldots, N$). In that case, it is said that there are three modes or ways, hence the name three-way data, arranged in a three-way data array or tensor. In the example, the first mode or way is the sample, the second one is excitation, and the third one is emission. In general, if the chemical data come from several ways, they will conform to a multi-way array, and it is already common to have more than three modes.

Similar to PCA, the objective of a multi-way analysis is to obtain the underlying structural model embedded in the data. Among the possible models for three-way chemical data, parallel factor analysis (PARAFAC) [13], [14], has gained popularity, in particular for the analysis of molecular fluorescence spectroscopic data [15]. This is a suitable manner of introducing the reader to the subject of multiway models.

7.2.1. *Fundamentals of PARAFAC*

Equation (7.14) shows that PCA is a bilinear model. According to Equation (7.17), when A PCs are retained, the structural PCA model of data matrix $\mathbf{X} = (x_{ij})$, a two-dimensional array, becomes

$$x_{ij} = \sum_{h=1}^{A} t_{ih} p_{hj} + e_{ij}, \quad i = 1, \ldots, N, \quad j = 1, \ldots, V. \qquad (7.21)$$

For the convenience of generalising to multi-dimensional arrays, the loadings $\mathbf{p}_1, \ldots, \mathbf{p}_A$ (rows in $\mathbf{P}$) are considered column vectors, so that Equation (7.21) becomes

$$x_{ij} = \sum_{h=1}^{A} t_{ih} p_{jh} + e_{ij}, \quad i = 1, \ldots, N, \quad j = 1, \ldots, V. \qquad (7.22)$$

Then, a PARAFAC model with A factors is a trilinear structural model for the three-way data $\underline{\mathbf{X}} = (x_{ijk})$, defined as in Equation (7.23):

$$x_{ijk} = \sum_{h=1}^{A} a_{ih} b_{jh} c_{kh} + e_{ijk},$$
$$i = 1, \ldots, N, \quad j = 1, \ldots, V, \quad k = 1, \ldots, W. \qquad (7.23)$$

The loading vectors are arranged as the columns of matrices $\mathbf{A}_{NA} = (a_{ih})$, $\mathbf{B}_{VA} = (b_{jh})$, and $\mathbf{C}_{WA} = (c_{kh})$, which are then the loadings of modes 1, 2, and 3, respectively. The residuals define a cube $\mathbf{E} = (e_{ijk})$. By comparing with the PCA decomposition into scores $\mathbf{T}_{NA}$ and loadings (by columns) $\mathbf{P}_{NA}$, the analogy is patent. That is the reason why, occasionally, the loadings on the sample mode (the first mode in the example of fluorescence data) are still called scores.

The loading matrices $\mathbf{A}$, $\mathbf{B}$, and $\mathbf{C}$ are estimated using least squares. To avoid trivial solutions, as in PCA, the columns of loadings in matrices $\mathbf{B}$ and $\mathbf{C}$ are normalised (unit length). However, contrary to PCA, the PARAFAC factors are not orthogonal, nor do they constitute "nested" models. That is, the PARAFAC model with $A+1$ factors is not the PARAFAC model that only consists of adding a new factor to the A previous ones.

On the other hand, PARAFAC is a unique structural model, which means that the estimated parameters (matrices $\mathbf{A}$, $\mathbf{B}$, and $\mathbf{C}$) constitute the unique solution to the problem, while any other solution represents a different fit to the data. Therefore, if the model is compatible with the experimental data $\underline{\mathbf{X}}$, the PARAFAC loadings are estimates of the true underlying parameters. Continuing with the example of three-way fluorescence data (sample, excitation, and emission modes), suppose that a PARAFAC model with three factors is appropriate for $\underline{\mathbf{X}}$. This immediately means the existence of three fluorophores in the analysed samples. Moreover, the uniqueness of the PARAFAC solution implies that $\mathbf{b}_1$ and $\mathbf{c}_1$ (the first columns in $\mathbf{B}$ and $\mathbf{C}$, that is, the loadings of the first factor in modes 2 and 3) are estimates of the excitation and emission spectra, respectively, of the fluorophore associated with Factor 1. Similarly, $\mathbf{b}_2$ and $\mathbf{c}_2$ are related to the corresponding spectra of the fluorophore modelled with Factor 2, and $\mathbf{b}_3$ and $\mathbf{c}_3$ are those of the third factor. If the emission and excitation spectra of various compounds are available (calibration standards, databases, etc.), it is possible to unequivocally identify each factor simply by comparing the spectra of the pure compounds with the loadings of the appropriate PARAFAC model.

Furthermore, the uniqueness of PARAFAC means that it is possible to calibrate analytes in test samples that contain any other compounds. In the mid-1990s, Booksh and Kowalski called this property the "second-order advantage" of an analytical method [16]. Its practical usefulness is so obvious that it needs no further comments.

The structure of a PARAFAC model is related to the principle of parallel proportional profiles introduced by Catell in 1944 [17]. According to this principle, if several data matrices are formed by the same set of profiles or loadings and the matrices differ from one another in the proportion in which these loadings intervene in each of them, then a model without rotational ambiguity will be generated. In the field of chemical analysis, it is obvious that data cubes generated by hyphenated instrumental techniques (e.g., GC-MS, HPLC-DAD, HPLC-FLD, and LC-MS) or excitation-emission molecular fluorescence spectroscopy obey, at least in theory, Catell's principle.

Based on the PARAFAC structural model of A factors in Equation (7.23), the factorisation of the individual $V \times W$ matrices corresponding to each "object" in the first mode, $\mathbf{X}_i$, $i = 1$, $2, \ldots, N$ (i.e., a "slice" or slab of the data cube $\underline{\mathbf{X}}$ taken in the first mode) is

$$\mathbf{X}_i = \mathbf{B}\mathbf{D}_i\mathbf{C}^T + \mathbf{E}_i, \quad i = 1, \ldots, N. \tag{7.24}$$

In Equation (7.24), $\mathbf{D}_i$ is a diagonal matrix whose diagonal elements correspond to the ith row of matrix $\mathbf{A}$, and $\mathbf{E}_i$ is the matrix of the residuals between the estimated ith sample (computed with the PARAFAC model with A factors) and $\mathbf{X}_i$. In terms of the example, Equation (7.24) would be the contribution of each fluorophore to construct the ith sample.

As discussed in Section 7.1.8, the structural model of A PCs for classical two-way matrices, unlike a structural model of A PARAFAC factors, does not possess the uniqueness property as a consequence of rotational ambiguity, and therefore it lacks the second-order advantage.

The use of methods (such as PARAFAC) that respect the inherent three-way structure of data results in more organised information that eventually leads to the disappearance of the rotational ambiguity.

7.2.2. *Analysis of a PARAFAC model*

The exploration of the validity of a PARAFAC model requires the analysis of several aspects, starting with the percentage of variance explained by the model, both total and per factor.

The projection spaces are defined using Equation (7.23) for each individual mode. Consider the loading matrix $\mathbf{A}$, with which the value of Hotelling's T^2 is calculated, i.e., the distance from each row of $\mathbf{A}$ to the vector of means of $\mathbf{A}$. This calculation is based on the covariance matrix $\mathbf{S_A}$, which is not diagonal because the columns of $\mathbf{A}$ are not orthogonal to one another. Assuming multivariate normality, the critical value is established at a confidence level of c. Analogously, the projection space for the other modes is defined from the loading matrices $\mathbf{B}$ and $\mathbf{C}$ and their covariance matrices $\mathbf{S_B}$ and $\mathbf{S_C}$, respectively.

According to the structural model in Equation (7.23), the residuals cube $\underline{\mathbf{E}}$ is the difference between the values estimated with the PARAFAC model of A factors and the experimental data cube. Consequently, the residual sum of squares can be calculated for each mode, for example, $SS_i = \sum_{j=1}^{V} \sum_{k=1}^{W} e_{ijk}^2$, $i = 1, \ldots, N$, for mode 1, so that it is possible to establish a critical value Q_c, and similarly for modes 2 and 3. Figure 7.7 corresponding to the case-study in Section 7.2.4.1 is an example of such a graph for the sample mode and will be discussed in more detail there.

As a consequence, the critical values for statistics T^2 and Q define a bounded region in the projection space for each mode in $\underline{\mathbf{X}}$ whose interpretation is the same as that of a PCA model. The inspection of the position of data points (chemical samples) in the three projection spaces is important to decide if, and in what way, a point deviates from the majority. In this regard, it should be remembered that

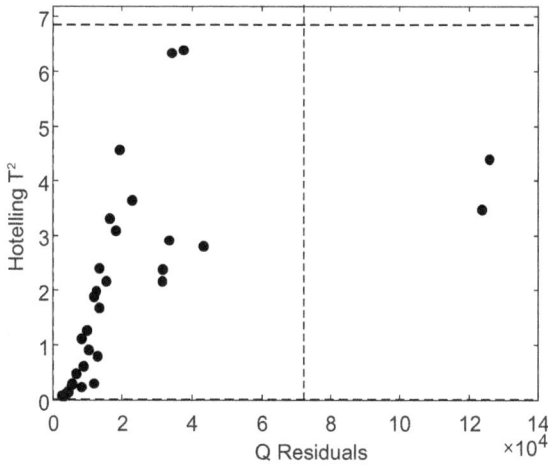

Figure 7.7. Values of Q residual and Hotelling's T^2 in the sample mode for the PARAFAC model with two factors after outlying samples were removed. Dashed lines mark the threshold values at 95% confidence.

PARAFAC is a least-squares method and, therefore, very sensitive to the presence of outliers. For this reason, this control is required when the model is applied to a test sample, that is, a new sample with data in a $V \times W$ matrix $\mathbf{X}_{\text{new}}$. Applying the model involves computing $\mathbf{D}_{\text{new}}$ in Equation (7.24) using the matrices $\mathbf{B}$ and $\mathbf{C}$ that have been obtained with the calibration samples in $\underline{\mathbf{X}}$. It is necessary to check that T^2_{new} and Q_{new} are in the projection space of mode 1 before accepting the contribution of each fluorophore in this new sample.

Similar to PCA, Q and T^2 statistics are essential tools. They help in detecting outliers while also identifying the mode in which the "anomaly" occurs.

The analysis of the loadings on each factor in each mode deserves special attention. When proposing a model with more factors than necessary, one or several of the profiles might degenerate (e.g., loadings of factors 1 and 2 in the excitation mode that have opposite signs). Nevertheless, in the field of chemical analysis, particularly in the calibration stage, it is possible to verify that the loadings obtained correspond to the signals of the analytes and that the sample loadings reproduce the structure of the composition known for the calibration samples.

7.2.3.   *Compatibility of a PARAFAC model: Trilinearity*

The key question in a multiway analysis is to decide if a structural model with A factors, a PARAFAC model in this case, is compatible with the experimental data cube $\underline{\mathbf{X}}$. If that is the case, the data are said to be trilinear. This can lead to confusion between the real characteristics of the data and those of the model with which they are to be described. For example, with small concentrations of fluorophores in the samples, the model for the fluorescent intensity of A fluorophores has a mathematical expression analogous to a PARAFAC model with A factors [18]. However, if the sample(s) contain(s) greater concentration(s) or experimental measurements corresponding to wavelengths with a Rayleigh behaviour, it is said frequently that "the trilinearity of the data has been lost", when instead, it should be said that the PARAFAC model is not valid for these data.

Specific procedures linked to the properties of PARAFAC have been designed, which generally work through proof by contradiction (*reductio ad absurdum*): if the data supports a PARAFAC model, then one of the properties seen above will be fulfilled; otherwise, the model is not adequate.

This is the case of the *split-half method*, which consists of dividing one of the modes of the $\underline{\mathbf{X}}$ cube into four parts (for example, splitting the sample mode is reasonable for excitation-emission data). Then, two cubes are constructed by joining, for example, "pieces" 1 and 3 on one side and pieces 2 and 4 on the other. Finally, for each cube, a PARAFAC model is built. The final step is to check the similarity of the loadings obtained in the remaining modes not used for the split. The comparisons are done among the two models and with that for the complete cube. The underlying idea is that if the loadings are different, there cannot be a common structure, and therefore the proposed model is not valid (in other words, the trilinear model is inadequate for the data).

Another criterion is to study the core consistency diagnostic (CORCONDIA) index, which measures the degree of possible rotability of the solution obtained while maintaining the same residual sum of squares [19]. This index varies from 0 to 100 (but sometimes it can even be negative, although in this case the trilinearity does not hold [20]), and whenever it approaches 100, the data are considered to have a trilinear structure. It is usual to explore the behaviour of CORCONDIA as a function of the number of factors. If there is a sudden decrease when moving from A to $A + 1$ factors, the model with A factors should be used. See Section 7.3.3 for an explanation about the computation of this index.

7.2.4. *Worked examples*

Contrary to Tucker3, explained in Section 7.3.3, PARAFAC is not yet implemented in CAT. However, the computations of the following worked examples can be done with free-access software, such as the algorithms in https://ucphchemometrics.com/186-2/algorithms/.

7.2.4.1. **Enrofloxacin in water from poultry farms**

Goal: To propose a cheap, fast, and accurate analytical procedure to determine enrofloxacin in feeding water from poultry farms. The method uses excitation emission fluorescence and three-way PARAFAC. The advantage of this proposal is that the procedure permits the identification and determination of the quantity of enrofloxacin present in water samples from poultry farms without the need to determine the possible interferents or separate them in a step prior to calibration [21].

Data: A total of 36 samples have been measured on a PerkinElmer LS50 B luminescence fluorimeter. The excitation-emission matrix (EEM) for each sample was obtained by recording the fluorescent intensity at emission wavelengths from 358 to 550 nm (every 1 nm) when the excitation wavelengths vary from 240 to 310 nm (every 5 nm). The data cube is a $36 \times 193 \times 15$ tensor. The FL WinLab software (PerkinElmer) was used for measurements, and the data were imported to Matlab using the INCA software [22] that avoids the non-trilinear parts of the spectrum by setting them to missing values.

Training set: Twenty standards, samples 1–20 in the data cube, were used for calibration and prepared at 10 levels of concentration, evenly distributed from 5 to $50\,\mu\text{g}\,\text{L}^{-1}$ with two replicates at each level.

Test set: It consists of standards with concentrations of 5, 15, 40, and $50\,\mu\text{g}\ \text{L}^{-1}$, which will be used to estimate the prediction capability of the model, plus three blanks that have been measured throughout the experiment to ensure the absence of experimental drift. There are also nine samples of water, the test samples, prepared with feeding water coming from five poultry farms, two replicates for every farm but for one of them, which has a single replicate.

The file "poultry.xlsx" contains the individual 193×15 EMM for the 32 samples used for PARAFAC, one per sheet, in the same order as in Figure 7.8(a).

Guided solution: A PARAFAC model has been carried out with 32 samples in the first mode because four of the 20 calibration standards had to be removed, as they surpass the threshold values established for Q and T^2 statistics (samples with 5, 15, and $40\,\mu\text{g}\,\text{L}^{-1}$ in the first replicate and $50\,\mu\text{g}\,\text{L}^{-1}$ in the second replicate). Two factors were selected for the PARAFAC model, with the CORCONDIA index equal to 100% and 86.68% of explained variance. The similarity index measured in splits and the overall model is 99.3% (*split-half* method). The values of statistics T^2 and Q are depicted in Figure 7.7, where it is seen that all the samples are below their corresponding limits at 95% confidence level (dashed lines), except for the two replicates of the last farm (samples 30 and 31 in Figure 7.8(a)) that surpass only the critical value for the Q statistic.

Figure 7.8 shows the profiles of the three modes: sample, emission, and excitation. On the abscissa of Figure 7.8(a), the first 16 samples are the calibration standards, the next six are the prediction

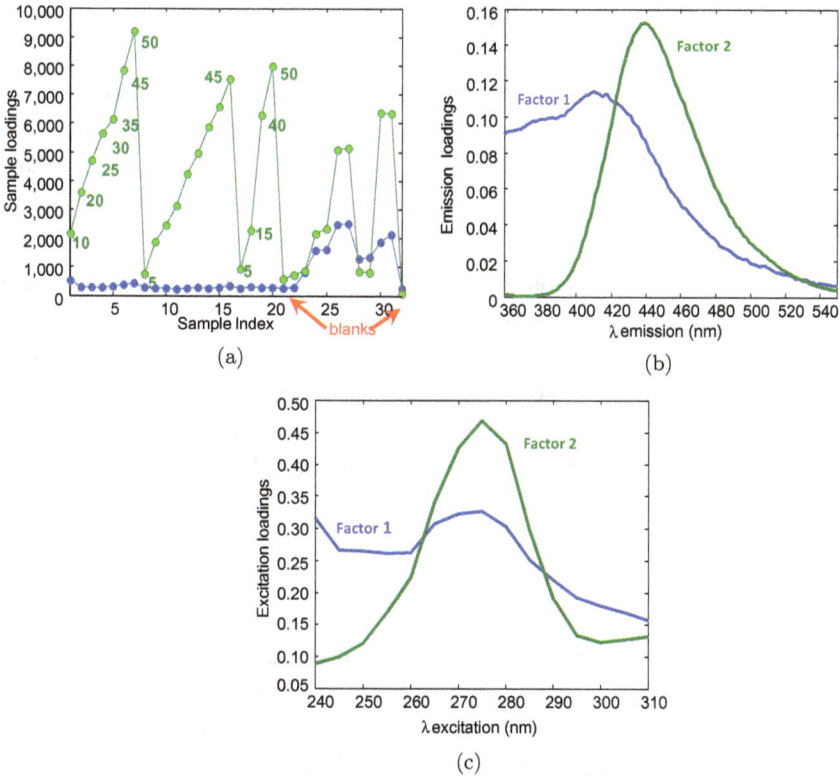

Figure 7.8. PARAFAC model. (a) Sample, (b) emission, and (c) excitation loadings. In (a), samples 1–16 are calibration standards, 17–20 are prediction samples, samples 21, 22, and 32 are blanks, and the test samples from the different farms are numbered from 23 to 31. Factor 1 in blue corresponds to the chemical sample matrix, and Factor 2 is related to enrofloxacin (in green).

samples and two blanks, the next nine are the farm samples, and the last one is another blank.

Figure 7.8(a) shows that, in the sample mode, the factor in blue, Factor 1, corresponds to a fluorophore not present in the laboratory-prepared samples. Note that the first 22 samples and sample number 32 (which is a blank) have loadings equal to practically zero, while loadings are non-null for samples 23–31, which are the test samples (feeding water from poultry farms). Thus, the presence of a fluorophore, that is, the matrix effect, is clearly observed. Factor 2 corresponds to enrofloxacin (in green in all graphs). The increasing pattern, according to the concentration in the samples prepared at the laboratory, is clear; additionally, the eight test samples from the farms all contain a quantity of enrofloxacin greater than zero.

In Figure 7.8(b), the emission profiles for both the matrix (Factor 1, in blue) and enrofloxacin (Factor 2, in green) are shown. Figure 7.8(c) depicts the excitation profiles obtained using PARAFAC for these two factors, with the same colouring code. For unequivocal identification, correlation is used. Precisely, we compute the correlation coefficients between the excitation and emission spectra obtained for a standard sample in the laboratory and the corresponding PARAFAC loadings of Factor 2. In both the excitation and emission profiles (modes 2 and 3), the correlation coefficients were greater than 0.99. Therefore, enrofloxacin has been identified in the feeding water.

After identifying the factor of the PARAFAC model related to enrofloxacin, a calibration model is fitted with the 16 standard samples. It is an ordinary least-squares (OLS) univariate regression line relating the sample loadings of enrofloxacin (those of Factor 2 at mode 1) as a function of the concentration. The fitted line is validated with the usual hypothesis tests of significance of the regression (in this case, a p value of less than 10^{-4}) and lack of fit (a p value equal to 0.53) and also checking that it has normal and independent residuals with the same variance.

The fitted equation is $loading_{enrofloxacin} = 22.43 + 173.09 \, C_{enrofloxacin}$ (correlation coefficient equal to 0.993). This calibration line is used to compute the concentration of the prediction samples prepared in the laboratory, $C_{calculated}$, from their PARAFAC sample loadings (mode 1). The accuracy line is $C_{calculated} = -0.0147 + 1.00054 \, C_{enrofloxacin}$, which implies that there is no bias because, at the 5% significance level, the slope equals one and the intercept equals zero. The mean of relative errors (in absolute value) is 6.8% for the calibration standards and 8.8% for the prediction samples.

The concentration of enrofloxacin in the test samples is determined in the same manner as that for prediction samples by applying the calibration line. This is supported by the fact that the PARAFAC decomposition obtains the true underlying profiles of the analyte, even in the presence of uncalibrated interferents. The nine water samples from the five farms contained the following amounts of enrofloxacin (two replicates except for the first one): 4.9, 12.3 and 13.2, 4.7 and 4.5, 29.2 and 29.5, and 36.3 and 36.4 μg L^{-1}, for farms 1–5, respectively.

7.2.4.2. Solving overestimation in migration of melamine from food contact materials

Goal: To determine melamine migration from food contact materials (FCM). It is obvious that any compound in the FCM can migrate to foodstuffs, a reason that explains why the European Union regulates some compounds that are potentially dangerous to human health. This is the case for melamine in plastic materials whose specific migration limit (SML) was fixed by European regulations at 2.5 mg kg^{-1} [23]. The goal of this example is to compare the results obtained with a univariate analysis and a multivariate approach based on combining a PARAFAC decomposition with HPLC-DAD signals.

The quantification of melamine through a conventional univariate data analysis leads to results that overestimate the quantity of melamine [24]. This can be caused by the presence of non-intentionally added substances (NIASs) in migration samples, substances that can share their retention time with that of melamine (or be very close to each other).

Data: Measurements recorded with HPLC-DAD were arranged in a single three-way tensor, **X**, with size $76 \times 151 \times 30$. The first number refers to the chromatographic profile (i.e., the number of retention times of the chromatographic region of interest), the second one to the spectral profile (number of wavelengths), and the third one to the sample profile (number of samples). The file "migration.xlsx" contains the individual 76×151 matrices for each sample, one per sheet, in the order shown in Figure 7.9(c).

Training set: It is formed by 14 samples of melamine standards to build the calibration line. Their concentrations vary from 0 to 10 mg L^{-1}, two of them with two replicates and with two additional

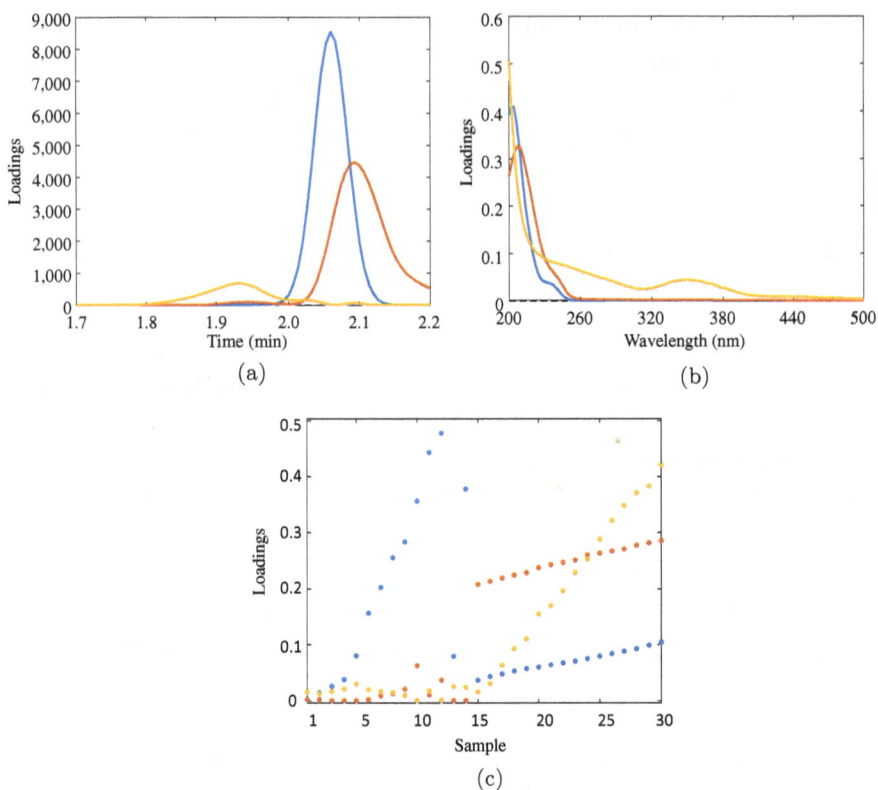

Figure 7.9. Loadings of the PARAFAC model: (a) chromatographic, (b) spectral, and (c) sample profiles (1–14 calibration standards, 15–30 kinetic migration cycles); blue factor is related to melamine, while orange and yellow factors are two interferents.

blanks. All samples were prepared in the same simulant, the so-called simulant B (3% acetic acid w/v in aqueous solution). The same simulant is used for the samples in the test set. More experimental details can be found in Ref. [24].

Test set: It contains 16 additional samples to build the kinetic curve. They were obtained by migration of melamine from a bowl also into simulant B during 16 cycles, each measured for 30 min. The accumulated quantity of melamine was considered in each cycle.

Guided solution: A PARAFAC model is carried out with 30 samples (14 standards plus 16 migration samples). In general, better results are obtained when the standard samples are combined with the test samples into a single cube to perform the decomposition [25]. Imposing nonnegativity on the three profiles, a model with three factors was built and validated. The model has a CORCONDIA index of 97% and explains 84.3% of the variance of the signals.

In addition, as explained in Section 7.2.2, it is possible to check that the model is correct by studying the consistency between what was prepared in the laboratory and what is obtained using PARAFAC because the composition of the standards is known by design.

The loadings of the three factors obtained using PARAFAC are shown in Figure 7.9. Figure 7.9(a), for the chromatographic mode, shows the profiles of three chromatographic peaks. The peak corresponding to melamine is in blue, while those that correspond to two overlapping interferents eluting before and after the retention time of the melamine peak are in yellow and orange, respectively. It is obvious that integrating the melamine peak without taking this fact into account will overestimate its peak area. Furthermore, as can be seen in Figure 7.9(b) for the spectral profile, the spectra of these three analytes overlap.

The loadings on the sample mode are presented in Figure 7.9(c), in which the first 14 correspond to standard samples, while the remaining 16 relate to the migration samples. Those for melamine, in blue, are the only ones that show increasing behaviour in the first 14 points, corresponding to the 10 standards (two of them replicated) and the two blanks, while the loadings of the other two factors, yellow and orange points, are practically constant, almost null. Comparing the melamine-related loadings to those from the other samples, the contribution of the interferents is not significant for the calibration standards; however, it increases in the last 16 samples, though at different ratios (different slopes). These last samples correspond to the migration cycles in which the amount of melamine (blue points) also increases since each slab in the data cube is a 76×151 matrix whose elements are the sum of the elements of all preceding matrices (recall that accumulated quantities were recorded in the migration cycles).

An important consideration in migration from FCM is the unequivocal identification of the migrant. In this case, the identification of melamine is carried out in two ways, as recommended in many regulations: (i) taking into account the coincidence in the retention time of a standard with that obtained with the chromatographic profile of the PARAFAC decomposition, and (ii) correlating the spectral profile of the melamine standard with that obtained using PARAFAC. The correlation coefficient in this case was 0.999 for the first 25 wavelengths of the standard spectrum after normalising (by dividing by its length) so that it can be compared with the profile defined by the PARAFAC loadings. Normalisation is necessary for the reasons explained in Section 7.2.1, that is, the spectral loadings are normalised by construction.

With the sample loadings of the standards for the blue factor, identified as melamine, the loadings versus concentration calibration curve is fitted. Solving the calibration equation for the loadings of the samples obtained in the migration experiment for the same factor, gives the quantity of migrated melamine in each cycle. After the validation of the model, the equation of the kinetic curve for the bowl for the 16 cycles of 30 min was $y = 0.301 + 0.044x$ ($R^2 = 99.53\%$).

Figure 7.10 shows the concentration of melamine obtained in the 16 cycles as black bars. It also shows, in grey bars, the concentration

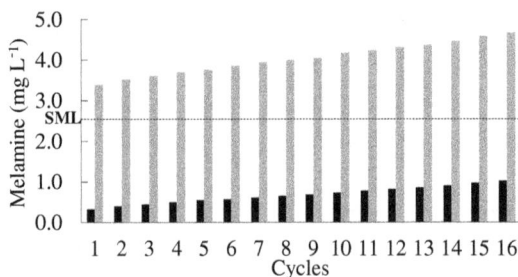

Figure 7.10. Calculated melamine concentration (in $mg\,L^{-1}$) with a univariate conventional analysis (grey bars) and with a PARAFAC decomposition (black bars) obtained for migration kinetics from a bowl exposed for 16 cycles. The horizontal dashed line indicates the melamine SML.

calculated with the univariate analysis by integrating the whole peak area (not shown here) interfered by the two concomitants. The dotted line marks the SML $(2.5\,\mathrm{mg\,L^{-1}})$.

The overestimation of the amount of melamine in all the cycles is obvious when the quantification is carried out with the usual integration of the peak area. The conclusion is that the SML is exceeded, while it is not the case when the concentration is obtained by means of a PARAFAC decomposition, correctly separating the melamine from the two interferents that practically co-elute at the same retention time as melamine. Moreover, the amount obtained by the univariate method would be 10 times greater than that obtained with PARAFAC, which would cause a false nonconformity in the analysis of this bowl.

7.3. Some Other Methods for Analysing Data Matrices or Cubes

The purpose of this section is to introduce methods for describing matrices or data cubes by constructing linear combinations of the original variables. Despite the popularity and usefulness of PCA and PARAFAC, the authors believe the reader should be aware of several alternative methods, which are also widely used.

This section can be considered to be of a higher level than the previous ones, though it lacks examples of application. Nevertheless, it can help the interested readers in directing their attention to alternative methods and more advanced techniques. In that sense, Section 7.3.1 draws attention to robust alternatives for dealing with (ubiquitous) outlier data, Section 7.3.2 provides some alternatives for bilinear decomposition, and Section 7.3.3 provides alternatives for trilinear decompositions.

Given a data matrix or cube containing the experimental information obtained during the research on a chemical problem, the preliminary question is: Which method of analysis may be suitable for these data? This question, which is critical since it conditions the resulting conclusions, cannot be even posed if only one method of analysis is known.

The following list of alternatives is obviously very limited; its only intention is to provide the reader with a simplified guideline to these methods.

7.3.1. *Robust PCA and PARAFAC*

In all the previous theoretical sections, it is assumed that no outliers existed. However, anomalous (outlier) data are ubiquitous, and this justifies the need for robust versions of PCA and PARAFAC.

It is known that the presence of anomalous data causes any estimation based on least squares to fail. As an illustrative exercise, the reader is asked to modify only the sign of value x_{12} in Table 7.1 to -2.34, then centre the data and calculate the two PCs. (*Hint*: You will observe the changes on eigenvalues and loadings and also that object O_1 has now a negative score in PC_2, while that of O_3 is positive. And more importantly, none of the new scores has a T^2 value greater than the limit at a 95% confidence level, so it would remain undetected.)

There are many strategies to detect outliers based on the analysis of their effects on an estimate. Unfortunately, they are generally not suitable for detecting the combined effect of several outliers, which can further mask their own behaviour, causing other points to appear as if the latter were the outliers. For this reason, it is much more efficient to carry out the estimation using a robust fitting criterion which is less sensitive to the presence of anomalous data. This is possible whenever the median of squares, rather than the mean of squares, is used. Thus, the outlying samples will be detected, and the analysis should be repeated by weighing their influence or even, when appropriate, eliminating them. A good, updated review of robust methods in the field of chemometrics for both PCA and PARAFAC can be found in Ref. [26].

7.3.2. *Methods for two-way data matrices*

The bilinear structure of PCA was described in Section 7.1.3. This structure is somehow truncated to model the idea that the data in matrix $\mathbf{X}_{NV}$ can be, in fact, described in an A-dimensional space, with $A \leq \min\{N, V\}$. To sum up, PCA is a solution of the bilinear model shown in Equation (7.25). Just compare it with Equation (7.17) and observe that $\mathbf{G}$ is $\mathbf{T}$ (the matrix whose columns are the PCA scores) and $\mathbf{F}$ is $\mathbf{P}$ (the matrix whose rows are the PCA loadings):

$$\mathbf{X}_{NV} = \mathbf{G}_{NA}\mathbf{F}_{AV} + \mathbf{E}_{NV}. \tag{7.25}$$

It is evident that the aim of a bilinear decomposition is to split the variability observed in the data matrix $\mathbf{X}$ into a structural variability attributable to chemical causes (bilinear model) and another residual one in $\mathbf{E}$. Anyhow, for any given $\mathbf{X}$ and A, it is clear that there are infinitely many bilinear models (pairs of $\mathbf{G}$ and $\mathbf{F}$ for Equation (7.25)), which imply the need to set a fitting criterion and/or specific characteristics of $\mathbf{G}$ and/or $\mathbf{F}$. In the case of PCA, the criterion is to minimise the residual sum of squares by imposing that the rows of $\mathbf{F}$ (loadings) are orthonormal and the columns of $\mathbf{G}$ (scores) are orthogonal. Countless procedures have been developed with the aim of linking $\mathbf{G}$ and $\mathbf{F}$ directly to the underlying structure of the data, especially using the *a priori* information about the problem. What follows is a list of the most usual ones [27].

Non-negative matrix factorisation (NMF) [28]: Continuing with the criterion of minimising the residual sum of squares, the method assumes that the elements of $\mathbf{X}$ are greater than or equal to zero, so data autoscaling (and other pretreatments such as column centring and SNV) is not applicable. NMF imposes the restriction that the values of $\mathbf{G}$ and $\mathbf{F}$ must be greater than or equal to zero. This nonnegativity constraint is supposed to link the calculated factors to the underlying chemical/physical model by preventing subtractions, that is, by assuming that the data in $\mathbf{X}$ is in fact the sum of several factors. There is a plethora of algorithms to obtain the solution, such as multiplicative updates of $\mathbf{G}$ and $\mathbf{F}$ or alternating least squares that allow incorporating constraints based on the knowledge of the problem and are less prone to local minima. The gradient of the residual sum of squares with respect to $\mathbf{G}$ and $\mathbf{F}$ has also been used to accelerate the convergence [29].

Multivariate curve resolution alternating least squares (MCR-ALS) [30]: Among the family of MCR methods, this is the most widely used method in chemometrics to iteratively solve the model of Equation (7.25) [31]. With the data matrix $\mathbf{X}$ and number of components A, which are supposed to describe the problem, $\hat{\mathbf{X}}$ is computed using Equation (7.17), and then the sum of squares of the elements of $\hat{\mathbf{X}} - \hat{\mathbf{G}}\hat{\mathbf{F}}$ are alternatively minimised in $\hat{\mathbf{G}}$ and $\hat{\mathbf{F}}$. The flexibility of the algorithm permits the incorporation at each iteration of the (scientist-imposed) constraints in the model so that it collects the characteristics of the data in $\mathbf{X}$ (for example, nonnegativity, unimodality, closure, selectivity, normalisation, a specific functional relation, peak shape, and trilinearity). This, together with a strategy to analyse data from different sources on the same objects (data matrices and/or cubes), makes the method extremely flexible [32]. Recently, the degree of rotational ambiguity [33], including the effect of instrumental noise [34], has been characterised and quantified. In addition, this ambiguity can be reduced considerably by using the physical-chemical knowledge about the structure of the data and incorporating it into MCR-ALS through the appropriate restrictions [35].

Independent component analysis (ICA) [36]: The purpose of this group of methods is to replace the restriction of uncorrelation between the rows of $\mathbf{F}$ by that of independence. Remember that for data with multivariate normal distribution, both concepts are equivalent, but this is not so in general. PCA constructs uncorrelated components, which implies that it minimises the sum of squares that constitutes the second-order moment of a probability distribution, while ICA maximises higher-order moments, such as kurtosis (a fourth-order moment). Therefore, PCA can extract independent loadings when the higher-order moments of the data are small, but not in other cases.

Another group of methods incorporates the uncertainty in the measurements in $\mathbf{X}$ when estimating $\mathbf{G}$ and $\mathbf{F}$ of the bilinear model, as well as the design that has been followed to obtain the samples, which will affect the correlation structure between the measured variables. This is the statistical version of the fit of a bilinear model to experimental data. As in any statistical estimation problem, there are two usual approaches based on the maximum likelihood and the Bayesian criteria. Three of the most often used techniques are:

Positive matrix factorisation (PMF) [37]: For this method, both the data matrix $\mathbf{X} = (x_{ij})$ and their individual standard deviations in $\mathbf{S} = (s_{ij})$ are known. The problem is solved in the weighted least-squares sense: $\mathbf{G}$ and $\mathbf{F}$ are determined so as to minimise the sum of squares divided (element-by-element) by s_{ij}^2. Furthermore, like in NMF, the solution is constrained so that all the elements of $\mathbf{G}$ and $\mathbf{F}$ are required to be non-negative.

Maximum likelihood factor analysis (MLFA) [38]: MLFA is also referred to simply as factor analysis, common factor analysis, or maximum likelihood common factor analysis. It involves finding $\mathbf{G}$ and $\mathbf{F}$ that maximise the probability of obtaining the existing matrix $\mathbf{X}$. In this context, the solution of PCA that minimises the residual sum of squares is the maximum likelihood estimation when measurement errors are independently and identically distributed with a normal distribution. Among the MLFA methods, some common ones are *principal axis factorisation* (PAF) and *maximum likelihood principal component analysis* (MLPCA) [39]. The latter depends on knowing not only $\mathbf{X}$ but also the covariance matrix of the multivariate normal distribution of the errors, which makes its use difficult. MLPCA can also be incorporated into MCR-ALS simply by using the $\hat{\mathbf{X}}$ estimated with MLPCA instead of PCA when minimising $\hat{\mathbf{X}} - \hat{\mathbf{G}}\hat{\mathbf{F}}$.

Bayesian non-negative factor analysis (BNFA) [40]: In BNFA, $\mathbf{G}$ and $\mathbf{F}$ are regarded as random parameters with their own distributions. Therefore, it is necessary to specify both the prior probabilities of $\mathbf{G}$ and $\mathbf{F}$ and the data distribution. Bayes' theorem gives the posterior probability, that is, the conditional probability of $\mathbf{G}$ and $\mathbf{F}$ given the current data matrix $\mathbf{X}$. BNFA needs to incorporate A and the constraints for factor identification. It has the advantage that it provides the uncertainty in the estimates of $\mathbf{G}$ and $\mathbf{F}$.

7.3.3. Methods for three-way data cubes

PARAFAC2 [41]: In chromatographic applications, the shift of chromatographic peaks from one run to another is usual. This means that the principle of parallel proportional profiles is not fulfilled, and hence, the PARAFAC structural model is not applicable. An alternative is PARAFAC2, which, unlike the model in Equation (7.24), decomposes each slab of the $\mathbf{X}$ cube as

$$\mathbf{X}_i = \mathbf{B}_i \mathbf{D}_i \mathbf{C}^T + \mathbf{E}_i, \quad i = 1, \ldots, N, \tag{7.26}$$

where $\mathbf{B}_i$ contains the elution profiles of the ith sample, which are different from sample to sample, $\mathbf{C}$ is the matrix with the spectral profiles common to all samples, and $\mathbf{D}_i$ is a diagonal matrix, as in Equation (7.24), made up of the contribution of each factor in the ith sample.

There are intermediate possibilities between assuming that for all samples $\mathbf{B}_i = \mathbf{B}$ (PARAFAC) and not imposing any constraints. The constraint imposed in PARAFAC2 is $\mathbf{B}_i^T \mathbf{B}_i = \mathbf{B}^T \mathbf{B}$, that is, a set of elution profiles $\mathbf{B}_i$ is estimated for each sample i with the constraint that the cross-product of the profiles is kept constant. In particular, this condition is fulfilled in the case of shifts in the chromatographic peaks (see annex in Ref. [42] for more details).

Tucker3 [43]: This is the most general structural model for three (or more) way data. The Tucker3 model assumes that each element of the $\mathbf{X} = (x_{ijk})$ data cube is defined as

$$x_{ijk} = \sum_{l=1}^{L} \sum_{m=1}^{M} \sum_{u=1}^{U} a_{il} b_{jm} c_{ku} g_{lmu} + e_{ijk},$$

$$i = 1, \ldots, N, \quad j = 1, \ldots, V, \quad k = 1, \ldots, W. \tag{7.27}$$

Contrary to Equation (7.23), the number of factors (L, M, and U) is not the same for each mode, while in PARAFAC all three modes consist of the same A factors. In addition, a g_{lmu} term appears, which links the loadings of one way (mode) with those of the others. For example, g_{112} indicates the relationship between the first factor of mode 1 (vector $\mathbf{a}_1$), the first factor of mode 2 ($\mathbf{b}_1$), and the second factor of mode 3 ($\mathbf{c}_2$). Note that PARAFAC is a Tucker3 model with $g_{lmu} = 1$ if $l = m = u$ and $g_{lmu} = 0$ otherwise.

The tensor $\mathbf{G} = (g_{tmu})$ is the *core* of a Tucker3 model for $\underline{\mathbf{X}}$, and it is used to compute the CORCONDIA index, which measures the trilinearity of the data (as introduced in Section 7.2.3). Let $\mathbf{A}$, $\mathbf{B}$, and $\mathbf{C}$ be the matrices whose columns are the loadings in Equation (7.23) of a PARAFAC model with A factors for tensor $\underline{\mathbf{X}}$. These matrices are used to compute, using least squares, the tensor $\underline{\mathbf{G}}$ via Equation (7.27). If $\underline{\mathbf{T}}$ denotes the core of the PARAFAC model seen as a Tucker3 model, then the CORCONDIA index is the Frobenius' distance between both tensors, written as the percentage of the variation in $\underline{\mathbf{G}}$, consistent with the variation in $\underline{\mathbf{T}}$, that is, CORCONDIA $= 100 \left(1 - \frac{\sum_t \sum_m \sum_u (g_{tmu} - t_{tmu})^2}{A} \right)$ %, which summarises in a single index the degree of similarity between both tensors.

The Tucker3 method is frequently used in environmental analyses, where the chemical compounds have different mutual relationships depending on the site. In this case, the core tensor describes these relations, but in a qualitative way, since there is no guarantee of a unique solution.

7.4. Lessons Learnt

Sections 7.1.1–7.1.3 presented a complete theory for constructing new variables, the PCs, by linear combinations of the experimental variables. These PCs have well-defined mathematical properties as the unique solution of a minimisation problem by successive orthogonal projections onto subspaces of one dimension.

Their possible chemical interpretation, through the identification of the subspace of chemical interest (Section 7.1.7), leads to the idea of the chemical rank of the problem as opposed to the mathematical rank of the data.

This *a priori* possibility of finding the chemical meaning for each individual PC is reduced because of the rotational ambiguity

(Section 7.1.8). However, the chemical structure remains intact, and thus, the interpretability of the PCs in the rotated subspace is maintained. The illustration on how to do that was presented in Case studies 7.1.10.1, "A mixture of two colourants", and 7.1.10.2, "Fraud detection".

They showed that the chemical interpretation is not strictly dependent on the mathematical properties of the PCA. In particular, the *rules* for deciding the number of PCs should be used flexibly. Note that, while in the example of Section 7.1.10.1 the two-dimensional space of PC_1 and PC_2 gathers the variability caused by the existence of two chemical compounds, in the example of Section 7.1.10.2 (despite the fact that "only" the tallow content has been modified in the samples), the design used makes it necessary the use of two PCs to describe the effect of added tallow on the chromatograms. Further, there is still a third component (uncorrelated with the other two) that is linked to the characteristics of the chromatographic columns used for the determinations.

The most striking case is the study presented in Section 7.1.10.3 ("Vinegars") because the answer to the problem about ageing lies in the third PC, which is not the greatest source of data variability (it only corresponds to 12.6% of the variability). Moreover, this third component also shows that the Rioja samples are distinguishable from those from Jerez, provided that they have been aged for over a year.

The view of PCA as a bilinear model with constraints in Section 7.1.3 is of great interest for new developments in the search for "loadings" with a more direct chemical interpretation. It turns out that rotational ambiguity can be removed if the chemical information is increased in a structured way, moving from a vector to a data matrix on a per-sample basis. This is the multi-way analysis developed in Section 7.2, focused on three-way data and PARAFAC, with its property of unequivocal identification of the studied compounds, even in the presence of non-calibrated interferents.

In chemical analysis, the use of PARAFAC is highly recommended because it allows for experimentally verifying the coherence between the PARAFAC factors and the experimental signals or samples. This reasoning was exemplified in the two case studies in Sections 7.2.4.1 ("Enrofloxacin in water from poultry farms") and 7.2.4.2 ("Solving overestimation in migration of melamine from food contact materials").

In Section 7.3.3, the PARAFAC2 method was introduced. It responds to the possibility of weakening the proportionality principle, which guarantees the trilinearity of PARAFAC while maintaining the uniqueness of the solution. In addition, the knowledge that PARAFAC is a particular case of the Tucker3 structural model allowed us to "generalise" the analysis of data cubes (for example, environmental data) in which there may be relationships among the factors.

Historically, the structural models described in this chapter began their mathematical journey more than 120 years ago, and the associated statistical methodology was formulated 60 years ago, that is, well before the advent of instrumental analytical techniques generating a vector or matrix of data for each sample was accessible. It is worth remembering, paraphrasing G.P.E. Box, that once the experimental data matrix or cube is obtained, all structural models are false, but some of them might be very useful.

References

[1] Pearson, K. (1901). On lines and planes of closest fit to systems of points in space. *Philosophical Magazine*, 2, 559–572.

[2] Hotelling, H. (1933). Analysis of a complex of statistical variables into principal components. *Journal of Educational Psychology*, 24, 417–441, 498–520.

[3] Anderson, T. W. (1963). Asymptotic theory for principal component analysis. *The Annals of Mathematical Statistics*, 34, 122–148.

[4] Jolliffe, I. T. (2002). *Principal Component Analysis*, 2nd edn. New York: Springer Verlag.

[5] Jackson, J. A. (1991). *User's Guide to Principal Components*. New York: Wiley.

[6] Esbensen, K. H. and Geladi, P. (2009). Principal component analysis: Concept, geometrical interpretation, mathematical background,

algorithms, history, practice. In Brown, S. D., Tauler, R., and Walczak, B. (eds.), *Comprehensive Chemometrics*. Oxford: Elsevier.

[7]   Geladi, P. and Linderholm, J. (2020). Principal component analysis. In Brown, S. D., Tauler, R., and Walczak, B. (eds.), *Comprehensive Chemometrics*. Oxford: Elsevier.

[8]   Stewart, G. (1993). On the early history of the singular value decomposition. *SIAM Review*, 35, 551–566.

[9]   Todeschini, R. (1997). Data correlation, number of significant principal components and shape of molecules. The K correlation index. *Analytica Chimica Acta*, 348, 419–430.

[10]   Kaiser, H. F. (1958). The varimax criterion for analytic rotation in factor analysis. *Phychometrika*, 23, 187–200.

[11]   Sarabia, L. A., Ortiz, M. C., and Checa, M. A. (1989). Pattern recognition for detection of tallow in lard. In *Agriculture, Food Chemistry and the Consumer*, Vol. 2. Paris: l'Institute Nationale de la Recherche Agronomique, pp. 602–606.

[12]   Benito, M. J., Ortiz, M. C., Sánchez, M. S., Sarabia, L. A., and Íñiguez, M. (1999). Typification of vinegars from Jerez and Rioja using classic chemometric techniques and neural network methods. *The Analyst*, 124, 547–552.

[13]   Leurgans, S., Ross, R. T., and Abel, R. B. (1993). A decomposition for three-way arrays. *SIAM Journal on Matrix Analysis and Applications*, 14, 1064.

[14]   Bro, R. PARAFAC. (1997). Tutorial and applications. *Chemometrics and Intelligent Laboratory Systems*, 38, 149–171.

[15]   Murphy, K. R., Stedmon, C. A., Graeberc, D., and Bro, R. (2013). Florescence spectroscopy and multi-way techniques. PARAFAC. *Analytical Methods*, 5, 6557–6566.

[16]   Booksh, K. S. and Kowalski, B. R. (1994). Theory of analytical chemistry. *Analytical Chemistry*, 66(15), 782A–791A.

[17]   Cattell, R. B. (1944). "Parallel proportional profiles" and other principles for determining the choice of factors by rotation. *Psychometrika*, 9, 267.

[18]   Leurgans, S. and Ross, R. T. (1992). Multilinear models: Applications in spectroscopy. *Statistical Science*, 7, 289–319.

[19]   Bro, R. and Kiers, H. A. L. (2003). A new efficient method for determining the number of components in PARAFAC models. *Journal of Chemometrics*, 17, 274–286.

[20]   Rubio, L., Ortiz, M. C., and Sarabia, L. A. (2014). Identification and quantification of carbamate pesticides in dried lime tree flowers by means of excitation-emission molecular fluorescence and parallel

factor analysis when quenching effect exists. *Analytica Chimica Acta*, 820, 9–22. doi: 10.1016/j.aca.2014.02.008.

[21]   Giménez, D., Sarabia, L. A., and Ortiz, M. C. (2005). Identification and quantification of enrofloxacin in poultry feeding water through excitation emission fluorescence and three-way PARAFAC calibration. *Analyst*, 130, 1639–1647.

[22]   Andersson, C. A. (2002). INCA 1.41, Department of Dairy and Food Science, Frederiksberg, Denmark. Available at: https://food.ku.dk/ english/research_at_food/sections/food-analytics-and-biotechnology /food-datasets---test/. Last visit: 12/10/2022.

[23]   Commission Regulation (EU). (2011). No 1282/2011 of 28 November 2011 amending and correcting Commission Regulation (EU) No 10/2011 on plastic materials and articles intended to come into contact with food. *Official Journal of the European Union Law*, 328, 22–29.

[24]   Arce, M. M., Ortiz, M. C., and Sanllorente, S. (2022). Univariate data analysis versus multivariate approach in liquid chromatography. An application for melamine migration from food contact materials. *Microchemical Journal*, 181, 107648.

[25]   Ortiz, M. C., Sarabia, L. A., García, I., Giménez, D., and Meléndez, E. (2006). Capability of detection and three-way data. *Analytica Chimica Acta*, 559, 124–136.

[26]   Hubert, M. (2020). Robust methods for high-dimensional data. In Brown, S. D., Tauler, R., and Walczak, B. (eds.), *Comprehensive Chemometrics*, Vol. 1. Oxford: Elsevier, pp. 149–171.

[27]   Tauler, R. (2022). Personal communication about plenary conference "Omidikia, N., Ghaffari, M., Jansen, J., Buydens, L., and Tauler, R. Bilinear model factor decomposition: a general mixture analysis tool", CAC2022, Rome, Italy.

[28]   Lee, D. D. and Seung, H. S. (1999). Learning the parts of objects by non-negative matrix factorization. *Nature*, 401, 788–791.

[29]   Chih-Jen, L. (2007). Projected gradient methods for nonnegative matrix factorization. *Neural Computation*, 19, 2756–2779.

[30]   Tauler, R. (1995). Multivariate curve resolution applied to second order data. *Chemometrics and Intelligent Laboratory Systems*, 30, 133–146.

[31]   de Juan, A., Rutan, S. C., and Tauler, R. (2020). Two-way data analysis: Multivariate curve resolution – Iterative resolution methods. In Brown, S. D., Tauler, R., and Walczak, B. (eds.), *Comprehensive Chemometrics*, Vol. 2. Oxford: Elsevier, pp. 153–171.

[32] de Juan, A. and Tauler, R. (2021). Multivariate curve resolution: 50 years addressing the mixture analysis problem – A review. *Analytica Chimica Acta*, 1145, 59e78.

[33] Jaumot, J. and Tauler, R. (2010). MCR-BANDS A user friendly MATLAB program for the evaluation of rotational ambiguities in multivariate curve resolution. *Chemometrics and Intelligent Laboratory Systems*, 103, 96–107.

[34] Olivieri, A. C. and Tauler, R. (2021). N-BANDS: A new algorithm for estimating the extension of feasible bands in multivariate curve resolution of multicomponent systems in the presence of noise and rotational ambiguity. *Journal of Chemometrics*, 35, e3317.

[35] Olivieri, A. C. (2022). Evaluation of the ambiguity in second-order analytical calibration based on multivariate curve resolution. A tutorial. *Microchemical Journal*, 179, e107455.

[36] Westad, F. and Kermit M. Independent component analysis. In Brown, S. D., Tauler, R., and Walczak, B. (eds.), *Comprehensive Chemometrics*, Vol. 2. Oxford: Elsevier, pp. 39–56.

[37] Paatero, P. and Tapper, U. (1994). Positive matrix factorization: A non-negative factor model with optimal utilization of error estimates of data values. *Environmetrics*, 5, 111–126.

[38] Wentzell, P. D., Giglio, C., and Kompany-Zareh, M. (2021). Beyond principal components: A critical comparison of factor analysis methods for subspace modelling in chemistry. *Analytical Methods*, 13, 4188–4219.

[39] Wentzell, P. D., Andrews, D. T., Hamilton, D. C., Faber, K., and Kowalski, B. R. (1997). Maximum likelihood principal component analysis. *Journal of Chemometrics*, 11, 339–366.

[40] Park, E. S., Lee, E.-K., and Oh, M.-S. (2021). Bayesian multivariate receptor modeling software: BNFA and bayesMRM. *Chemometrics and Intelligent Laboratory Systems*, 11, 10428.

[41] Bro, R., Andersson, C. A., and Kiers, H. A. L. (1999). PARAFAC2-Part II. Modeling chromatographic data with retention time shifts. *Journal of Chemometrics*, 13, 295–309.

[42] Valverde-Som, L., Herrero, A., Reguera, C., Sarabia, L. A., and Ortiz, M. C. (2022). A new multi-factor multi-objective strategy based on a factorial presence-absence design to determine polymer additive residues by means of head space-solid phase microextraction-gas chromatography-mass spectrometry. *Talanta*, 124021. doi: 10.1016/j.talanta.2022.124021.

[43] Tucker, L. R. (1966). Some mathematical notes on three-mode factor analysis. *Psychometrika*, 31, 279.

Chapter 8

Introducing Predictive Regression Techniques

Luis Antonio Sarabia, María de la Cruz Ortiz,
and María Sagrario Sánchez

Objectives and Scope

Regression analysis consists of obtaining, in the form of a mathematical model, reliable predictions of the responses from new predictors. The chapter begins with a summary of the fundamental concepts of multiple linear regression by least squares (OLS), with special emphasis on prediction and its precision. Also, the problems in least-squares models generated by multicollinearity and highly correlated predictors are pointed out. Both issues are of interest in chemometrics because they occur in almost all multivariate calibrations, whose purpose is to make a prediction as accurately as possible.

It shows the usefulness of biasing OLS to achieve less variance in predictions and, therefore biased linear regression methods must be used to overcome the serious problems caused by multicollinearity. One of these methods is principal component regression, which is discussed in detail along with the criteria being optimised with the regression and the advantages and disadvantages of using it discussed in depth.

A different criterion gives the partial least-squares (PLS) regression method, another biased regression method also based on latent variables. The general model with more than one response is presented, as well as the two customary algorithmic implementations.

In summary, the regression methods are a set of tools that must be used flexibly and judiciously, depending on the characteristics of the training set. The methodology required for regression originated with the "scientific method" and has not stopped growing following the needs of researchers and the development of instrumentation. A paradigmatic case is PLS, motivated by multivariate calibrations in chemometrics and developed during the last quarter of the 20th century with enormous diffusion in many scientific fields beyond chemistry.

The use of statistical software is essential if the regression is to be thoroughly analysed. However, the bold application of regression methods can produce poor results, so it is important for the reader to understand the most widely used methods in chemometrics, their limitations, and when they should be used. But it is almost impossible to establish a "gold standard" – an exact sequence of steps to perform a regression analysis, which, in practice, is a very meticulous task.

Some practical "hints" and ideas are discussed, along with several case studies that revise, explain, and illustrate the theoretical aspects.

8.1. A Brief Introduction to Regression

In its simplest version, *regression analysis* refers to a set of procedures to estimate the relation between some predictor variables $X_1, X_2, \ldots, X_V$ (also named covariates, independent variables, or input variables in machine learning) and a response variable Y (dependent or output variable). In this chapter, the predictor and response variables are all real-valued and are measured for N objects, forming a $N \times V$ data matrix $\mathbf{X} = (x_{ij})$ and an N-dimensional vector $\mathbf{y} = (y_i)$, $i = 1, \ldots, N$, $j = 1, \ldots, V$. The rows of $\mathbf{X}$ are thus V-dimensional row vectors $\mathbf{x}_i$ representing objects (chemical specimens or "samples"), whose corresponding response value is y_i. Sometimes, to emphasise the size of the matrices, they can also be denoted as $\mathbf{X}_{NV}$, $\mathbf{y}_N$. In any case, $\mathbf{X}$ and $\mathbf{y}$ constitute the *training set* to build the prediction rule (regression model).

The purpose of a regression analysis is to compute, by using the information contained in the training set, a function $\hat{f}$ that estimates the unknown relation between predictors and response. There are two different conceptual frameworks for obtaining a regression model: supervised learning and function approximation. To describe them briefly, assume that the true relationship between response and predictors is a sum of a deterministic function f and a random variable ε, formally, $Y = f(X_1, X_2, \ldots, X_V) + \varepsilon$, and that a representative training set is available.

In the case of supervised learning, there will be a learning algorithm (a computer program) that calculates $\hat{f}(\mathbf{x}_i)$, $i = 1, \ldots, N$. The learning algorithm [1–4] has the ability to modify the estimated function $\hat{f}$ based on the differences between the original value of the response and the one generated by the algorithm, $y_i - \hat{f}(\mathbf{x}_i)$. This step of the procedure is named learning by example or learner training. At the end of the learning step, the generated and true responses are expected to be close enough to guarantee the goodness of the response prediction for a new object (specimen or sample). Supervised learning has seen enormous development in the fields of machine learning and neural networks.

The other case is the function approximation and estimation scheme, which is the one followed in this chapter. It considers the rows of the training set $\{\mathbf{X}, \mathbf{y}\}$ as points in the $(V + 1)$-dimensional Euclidean space and a single function f, defined in a subspace in V dimensions, such that $y_i = f(\mathbf{x}_i) + \varepsilon_i$. The purpose is now to obtain a useful approximation of the unknown function f by estimating the parameters that precisely define it. The regression analysis can be seen as a form of function approximation that can be solved using geometric concepts pertaining to Euclidean spaces and mathematical concepts of statistical inference, with which an undergraduate student is likely already familiar.

Some key elements of the regression analysis:

✓ There is a training set $\{X, \mathbf{y}\}$ with samples (the rows of $\mathbf{X}$) representative of the phenomenon under study and whose responses are known (the rows of $\mathbf{y}$).

✓ Each response value y_i is supposed to be the result of some deterministic parametric function f responding to x_i and some random "error", e_i (a realisation of a random variable).

✓ The goal is to estimate the parameters that define f and those that define the probability distribution of the random variable.

A distinction can be made in the strategy followed in the context of function approximation, depending on how the training set is generated:

(i) *Designed experiments*: In this case, the experimenter can choose the experiments (the experimental conditions under which they are performed), i.e., the points of the design $(x_{i1}, \ldots, x_{iV})$, $i = 1, \ldots, N$, can be set without error. Then, the response is measured in them. The design of experiments (DOE, Chapter 4) is a methodological strategy for selecting the points that give the most accurate estimates of the functional dependence of Y on $X_1, X_2, \ldots, X_V$ by using the minimum number of experiments.

(ii) *Experiments with a fixed design*: $\mathbf{X}$ and $\mathbf{y}$ are specified in advance, but $(x_{i1}, \ldots, x_{iV})$, $i = 1, \ldots, N$, are fixed and measured without error (they are not random variables). The response Y is the one that "carries" the measurement error (random variable), that is, if the experiment is repeated, the same $(x_{i1}, \ldots, x_{iV})$ values will be obtained, but y_i in vector $\mathbf{y}$ may be different. An example of this situation is multivariate calibration using spectral data, e.g., NIR spectra. Thus, the researcher selects the training samples and then measures the spectrum (predictor variables) of each sample; the concentration of the compound is known in advance as per the preparation of the training samples. This fixed design is the most commonly used type of design in regression analysis and determines its success.

(iii) *Experiments with a random design*: In this case, $X_1, X_2, \ldots, X_V$ and Y are random variables with a joint probability distribution (which is unknown). The data in the training set are a random sample of the population. If the experiment is repeated, both the responses and the design points will be different. They are also called observational or descriptive studies.

The conceptual structure to which experiments with a random design or experiments with a fixed design respond is very different and related to the purpose of the experiment. By means of $\hat{f}$, one can achieve the following:

(i) Understand the nature of the relationship between response and predictors by analysing the form of function $\hat{f}$, for example, to decide which variable X_i has the greatest effect on the response. This is a difficult task with observational data.

(ii) Make a prediction. In this case, model $\hat{f}$ is used to predict the values of the response (Y) in future observations (test samples) for which the vector of the predictors $(x_1, \ldots, x_V)$ is known but not that of the response. Therefore, $\hat{y} = \hat{f}(x_1, \ldots, x_V)$ is the estimated value of the response. A good future prediction is possible without the need of achieving objective (i), for example, when there is collinearity or high correlation between predictors, in which case the effect of one variable can be observed through others. It should be emphasised that the purpose of a regression model is to obtain good predictions of test samples, not necessarily to make excellent predictions for the samples of the training set. This issue is the key to developing biased regression methods, which is the goal of this chapter.

The analytical techniques that provide a vector of predictors for each sample (e.g., GC-MS, NIR spectroscopy, fluorescence, and UV-visible) are ubiquitous, and, clearly, they require multivariate regression for proper calibrations. That means a training set, formed by the N spectra of the calibration samples, $\mathbf{X}$, and the corresponding concentrations of a compound, $\mathbf{y}$, is needed. Obviously, the estimated predictive model is the calibration model for the specific analytical procedure that will be used to estimate the concentration of the compound in a test sample, as measured in the same way as those of the training set. In this type of data, the predictors are highly correlated, and in principle, the model will not be used to understand the nature of the relationship between the predictors (e.g., the spectrum) and the concentration, although information about this type can be sometimes obtained when several compounds are calibrated together.

The use of multivariate signals is not limited to laboratory calibration tasks; rather, their applicability extends to many other areas. It allows analytical determinations to be made outside the laboratory using online sensors for industrial, environmental, or agricultural control. It also helps in improving the identification and assessment of the origin of food products, and it can also be used to detect frauds and adulterations.

8.2. Basics of Linear Regression and Least Squares

The subject of regression has been extensively discussed in the literature, and there are several monographs adapted to an undergraduate level, such as Refs. [5] and [6] for the case of linear regression models. The purpose of this chapter is to compile the basic results of multivariate linear regression by least squares (OLS), which acts as a reference to understand the development and advantages of using principal component regression (PCR) or partial least squares (PLS) regression. Both PCR and PLS have become key in calibrating first-order signals, that is, the calibration of analytical methods that provide a data vector for each analysed sample.

The generic regression model can be written as the sum of a deterministic function, f, and a random variable, ε, as in Equation (8.1):

$$Y = f(X_1, X_2, \ldots, X_V) + \varepsilon. \tag{8.1}$$

ε is a random variable, and therefore Y is also a random variable. However, the expected value of Y depends only on f. In other words, the mathematical expectation of the random variable ε is zero, formally $E(\varepsilon) = 0$, and thus, the conditional expectation $E(Y|(x_1, \ldots, x_V)) = f(x_1, \ldots, x_V)$ for any new values of the predictor variables $(x_1, \ldots, x_V)$.

Equation (8.1) implies that, in addition to the experimental data in the training set, $\mathbf{X}$ and $\mathbf{y}$, performing a regression analysis requires specifying:

- the functional form of f (linear, quadratic, exponential, etc.);
- the statistical distribution of the random variable ε;
- the fitting criterion, which defines how to quantify the closeness between the experimental values in the training set, y_i, and those computed with the fitted model $\hat{y}_i = \hat{f}(\mathbf{x}_i)$;
- the type of the design (fixed or random).

The multilinear regression (MLR) method uses linear models with normal random error and the least-squares criterion – the reason why sometimes it is also referred to as ordinary least-squares (OLS) regression.

In the particular case of a linear regression, the functional term in Equation (8.1) is a linear function of the predictors, as expressed in Equation (8.2):

$$f(X_1, \ldots, X_V) = b_0 + b_1 X_1 + \cdots + b_V X_V. \tag{8.2}$$

The least-squares theory is concerned with estimating its parameters, namely the coefficients b_i in Equation (8.2). The foundations of the theory were laid by Gauss (1809) and Markoff (1900), as can be consulted in the historical notes about the subject in Chapter 4 of Ref. [7].

Throughout this chapter, the distribution of ε (the random component of the model in Equation (8.1)) is supposed to be normal with mean zero and standard deviation σ, independent and identically distributed (i.i.d.) with respect to the vector of predictor variables $(x_1, \ldots, x_V)$, that is:

$$\varepsilon = N(0, \sigma) \ i.i.d. \tag{8.3}$$

In matrix form, the training set with predictors $\mathbf{X}$ and responses $\mathbf{y}$, as well as the elements of the regression model are written in Equation (8.4):

$$\mathbf{X} = \begin{pmatrix} 1 & x_{11} & x_{12} & \cdots & x_{1V} \\ 1 & x_{21} & x_{22} & \cdots & x_{2V} \\ \vdots & \vdots & \vdots & \ddots & \vdots \\ 1 & x_{N1} & x_{N2} & \cdots & x_{NV} \end{pmatrix}$$

$$\mathbf{y} = \begin{pmatrix} y_1 \\ y_2 \\ \vdots \\ y_N \end{pmatrix} \quad \mathbf{b} = \begin{pmatrix} b_0 \\ b_1 \\ b_2 \\ \vdots \\ b_V \end{pmatrix} \quad \varepsilon = \begin{pmatrix} \varepsilon_1 \\ \varepsilon_2 \\ \vdots \\ \varepsilon_N \end{pmatrix}. \tag{8.4}$$

With this notation, Equations (8.1) and (8.2) result in the linear model in Equation (8.5):

$$\mathbf{y} = \mathbf{X}\mathbf{b} + \varepsilon. \tag{8.5}$$

Note that matrix $\mathbf{X}$ now contains a column of ones added to the predictor variables (which is then a $N \times (V+1)$ matrix). The reason is that the model in Equation (8.2) includes the constant b_0. Additionally, $\mathbf{X}$ is supposed to be of full rank (recall here what has been explained in Section 2.1.3 in Chapter 2).

The least-squares criterion dictates that the estimates of the coefficients, $\hat{\mathbf{b}} = (\hat{b}_0, \ldots, \hat{b}_V)^T$, are the ones that minimise the sum of the squared differences between the experimental values and those predicted with the model in Equation (8.2), which is a function of the coefficients vector. Formally,

$$\hat{\mathbf{b}} = (\hat{b}_0, \dots, \hat{b}_V)^{\mathrm{T}}$$

$$= \arg\min_{(b_0,\dots,b_V)} \left\{ \sum_{i=1}^{N} [y_i - (b_0 + b_1 x_{i1} + \dots + b_V x_{iV})]^2 \right\}, \quad (8.6)$$

where $(1, x_{i1}, \dots, x_{iV})$ is the ith row of the $\mathbf{X}$ matrix in Equation (8.4) and y_i is the corresponding row of the $\mathbf{y}$ vector of responses. As usual, the upper T denotes transposing.

The solution of the least-squares problem in Equation (8.6) is unique for each $\{\mathbf{X}, \mathbf{y}\}$; it can be explicitly written as in Equation (8.7) and does not depend on the probability distribution of the random variable ε of the model in Equation (8.5):

$$\hat{\mathbf{b}} = (\mathbf{X}^{\mathrm{T}}\mathbf{X})^{-1}\mathbf{X}^{\mathrm{T}}\mathbf{y}. \quad (8.7)$$

The $(V + 1)$-dimensional vector of the least-squares estimate $\hat{\mathbf{b}}$ in Equation (8.7) has the properties summarised in the following paragraphs.

Equation (8.7) shows that each coordinate (component or element) of $\hat{\mathbf{b}}$ is a linear function of the values in $\mathbf{y}$, which are random variables. Therefore, $\hat{\mathbf{b}}$ is a random vector, with each of its components (coordinates) $\hat{b}_i$ ($i = 0, 1, \dots, V$) being a random variable. Its individual expected (mean) value coincides with the theoretical value b_i, as formally expressed in Equation (8.8), so the estimates $\hat{b}_i$ are unbiased:

$$\mathrm{E}(\hat{\mathbf{b}}) = \mathbf{b}. \quad (8.8)$$

Their covariance matrix is a $(V + 1) \times (V + 1)$ square symmetric matrix that can also be written as a function of $\mathbf{X}$, as in Equation (8.9):

$$\mathrm{cov}(\hat{\mathbf{b}}) = \sigma^2 (\mathbf{X}^{\mathrm{T}}\mathbf{X})^{-1} = \sigma^2 (c_{ij}), \quad (8.9)$$

where c_{ij} denotes the generic element of the $(\mathbf{X}^{\mathrm{T}}\mathbf{X})^{-1}$ matrix. Therefore, the diagonal elements $\sigma^2 c_{ii}$ correspond to the variance of each $\hat{b}_i$. $(\mathbf{X}^{\mathrm{T}}\mathbf{X})^{-1}$ is known as the *dispersion* matrix because it provides the variance factors depending on the design matrix $\mathbf{X}$.

The estimate of the response, computed with $\hat{f}$ for a test vector $\mathbf{x}_t = (1, x_{t1}, \ldots, x_{tV})$ is given in Equation (8.10), and it is also an unbiased estimate (i.e., it is the mean of Y, which is a random variable according to Equation (8.1)). Note that $\mathbf{x}_t$ is written as a row vector because the objects in the training set are rows in the $\mathbf{X}$ matrix:

$$\hat{y}_t = \mathbf{x}_t \hat{\mathbf{b}}. \tag{8.10}$$

> Be aware that there is a random variable in the regression model. Consequently, *all* the computed parameters and predictions made with the constructed model are *not* just "simple" numbers, but mean values of the corresponding random variable.

Usually, the predictor variables are supposed to be mean centred. In that case, the $\mathbf{X}$ matrix in Equation (8.4) becomes $\mathbf{X}_c$ in Equation (8.11), where the vertical line in the final expression of the matrix indicates its separation (partition) into two blocks: the column of ones and matrix $\tilde{\mathbf{X}}$ with the centred values of the predictor variables:

$$\mathbf{X}_c = \begin{pmatrix} 1 & x_{11} - \bar{x}_1 & x_{12} - \bar{x}_2 & \cdots & x_{1V} - \bar{x}_V \\ 1 & x_{21} - \bar{x}_1 & x_{22} - \bar{x}_2 & \cdots & x_{2V} - \bar{x}_V \\ \vdots & \vdots & \vdots & \ddots & \vdots \\ 1 & x_{N1} - \bar{x}_1 & x_{N2} - \bar{x}_2 & \cdots & x_{NV} - \bar{x}_V \end{pmatrix} = (\mathbf{1}_N | \tilde{\mathbf{X}}). \tag{8.11}$$

Consequently, $(\mathbf{X}_c^{\mathrm{T}} \mathbf{X}_c)^{-1} = \begin{pmatrix} 1/N & \mathbf{0}^{\mathrm{T}} \\ \mathbf{0} & (\tilde{\mathbf{X}}^{\mathrm{T}} \tilde{\mathbf{X}})^{-1} \end{pmatrix}$.

The estimate of the coefficients vector, $\mathbf{a}$, for the new model with centred predictors has $\bar{y}$ as its first element and then $(\tilde{\mathbf{X}}^{\mathrm{T}} \tilde{\mathbf{X}})^{-1} \tilde{\mathbf{X}}^{\mathrm{T}} \mathbf{y}$, as expressed in Equation (8.12), where again the horizontal line separates the two blocks being distinguished in the partitioned vector (remember, this is not a division):

$$\hat{\mathbf{a}} = \begin{pmatrix} \bar{y} \\ \hline (\tilde{\mathbf{X}}^{\mathrm{T}} \tilde{\mathbf{X}})^{-1} \tilde{\mathbf{X}}^{\mathrm{T}} \mathbf{y} \end{pmatrix}. \tag{8.12}$$

The centring of the predictors in the training set and the consequent change of parameters, from **b** to **a**, in the linear model does not affect the estimated value $\hat{y}$, so the residuals of both models (with centred and uncentred variables) are the same.

With the raw data, the variance of the estimated response for a test vector $\mathbf{x}_t$ is given by

$$\text{var}(\hat{y}_t) = \mathbf{x}_t \text{cov}(\hat{\mathbf{b}})\mathbf{x}_t^\text{T}\sigma^2 + \sigma^2 = [\mathbf{x}_t \left(\mathbf{X}^\text{T}\mathbf{X}\right)^{-1}\mathbf{x}_t^\text{T} + 1]\sigma^2, \quad (8.13)$$

which, with centred data, becomes

$$\text{var}(\hat{y}_t) = \tilde{\mathbf{x}}_t \text{cov}(\hat{\mathbf{a}})\tilde{\mathbf{x}}_t^\text{T}\sigma^2 + \text{var}(\bar{y}) + \sigma^2$$

$$= \left[\tilde{\mathbf{x}}_t(\tilde{\mathbf{X}}^\text{T}\tilde{\mathbf{X}})^{-1}\tilde{\mathbf{x}}_t^\text{T} + \frac{1}{N} + 1\right]\sigma^2. \quad (8.14)$$

In Equation (8.13), $\mathbf{x}_t = (1, x_{t1}, \dots, x_{tV})$, and in Equation (8.14), $\tilde{\mathbf{x}}_t = (x_{t1} - \bar{x}_1, \dots, x_{tV} - \bar{x}_V)$ for the mean values $\bar{x}_i$ obtained with the original $\mathbf{X}$ in the training set.

Applying Equation (8.10) for the particular case of the ith row of $\mathbf{X}$, $\mathbf{x}_i = (1, x_{i1}, \dots, x_{iV})$, and replacing the estimated coefficients according to Equation (8.7), one obtains

$$\hat{y}_i = \mathbf{x}_i\hat{\mathbf{b}} = \mathbf{x}_i(\mathbf{X}^\text{T}\mathbf{X})^{-1}\mathbf{X}^\text{T}y_i = \mathbf{H}y_i. \quad (8.15)$$

$\mathbf{H}$ in Equation (8.15) is usually known as the *hat* matrix (puts the hat on y_i) and contains information about the regression model constructed in relation to the training set used, as it describes the transformation that is applied to each value of the response y_i to convert it into its estimate. Equation (8.15) also shows that this transformation only depends on the design matrix $\mathbf{X}$.

The vector $\mathbf{e} = (e_i) = \mathbf{y} - \hat{\mathbf{y}}$ of differences (the residuals) can be seen as N independent realisations of the random variable ε in the regression model in Equation (8.1). To characterise its mean and variance, the least-squares criterion in Equation (8.6) guarantees that the sum of the residuals is zero:

$$\sum_{i=1}^{N} e_i = 0. \quad (8.16)$$

Therefore, the mean of the residuals is zero. On the other hand, the variance σ^2 of ε can be obtained, independently of the training set, from replicated measurements (different values of Y obtained with the same $\mathbf{x}$, other than the ones in the training set). However, it is usually estimated from the $\mathbf{e}$ vector by using the residuals or errors mean square, MS_E, as defined in Equation (8.17):

$$\hat{\sigma}^2 = MS_\mathrm{E} = \frac{\mathrm{SS_E}}{N - (V+1)} = \frac{\sum_{i=1}^{N} e_i^2}{N - (V+1)}. \qquad (8.17)$$

The expected value of MS_E is precisely σ^2, and thus, $\hat{\sigma}^2$ is an unbiased estimator of σ^2. This estimate is used in Equation (8.9) and Equation (8.13) (or Equation (8.14)) to estimate the variances of $\hat{\mathbf{b}}$ and the predicted $\hat{y}_t$, respectively.

Due to the normal distribution of ε, each $\hat{b}_i$ and any estimated response $\hat{y}_t$ follow a Student's t distribution with $N - V - 1$ degrees of freedom (those of the estimate $\hat{\sigma}^2$). Therefore, confidence intervals can be computed.

For each individual coefficient b_i, at a $100(1 - \alpha)\%$ confidence level, the interval is

$$\hat{b}_i - t_{\alpha/2, N-V-1}\sqrt{\hat{\sigma}^2\, c_{ii}}$$
$$\leq b_i \leq \hat{b}_i + t_{\alpha/2, N-V-1}\sqrt{\hat{\sigma}^2\, c_{ii}}, \quad i = 0, 1, \ldots, V, \qquad (8.18)$$

where $t_{\alpha/2, N-V-1}$ is the $1 - \alpha/2$ percentile of a Student's t distribution with $N - V - 1$ degrees of freedom, and c_{ii} is the ith term of the diagonal of the $(\mathbf{X}^\mathrm{T}\mathbf{X})^{-1}$ matrix, see Equation (8.9).

At the same $100(1 - \alpha)\%$ confidence level, the joint confidence region for the $\mathbf{b}$ vector is an hyperellipsoid in the $(V+1)$-dimensional space, the one defined in Equation (8.19):

$$\frac{(\hat{\mathbf{b}} - \mathbf{b})^\mathrm{T}\mathbf{X}^\mathrm{T}\mathbf{X}(\hat{\mathbf{b}} - \mathbf{b})}{(V+1)\hat{\sigma}^2} < F_{\alpha, V+1, N-V-1}, \qquad (8.19)$$

where $F_{\alpha, V+1, N-V-1}$ is the $1 - \alpha$ percentile of a Snedecor's F distribution with $(V + 1)$ degrees of freedom in the numerator and $(N - V - 1)$ degrees of freedom in the denominator.

As for an estimated response, Equation (8.20) applies at a $100(1 - \alpha)\%$ confidence level:

$$\hat{y}_t - t_{\alpha/2,N-V-1}\sqrt{\text{var}(\hat{y}_t)} < y < \hat{y}_t + t_{\alpha/2,N-V-1}\sqrt{\text{var}(\hat{y}_t)}, \quad (8.20)$$

where $\text{var}(\hat{y}_t)$ is computed by substituting σ^2 in Equation (8.13) by its estimate $\hat{\sigma}^2$ in Equation (8.17).

It is also possible to apply hypothesis testing to make decisions regarding the estimated values. Particularly useful is the significance test on each individual coefficient (H_0: $b_i = 0$ and H_1: $b_i \neq 0$). If H_0 is retained, the corresponding variable X_i should not be included in the model to make predictions. The critical region of the two-sided test at a significance level of α is related to Equation (8.21). In other words, the nonzero coefficients are those that satisfy this equation.

$$\frac{|\hat{b}_i|}{\sqrt{\hat{\sigma}^2\, c_{ii}}} > t_{\alpha/2,N-V-1}. \quad (8.21)$$

When ε has a normal distribution, an important property is that the least-squares estimator $\hat{\mathbf{b}}$ has the property of being the maximum likelihood estimator, that is, it is the estimator that maximises the probability of having the experimental data in $\mathbf{X}$ and $\mathbf{y}$.

Given that the purpose of a regression model is to make predictions, estimators that are both unbiased (i.e., they provide, on average, the true value) and with the least possible variance are of interest, regardless of how they have been calculated. In this respect, the Gauss–Markoff theorem states that the estimator $\hat{\mathbf{b}}$ obtained through least squares using Equation (8.7) is the best linear unbiased estimator of $\mathbf{b}$.

This means that for any linear combination of $\mathbf{b}$, $\mathbf{w}^{\mathrm{T}}\mathbf{b}$, among the estimators that are linear in $\mathbf{y}$ and unbiased, $\mathbf{w}^{\mathrm{T}}\hat{\mathbf{b}}$ has minimum variance. If the assumption of normality for ε is added, then $\mathbf{w}^{\mathrm{T}}\hat{\mathbf{b}}$ is the minimum variance unbiased estimator of $\mathbf{w}^{\mathrm{T}}\mathbf{b}$, that is, the restriction is still unbiasedness, but not linearity in $\mathbf{y}$. Then, by only considering the linear combination with $\mathbf{w}$ having 1 in coordinate i and 0 elsewhere, $\mathbf{w}^{\mathrm{T}}\hat{\mathbf{b}} = \hat{b}_i$. Therefore, the Gauss–Markoff theorem implies that, for all i, $\text{var}(\hat{b}_i)$ is the minimum value out of all unbiased estimators if ε has a normal distribution, or of the linear unbiased estimators in the other case. The same holds for $\mathrm{var}(\hat{y}_t)$.

> The assumption of normality (that must be checked) provides a competitive advantage in unbiased linear regression: the support for all the inferential aspects for validation, hypothesis testing, and/or confidence intervals.

The statistical properties of least-squares estimation follow from the fact that ε has an i.i.d. $N(0, \sigma)$ distribution. Therefore, it is necessary to verify the hypotheses of normality, independence, and homoscedasticity with the residuals $\mathbf{e} = (e_i)$ obtained in the fit.

Regarding the functional part, it is also necessary to evaluate if the assumed linear model is compatible with the experimental data of the training set. To do that, hypothesis tests are used for the significance of the regression model (null hypothesis H_0: the regression model does not explain the variability of the response, against the alternative H_1: the model does explain it) and for lack of fit (H_0: no lack of fit, that is, the assumed regression model is compatible with the experimental data, H_1: there is lack of fit). It should be remembered that for the lack-of-fit test, replicates are needed, that is, different response measures obtained with the same vector of predictors. This fact must be taken into account when designing the experiment to obtain the training set.

Details of the methodology to validate the functional part f and the random part ε in Equation (8.1) can be found in Ref. [5].

8.3.   Bias versus Prediction

This section introduces – in a very elementary, almost intuitive way – the key concept of the balance between bias and variance of an estimator and its impact on linear regression models. A broad and detailed analysis of the bias–variance dilemma and the complexity of a model as well as its importance for making accurate estimations can be found in Chapter 7 of Ref [8], an advanced-level book.

The idea that an estimator is "good" when it is unbiased and precise (with less variance) has been introduced implicitly in the previous section. In fact, both concepts are also used to define the quality of an analytical measurement, and as such, they are familiar to any chemist and are also common in chemometrics.

However, from a statistical point of view, it is necessary to clarify the idea of what constitutes "good". This is done by specifying a loss function to quantify the notion. The most common specification is via the squared error loss, L:

$$L(b, \hat{b}) = (b - \hat{b})^2, \tag{8.22}$$

where b is the real value of a parameter that has been estimated as $\hat{b}$ using some experimental data. The fact is that L cannot be computed from a particular training set because it is a random variable itself, as a direct consequence of the random character of $\hat{b}$. The best scenario is to determine the distribution of $\hat{L}(b, \hat{b})$ over repeated experimental designs, although in that case, often only the location of that distribution is obtained, that is, the expected squared error (ESE):

$$\text{ESE} = E(\hat{L}(b, \hat{b})) = E((b - \hat{b})^2). \tag{8.23}$$

Therefore, statistical procedures that tend to minimise ESE over (conceptually) repeated experiments should be used or designed. In the case at hand, the repeated experiments would consist of several training datasets $\{\mathbf{X}_i, \mathbf{y}_i\}$ drawn from a usually unknown population.

If the term $E(\hat{b})$ is added and subtracted in Equation (8.23), when expanding the squares, the cross term becomes null and ESE can be decomposed into a sum of two addends, namely the square of the bias and the variance of the estimate, as expressed in Equation (8.24):

$$\text{ESE} = E((b - \hat{b})^2) = (b - E(\hat{b}))^2 + E(\hat{b} - E(\hat{b}))^2$$
$$= (\text{bias}(\hat{b}))^2 + \text{var}(\hat{b}). \tag{8.24}$$

Zero variance estimators are rarely good unless the researcher is very lucky. For example, if $\hat{b} = 3$, independent of the data, its bias will be $b - 3$ with $\text{var}(\hat{b}) = 0$. But clearly, $\hat{b}$ will be a good estimate only if the true value for b happens to be close to 3.

Unbiased estimators are seldom the best in terms of ESE. Although they are often very good, sometimes they are poor. Following the Gauss–Markoff theorem in Section 8.2, the linear least-squares estimate $\hat{\mathbf{b}}$ for a linear additive model is the minimum ESE estimate of $\mathbf{b}$ and $\hat{Y} = \hat{b}_0 + \sum_{i=1}^{V} \hat{b}_i X_i$ is an unbiased estimate of $Y = b_0 + \sum_{i=1}^{V} b_i X_i$.

The following paragraphs will show that the ESE of least-squares estimates can be reduced using biased estimators with less variance, even when ε is distributed as i.i.d. $N(0, \sigma)$. This is illustrated with the following example, adapted from Ref. [9].

8.3.1. *Example*

Consider a highly simplified regression situation in which the values of the predictor variables X_i in the design are uncorrelated ($\mathbf{X}^T\mathbf{X}$ is diagonal); suppose also that both predictors and response Y have a null mean, and assume that the variance σ^2 has been estimated with replicates and not with the residuals of the fit (that is, not with the data from the training set).

The least-squares estimate $\hat{y}_{LS} = \sum_{i=1}^{V} \hat{b}_i X_i$ is unbiased, but if the last addend is removed (emulating $b_V = 0$), the new estimate $\hat{y}_{BS} = \sum_{i=1}^{V-1} \hat{b}_i X_i$ is biased. Note that the estimates $\hat{b}_i, i = 1, \ldots, V - 1$, in both cases are the same because the variables X_i are uncorrelated.

Let us compare the bias and variance of both estimators when predicting with the two regressions a test vector $\mathbf{x}_t = (x_{t1}, \ldots, x_{tV})$, e.g., the spectrum of a test sample:

$$\text{var}(\hat{y}_{LS}(\mathbf{x}_t)) = \text{var}\left(\sum_{j=1}^{V} \hat{b}_j x_{tj}\right) = \sum_{j=1}^{V} x_{tj}^2 \text{var}(\hat{b}_j)$$

$$= \sum_{j=1}^{V} \frac{\sigma^2 x_{tj}^2}{SS(\mathbf{x}_j)} = \sigma^2 \sum_{j=1}^{V} \frac{x_{tj}^2}{SS(\mathbf{x}_j)}, \qquad (8.25)$$

where $\mathbf{x}_j$ is the jth column of $\mathbf{X}$, corresponding to predictor X_j, and $SS(\mathbf{x}_j)$ is the sum of squares of its elements, $SS(\mathbf{x}_j) = \sum_{i=1}^{N} x_{ij}^2$.

Since $\hat{y}_{LS}$ is unbiased, the application of Equation (8.24) with the variance in Equation (8.25) gives

$$\text{ESE}(\hat{y}_{LS}(\mathbf{x_t})) = \sigma^2 \sum_{j=1}^{V} \frac{x_{tj}^2}{SS(\mathbf{x}_j)}. \tag{8.26}$$

Similar computations with the other estimator yield a variance:

$$\text{var}(\hat{y}_{BS}(\mathbf{x_t})) = \text{var}\left(\sum_{j=1}^{V-1} \hat{b}_j x_{tj}\right) = \sigma^2 \sum_{j=1}^{V-1} \frac{x_{tj}^2}{SS(\mathbf{x}_j)}, \tag{8.27}$$

which is less than the one in Equation (8.25) because it has fewer terms (the last positive addend is missing).

Regarding the bias, taking into account that the $\hat{b}_i$ are uncorrelated (as X_i themselves are uncorrelated) and unbiased (that is, $E(\hat{b}_i) = b_i$), we get

$$\text{bias}(\hat{y}_{BS}(\mathbf{x_t})) = f(\mathbf{x_t}) - E(\hat{f}(\mathbf{x_t}))$$
$$= \sum_{j=1}^{V} b_j x_{tj} - E\left(\sum_{j=1}^{V-1} \hat{b}_j x_{tj}\right) = b_V x_{tV}. \tag{8.28}$$

The expected loss can be computed by again applying Equation (8.24) and substituting the bias and variance for those in Equations (8.27) and (8.28) to obtain

$$\text{ESE}(\hat{y}_{BS}(\mathbf{x_t})) = (b_V x_{tV})^2 + \sigma^2 \sum_{j=1}^{V-1} \frac{x_{tj}^2}{SS(\mathbf{x}_j)}. \tag{8.29}$$

Consequently, the difference between both estimators is

$$\text{ESE}(\hat{y}_{BS}(\mathbf{x_t})) - \text{ESE}(\hat{y}_{LS}(\mathbf{x_t}))$$
$$= (b_V x_{tV})^2 - \sigma^2 \frac{x_{tV}^2}{SS(\mathbf{x}_V)} = x_{tV}^2\left(b_V^2 - \frac{\sigma^2}{SS(\mathbf{x}_V)}\right). \tag{8.30}$$

Therefore, when $b_V^2 < \sigma^2/SS(\mathbf{x}_V)$, the biased estimate $\hat{y}_{BS}$ wins over the unbiased estimate $\hat{y}_{LS}$ for all $\mathbf{x}_t$.

8.4. Multicollinearity

The current routine use of analytical techniques that provide data vectors with up to thousands of components (e.g., IR spectroscopy and proton magnetic resonance) makes it absolutely necessary to use biased estimators. This is so because the huge quantity of variables, in general with a few samples, aggravates the poor condition of the $\mathbf{X}^T\mathbf{X}$ matrix. This fact introduces a high correlation between predictor variables and a very low ratio between the number of objects and variables, N/V.

For example, when handling UV–vis absorbance spectra, the signal recorded at the ith wavelength (predictor variable X_i) is highly correlated with that recorded at close wavelengths, and this is unavoidable even if the calibration samples were selected so that analyte concentrations follow an orthogonal design of experiments.

Multicollinearity among the V predictors refers to the situation in which some of them take values that are linearly related to one another. For example, consider N ternary mixtures of solvents for a mobile phase in HPLC. They are defined by the percentage of each solvent in the mixture, which is a vector of three coordinates (x_1, x_2, x_3), that is, there are three predictor variables X_1, X_2, and X_3. However, as $X_1 + X_2 + X_3 = 100$, all the three-dimensional points are in fact in a two-dimensional subspace. In that situation, it is impossible to fit a model through least squares to predict a response Y as a linear function of X_1, X_2, and X_3, because $\mathbf{X}^T\mathbf{X}$ is a 3×3 matrix with a rank of two and, consequently, it does not have an inverse (see the shaded box at the end of Section 2.2.1 in Chapter 2).

The situation described in the previous paragraph is not always so clear. In many cases, multicollinearity causes the $\mathbf{X}^T\mathbf{X}$ matrix to be almost singular, and although the least-squares linear estimate can still be obtained using Equation (8.7), it will be highly imprecise. This is so because the variance of the coefficients estimate, Equation (8.9), and that of the estimated response, Equation (8.13) or (8.14), depend on the inverse of the mentioned matrix, namely $(\mathbf{X}^T\mathbf{X})^{-1}$, whose elements are very large. Therefore, the capacity for inferential analysis of the results is lost. For example, if c_{ii} are large in Equation

(8.21), the test will tend not to reject the null hypothesis, making all the coefficients statistically equal to zero. Sometimes, there is the paradox of having a significant linear regression model but with no significant coefficients, that is, a model that explains the response Y but does not depend on the predictors X_i.

In practice, the problem is hard to diagnose because it is difficult to specify the dimension of the subspace that collects the relevant information for the response. There are several techniques to detect collinearity, which are generally based on the analysis of the correlation matrix between the predictors, $\mathbf{CR} = \text{corr}(X_1, X_2, \ldots, X_V)$. Among the evidences that point to collinearity, there are two criteria that are easy to check:

- If $\mathbf{X}^T\mathbf{X}$ is almost singular, so is $\mathbf{CR}$ with very small eigenvalues. Thus, if the quotient between the largest and smallest eigenvalues is greater than 10, it can be accepted that there is multicollinearity.
- Off-diagonal elements of $\mathbf{CR}$ close to 1 imply that the pair of involved variables behave like a single one, reducing the dimension of the space. On the other hand, note that the absence of bivariate high correlations does not imply the absence of multicollinearity because several predictors can be collinear without any two of them having a large correlation coefficient.

Finally, when $\mathbf{XX}$ is ill-conditioned, $\text{var}(\hat{y}_{LS}(\mathbf{x}_t))$ will be large (see Equation (8.13)), and accordingly the ESE of unbiased least-squares estimates will become hopelessly large. The only hope for obtaining reasonable estimates of the property of interest lies in the use of biased estimators. Unfortunately, the difficulty lies in introducing small squared bias while, at the same time, greatly reducing the variance for a particular problem.

> MLR models are, at best, imprecise for highly correlated and/or multicollinear predictor variables, which is a common situation with first-order chemical measurements. An alternative is to wisely bias a regression model to improve prediction.

8.5. Principal Component Regression

Early in the 1940s, Stone [10] proposed the orthogonalisation of the predictor variables as a method for backpropagating the regression coefficients up to the original variables to search for causal relationships. Subsequently, Kendall [11] suggested the idea of considering descriptive relationships between predictors and the response that are not necessarily causal. It was Massy [12] who articulated both concepts and published PCR in 1965 as a procedure for exploratory statistical analysis. It was a consequence of the idea of building statistical prediction models based not on a causal relationship between predictor variables and response but on a merely descriptive one.

The main underlying idea is that, when there are many predictors and high correlations/collinearities among several of them, there are numerous directions within the V-dimensional space along which the data show small variance. When computing a regression model, it is reasonable to bias the solution away from those directions to obtain a better ESE estimate. If the low-variance directions are primarily noise, this procedure will succeed in reducing the variance without unduly increasing the bias. However, this argument is valid as long as the training set is representative of the population. The argument will not be valid, and the bias could be very detrimental if the correlations in the design matrix $\mathbf{X}$ are caused by the sampling procedure that was followed to obtain the training set instead of by the underlying phenomenon in the population, which is the very aim of any study with a regression model.

In what follows, both the predictor and response variables are supposed to be mean centred. In Chapter 7, a change of coordinates (an orthogonal rotation) in the space of the variables $X_1, X_2, \ldots, X_V$ was studied. The orthogonal rotation generates new variables, named the principal components (PCs) $PC_1, PC_2, \ldots, PC_V$, which are uncorrelated to one another and with decreasing variance. Therefore, the PCs do not have multicollinearity and directions of low variability, which facilitates the application of the strategy mentioned above of biasing the model to directions far from those of low variance.

The idea is then to consider the PCs as the predictor variables to fit a linear regression model. Since $\mathbf{y}$ has a zero mean, the linear regression model has no independent term; therefore, the design matrix does not have the first column of ones written in Equation (8.4). Thus, the coefficients of the linear model are $\mathbf{b} = (b_1, \ldots, b_V)$, and to estimate them with a training set $\{\mathbf{X}, \mathbf{y}\}$, the predictor variables for the regression model are no longer those from $\mathbf{X}$ but the scores of the N objects on the PCs of $\mathbf{X}$. According to Equation (7.11) in Chapter 7, these scores are given by

$$\mathbf{T}_{NV} = \mathbf{X}_{NV} \mathbf{P}_{VV}^T, \tag{8.31}$$

where the sub-indexes refer to the size of the corresponding matrix. $\mathbf{T}$ is the matrix with the scores on each PC, in columns, and the jth row of $\mathbf{P}$ contains the loadings on PC_j.

To facilitate the reading of the rest of this chapter, hereinafter, $\mathbf{P}$ will be the matrix whose columns store the loadings of the PCs. Therefore, Equation (8.31) will become

$$\mathbf{T}_{NV} = \mathbf{X}_{NV} \mathbf{P}_{VV}. \tag{8.32}$$

With this notation, the training set for the PCR model is $\{\mathbf{T}, \mathbf{y}\}$ with the properties of $\mathbf{T}$ described in Chapter 7. In particular, $\mathbf{T}$ is mean centred and its covariance matrix is the diagonal matrix $\mathbf{S_T} = \mathrm{diag}(l_1, l_2, \ldots, l_V)$, with $l_i = \mathrm{var}(PC_i)$, $i = 1, \ldots, V$. Consequently,

$$(\mathbf{T}^T \mathbf{T})^{-1} = \frac{1}{N-1} \mathrm{diag}(1/l_1, \ldots, 1/l_V). \tag{8.33}$$

Then, applying Equation (8.7) and operating, each coefficient estimate becomes

$$\hat{b}_i = \frac{\mathbf{t}_i^T \mathbf{y}}{(N-1)l_i}, \quad i = 1, \ldots, V, \tag{8.34}$$

where $\mathbf{t}_i$ refers to the ith column of matrix $\mathbf{T}$.

It is clear that to obtain $\hat{b}_i$, only the scores of the ith PC are needed, that is, the model is a succession of univariate regressions of Y on each PC.

The variances of the estimated coefficients and predictions are computed with the estimate of the residual variance $\hat{\sigma}^2$ simply by substituting $(\mathbf{T}^T\mathbf{T})^{-1}$ in Equations (8.9) and (8.14), respectively, resulting in Equations (8.35) and (8.36):

$$\text{cov}(\hat{\mathbf{b}}) = \hat{\sigma}^2 \text{diag}\left(\frac{1}{(N-1)l_1}, \ldots, \frac{1}{(N-1)l_V}\right), \qquad (8.35)$$

$$\text{var}(\hat{y}_t) = \hat{\sigma}^2\left[1 + \frac{1}{N} + \mathbf{t}_t\text{diag}\left(\frac{1}{(N-1)l_1}, \ldots, \frac{1}{(N-1)l_V}\right)\mathbf{t}_t^T\right]$$

$$= \hat{\sigma}^2\left[1 + \frac{1}{N} + \sum_{i=1}^{V}\frac{t_{ti}^2}{(N-1)l_i}\right], \qquad (8.36)$$

where the row vector $\mathbf{t}_t = \tilde{\mathbf{x}}_t\mathbf{P}$ contains the scores obtained with $\tilde{\mathbf{x}}_t$, after centring a test vector $\mathbf{x}_t$, and $\hat{y}_t$ is the predicted response for it.

The variances in Equations (8.35) and (8.36) depend on the eigenvalues l_i that are decreasing when increasing i from the first to the last component. As a consequence, the confidence intervals on the estimated coefficients $\hat{b}_i$ become longer as i increases. A similar effect is exerted on the joint confidence region on the vector of coefficients $\hat{\mathbf{b}}$, with increasing volumes when successive PCs are added. Finally, longer confidence intervals on the predicted response are also obtained whenever more PCs are used. Furthermore, in this case, a positive term is added to the sum with each additional PC.

Of real importance here is the effect on the decisions that can be derived from them, for example, about the significance of the coefficients. The corresponding decision when testing whether b_i is zero or not at a significance level α in Equation (8.21) becomes

$$\frac{|\hat{b}_i|}{\sqrt{\hat{\sigma}^2/(N-1)l_i}} = \frac{|\hat{b}_i|\sqrt{(N-1)l_i}}{\hat{\sigma}} > t_{\alpha/2,N-V}. \qquad (8.37)$$

Again, when moving to later PCs (that is, when i increases from 1 to V), l_i becomes smaller, and so does the left-hand side of the inequality of Equation (8.37), making it increasingly difficult to surpass the critical value of $t_{\alpha/2,N-V}$. In other words, the test becomes more conservative and b_i will be considered null.

PCR models are MLR models for the scores of the PCs.

As such, under normal distribution, statistical inference can be applied to validate the model, test hypotheses, and calculate confidence intervals on the predictions.

If the regression is performed using all the PCs, the resulting model is exactly the same as when using the original variables. The idea is to bias this model by removing one or several components of the predictors in **T**. According to Equation (8.34), this does not change the estimated values for the coefficients of the retained components.

This situation is similar to that established in the example in Section 8.3.1. From the way the PCs are built, it is expected that ESE will improve by removing the last components from the model because they contribute to the prediction variance more than the first PCs. Furthermore, the directions of the last components explain the lowest variance of the data; therefore, it is adequate to restrict the model orthogonally by removing these components (the others being orthogonal to them by construction).

As more terms are added to the regression model, the restriction is removed away from low-variance directions. The degree to which the restriction is enforced (i.e., low-variance directions are removed) is controlled by the number of terms A (smaller A means a stricter restriction). The question then becomes how to determine A.

8.5.1. *Selection of principal components*

There are two approaches to selecting the "relevant" PCs, and thus, the bias of the model. They differ in the way in which the model's predictive quality is assessed, namely: (i) by validation (cross-validation or with an external set) using the residual sum of squares of the least-squares fit, or (ii) by selection of the significant components.

8.5.1.1. Use of the residual sum of squares

The purpose is to compute an estimate of the predictive capacity of a regression model with A PCs using the data of the training set $\{\mathbf{T}, \mathbf{y}\}$ formed by N objects and V PCs (obtained via Equation (8.32)). This can be done by following two major approaches:

(a) *Cross-validation*: This is a sequential method where, initially, all objects (rows of $\mathbf{T}$ and $\mathbf{y}$) are partitioned into G disjoint groups, called cancellation groups. In each iteration, the objects in the corresponding cancellation group are separated from $\{\mathbf{T}, \mathbf{y}\}$ in a test set $\{\mathbf{T}_t, \mathbf{y}_t\}$, and the remaining objects form a reduced training set $\{\mathbf{T}_{tr}, \mathbf{y}_{tr}\}$. Next, the training set $\{\mathbf{T}_{tr}, \mathbf{y}_{tr}\}$ is used to compute the regression model with the first PCs, which is then applied to predict the response for the objects in the test set $\mathbf{T}_t$.

Repeating this process with all cancellation groups gives an estimated value of the response to each object in $\mathbf{T}$. Note that each prediction $\hat{y}_{i-tr}$ is always computed with the model fitted with the $\{\mathbf{T}_{tr}, \mathbf{y}_{tr}\}$ such that $y_i \notin \mathbf{y}_{tr}$, i.e., when the object to be predicted was not part of the related training set $\mathbf{T}_{tr}$. The sum of squares between the values of y_i and its estimate $\hat{y}_{i-tr}(A)$ using the first A PCs is

$$PRESS(A) = \sum_{i=1}^{N}[y_i - \hat{y}_{i-tr}(A)]^2. \qquad (8.38)$$

The notation *PRESS* stands for "PREdiction Sum of Squares". It is also usual to express *PRESS* as root-mean-squared error in cross-validation (*RMSECV*), that is,

$$RMSECV(A) = \sqrt{\frac{PRESS(A)}{N}}. \qquad (8.39)$$

RMSECV (or *PRESS*) is, in a certain sense, an estimate of the model's prediction ability. The criterion is to choose the value of A in which *RMSECV* has a minimum, which, in this case, is equivalent to the optimal biasing of the least-squares linear model. In practice, only the components that improve *RMSECV*, e.g., by a specific percentage with respect to the previous one, are usually used. Also, a one-sided confidence interval on *RMSECV* has been proposed.

In the extreme case where each cancellation group is formed by a single object $(G = N)$, the so-called leave-one-out procedure, the properties of the **T** matrix reduce Equation (8.15) to a very simple expression for the calculation of *RMSECV*. Often, however, leave-one-out provides an underestimation of the prediction error. Although there are almost infinite rules to choose the structure of the cancellation groups, the usual recommendation is for G to be greater than or equal to 3.

In contrast to *RMSECV* in Equation (8.39), the root-mean squared error (*RMSE*) for a model with A components is related to the fit of the model with the whole training set $\{\mathbf{T}, \mathbf{y}\}$ but only using the first A components. It is precisely defined in the following Equation (8.40):

$$RMSE(A) = \sqrt{\frac{\sum_{i=1}^{N} [y_i - \hat{y}_i(A)]^2}{N}}. \qquad (8.40)$$

In least-squares linear regression, *RMSE* decreases as the number of components increases since, except for one factor, it coincides with the residual sum of squares, which is what is minimised (least-squares criterion) when all components are considered.

This explains why it is usual to represent both *RMSECV* and *RMSE* together as a function of the number of PCs to evaluate the effect of increasing the complexity of the model (via the number of components) on the prediction errors, especially on *RMSECV*.

Despite its widespread use, there is considerable debate about the use of cross-validation [13], particularly for PCR [14], that has led to a deep analysis of the problem of estimation of the prediction error with linear models by using cross-validation [15]. Section 1.2 in Ref. [15] presents an excellent analysis of the development of the topic since its proposal in 1974.

(b) *External test set*: When the number of objects N in the available training set $\{\mathbf{T}, \mathbf{y}\}$ is large enough, the whole set can be split into two (sub)sets: one is used for training, whereas the other is kept apart as an external test set. Hence the name of this alternative.

Note that cross-validation emulates this method, with each cancellation group acting as an external test set. The difference is that, at the end of the cross-validation iterations, the model is built with the whole training set. Therefore, the advantage of using an external dataset, thus building the model with only the reduced training set, is that the *RMSECV* (sometimes denoted as *RMSEP*) does not include the variability due to model change as the cancellation group changes.

8.5.1.2.   Selection of significant components

The idea is to select not the first A PCs but rather the PCs necessary to explain the response based on the estimated coefficients. The example in Section 8.3.1 shows that, under certain conditions, the removal of a variable leads to a lower ESE. According to Equation (8.30), the condition to reduce ESE when variable V is removed from the model is that $b_V{}^2 < \sigma^2/SS(\mathbf{x}_V)$ or, equivalently,

$$\frac{|b_V|}{\dfrac{\sigma}{\sqrt{SS(\mathbf{x}_V)}}} < 1. \tag{8.41}$$

The PCs verify the conditions stated in the example in Section 8.3.1, and the left-hand term in Equation (8.41) is the statistic of the left-hand term of Equation (8.37) applied to the estimated coefficient $\hat{b}_V$. The statistic is the one used to decide whether the coefficient is null or not. Therefore, if the inequality in Equation (8.41) holds, Equation (8.37) does not because $|b_V|/\left(\sigma/\sqrt{SS(\mathbf{x}_V)}\right) < 1 < t_{\alpha/2, N-V}$ for the usual significance levels, and thus the coefficient of PC_V is significantly null, and the contribution of PC_V to ESE is negligible. Accordingly, the variables removed to reduce ESE are among those with nonsignificant (statistically null) coefficients.

Consequently, a biased model with less ESE will be achieved by selecting the components whose estimated coefficient $\hat{b}_i$ is statistically different from zero, a decision that has to be made using the variance estimated with ordinary least squares (that is, with all the components). The advantage of this procedure, compared to cross-validation or an external set, is that it does not necessarily require selecting all the first A components. In that way, the researcher is able to explore the need to include all PCs or some of them, not necessarily consecutive, even selecting some *posterior* PC_i having very little variance yet highly correlated to the response.

An alternative to selecting variables with significant coefficients while taking into account the changes that occur in the estimation of the residual variance when modifying the model is to apply forward stepwise regression instead of just ordinary least squares, of course, on the PCs. In general, both selection procedures lead to the same result.

Although PCR uses PCA as a first step before regressing **y** on the scores, they are different because of their goals. In PCR, the prior PCA is used to stabilise the predictor matrix, but then a bias is introduced in the posterior regression model, related to the way the relevant PCs are selected.

When using cross-validation, only the first A components can be selected, but there are other methods that allow choosing nonsequential PCs on the basis of their relevant effect on **y** rather than the order they occupy in the PCA decomposition (which is only induced by the explained variance of the **X**-block).

8.5.2. *Analysis of the PCR model*

In PCR analysis, with A PCs, there are two key issues to ensure the validity of the model, namely the projection space defined by the selected components and the residuals of the fit:

1. The projection space is defined by the critical values of Q and T^2 statistics (see Section 7.1.9 of Chapter 7) at a given confidence level c. Thus, for any vector of original variables $\mathbf{x} = (x_1, \ldots, x_V)$, compute the scores $\mathbf{t} = (t_1, \ldots, t_A)$ and check that the value of its statistic T_t^2 is less than the defined limit T_c^2. Also, check with the corresponding residuals vector $\mathbf{e} = (e_1, \ldots, e_V)$ that its sum of squares $Q_t = \sum_i e_i^2$ is less than the threshold value Q_c. If a vector $\mathbf{x}$ of the calibration set exceeds both critical values, the regression model must be rebuilt without it, whereas if $\mathbf{x}$ is a test vector that surpasses both limits, then the PCR model should not be applied to it. This is a great advantage of multivariate calibration models over univariate ones: the former have some means to prevent their inappropriate use.

2. Regarding the residuals of the fit, $\mathbf{r} = \hat{\mathbf{y}}_{\text{PCR}} - \mathbf{y}$, it is customary to check its normality and compute a confidence interval to detect those that are uncharacteristically large. After their identification and removal, the PCR model is rebuilt, checking again that the remaining data also meet the criteria on Q and T^2 statistics.

   This iterative process must be done with great caution; sometimes it is necessary to modify the number of PCs after removing anomalous data because their presence modifies the values of *RMSE* and *RMSECV* and perhaps the dimension A of the PCR model. When it comes to multivariate calibrations with a training set $\{\mathbf{X}, \mathbf{y}\}$, the latent structure of the spectra is known; therefore, it is easier to diagnose if the anomaly lies in the predictors (spectrum not similar to those of the calibration standards) or the response (concentration measured by the reference method).

   In general, the normality of residuals can be accepted; otherwise, the response can be transformed to achieve it. Among the rich methodologies concerning this issue, Box–Cox transformations [5] are quite common. However, to check for homoscedasticity and independence of the residuals, replicates must be available. With these assumptions fulfilled (i.e., residuals approximately i.i.d. $N(0, \sigma)$), the significance of the model and lack of fit (again, if replicates are available) can be tested. In addition, a joint confidence region can be computed for the vector of coefficients (Equation (8.19)), as well as the confidence intervals for each individual coefficient (Equation (8.18)) and the calculated response (Equation (8.20)).

If the normality of the residuals cannot be accepted, non-parametric and/or computational methods are still available to establish confidence in the calculated response and significance of the model (for example, see the permutation test in the following section about PL S regression). In any case, their details are outside the scope of the current chapter.

In relation to the computed PCR model, some relevant aspects are as follows:

- The variability of the training set $\{\mathbf{X}, \mathbf{y}\}$ used by the PCR model is expressed by the percentage of variance of the $\mathbf{X}$ block explained by each PC used in the model, as well as the percentage of variance of the $\mathbf{y}$ block, that is, the variance of the response explained by the univariate linear model associated with that component.

  It is possible to know the total amount of information on the predictors (in percentage of variance) that the model uses to explain the information contained within the response (also in percentage of variance). For this, we must include the percentages of variance from $\mathbf{X}$ explained by each PC and, on the other hand, the percentage of variance explained for the response.
- It is very common to graphically represent scores in $\mathbf{T}$ and loadings in $\mathbf{P}$ on a per-component basis. However, it should not be forgotten that the predictors are PCs and caution must be exercised in their interpretation due to the rotational ambiguity (see Section 7.1.8 of Chapter 7).

Finally, note that PCR is a highly studied statistical tool (e.g., in December 2022, Scopus returned 4,097 documents with the keyword "Principal Component Regression") and with multiple adaptations, for example, in the case of multivariate calibration, to weight the least-squares criterion for considering the relative error of the calculated concentration [16].

8.6.   Partial Least-Squares Regression

Partial least-squares (PLS) regression was introduced by Wold in the late 1970s [17] and has been subsequently improved by other authors [18–20]. Currently, it is a widely used statistical tool; a search in Scopus (December 2022) with the keyword "Partial Least Squares" gives 59,633 documents, of which 28,932 were published from 2018 to 2022.

In PCR, a sequence of directions is chosen (the PCs defined by the columns of the $\mathbf{P}$ matrix in Equation (8.32)) and then a regression model is built using these components as predictors, which in fact is a sequence of univariate regression models, one with each component. The construction of the PCs considers only the $\mathbf{X}$ matrix of design, without the intervention of the response $\mathbf{y}$ anywhere in the process. The response is considered later, during the regression step.

This means that PCR starts by looking for directions $\mathbf{b}$ in $\mathbf{X}$ that sequentially maximise $\mathrm{var}(\mathbf{Xb})$, without considering the response. In particular, for the ith PC, $\mathbf{b}_i$, is the solution to the constrained maximisation problem in Equation (8.42), and as such, there is a high risk of underfitting.

$$\max_{\mathbf{b}}\{\mathrm{var}(\mathbf{Xb})\}$$
$$\text{subject to } \|\mathbf{b}\| = 1, \quad \text{and } \mathrm{corr}(\mathbf{b}_i, \mathbf{b}_j) = 0, \quad j = 1, \ldots, i-1.$$
$$(8.42)$$

The least-squares regression gives the direction $\mathbf{b}$ in the space of the predictors that is the most correlated with the response $\mathbf{y}$. In fact, the ordinary least-squares estimator $\hat{\mathbf{b}}_{\mathrm{OLS}}$ in Equation (8.7) is also the solution of the maximisation problem in Equation (8.43), and in this case, there is a high risk of overfitting:

$$\max_{\mathbf{b}}\{[\mathrm{corr}(\mathbf{Xb}, \mathbf{y})]^2\}. \qquad (8.43)$$

Regardless of the over(under)fitting risk, in cases with several responses, both OLS and PCR build as many regression models as there are responses, one for each response Y_i. Therefore, the regression model does not capture the implicit structure or the relations among the different responses.

On the contrary, PLS regression considers both aspects and has been highly successful in chemometrics, particularly in the simultaneous calibration of several analytes with multivariate signals.

To describe PLS conceptually, suppose a training set is available $\{\mathbf{X}_{NV_X}, \mathbf{Y}_{NV_Y}\}$ with centred data. That is, N objects in which V_X predictor variables and V_Y response variables have been determined. In the literature, PLS2 is occasionally used to denote the multi-response case (two or more responses). To simplify the notation, in this chapter, this distinction will not be made because the context will clearly define whether the space of the responses is univariate or multivariate.

The mathematical development of PLS for modelling one or several responses is beyond the scope of the chapter. However, a description of the most important steps will be given to understand its use in practical examples.

PLS regression is a compromise between the criteria of Equations (8.42) and (8.43) that is obtained by maximising their geometric mean or, equivalently, their product.

The procedure is sequential. To obtain the first latent variable, weights $\mathbf{r}_1$ and $\mathbf{q}_1$ are determined. These are the coefficients of the two linear combinations (one in $\mathbf{X}$ predictor matrix, and one in $\mathbf{Y}$ response matrix) that solve

$$
\begin{aligned}
&\max_{\mathbf{r},\mathbf{q}}\{\mathrm{var}(\mathbf{Xr})[\mathrm{corr}(\mathbf{Xr},\mathbf{Yq})]^2\mathrm{var}(\mathbf{Yq})\} \\
&= \max_{\mathbf{r},\mathbf{q}}\{[\mathrm{cov}(\mathbf{Xr},\mathbf{Yq})]^2\} \\
&= \max_{\mathbf{r},\mathbf{q}}\{[(\mathbf{Xr})^{\mathrm{T}}\mathbf{Yq}]^2\} \\
&\quad\text{subject to } \|\mathbf{r}\| = \|\mathbf{q}\| = 1
\end{aligned}
\tag{8.44}
$$

The maximisation of the product $\mathrm{var}(\mathbf{Xr})[\mathrm{corr}(\mathbf{Xr},\mathbf{Yq})]^2\mathrm{var}(\mathbf{Yq})$ tends to look for directions of large variance in both the $\mathbf{X}$- and $\mathbf{Y}$-blocks, avoiding those of small variance (although they do not have to be balanced), which leads to biasing the linear regression to reduce ESE. Besides, the criterion includes the term $\mathrm{corr}(\mathbf{Xr},\mathbf{Yq})$ that helps in avoiding directions in the predictors (the $\mathbf{X}$-block) with small correlation with the responses (the $\mathbf{Y}$-block), contrary to what happens with PCR.

> PLS represents a compromise between OLS and PCR, looking for the linear combinations of greatest variance in the predictors block and in the responses block, so that they have the maximum correlation between them, that is, discarding those directions that do not have sufficient predictive capacity.

Some further notation is required; thus, $\mathbf{t} = \mathbf{X}\mathbf{r}$ and $\mathbf{u} = \mathbf{Y}\mathbf{q}$ are the *scores* for the N objects in the $\mathbf{X}$-block and $\mathbf{Y}$-block, respectively. These scores define an inner linear relation according to Equation (8.44), precisely that of the regression coefficient $b_1 = \mathbf{u}_1^{\mathrm{T}}\mathbf{t}_1/\|\mathbf{t}_1\|^2$.

On the other hand, the coefficients obtained when regressing the $\mathbf{t}$ scores on $\mathbf{X}$ define the direction in the $\mathbf{X}$ space in which the $\mathbf{t}_1$ scores are projected, that is, the loading vector $\mathbf{p}_1 = \mathbf{X}^{\mathrm{T}}\mathbf{t}_1/\|\mathbf{t}_1\|^2$. Similarly, the coefficients generated when regressing the $\mathbf{t}$ scores on $\mathbf{Y}$ constitute the loading vector $\mathbf{q}_1 = \mathbf{Y}^{\mathrm{T}}\mathbf{t}_1/\|\mathbf{t}_1\|^2$, that is, the direction in the $\mathbf{Y}$ space in which the $\mathbf{t}_1$ scores are projected.

Next, successive linear combinations are built in the X-block, orthogonal to each other, that is, $\mathbf{t}_i^{\mathrm{T}}\mathbf{t}_j = 0, i \neq j$. The way the orthogonality is imposed gives rise to two different algorithmic approaches, namely, non-iterative partial squares (NIPALS) and straightforward implementation of a statistically inspired modification of the PLS (SIMPLS).

With NIPALS, $\mathbf{X}$ and $\mathbf{Y}$ are deflacted to obtain the matrices with the residuals of the first latent variable: $\mathbf{X}_1 = \mathbf{X} - \mathbf{t}_1\mathbf{p}_1^{\mathrm{T}}$ and $\mathbf{Y}_1 = \mathbf{Y} - \mathbf{u}_1\mathbf{q}_1^{\mathrm{T}} = \mathbf{Y} - b_1\mathbf{t}_1\mathbf{q}_1^{\mathrm{T}}$. Then, weights, scores, and loadings are calculated using $\mathbf{X}_1$ and $\mathbf{Y}_1$.

This process guarantees the orthogonality of the successive $\mathbf{t}_i$ scores, but it does not satisfy the criterion of Equation (8.44). What is more, the iteratively computed $\mathbf{t}_i$ and weights are not referred to the original variables, so they need to be "transformed" to obtain new ones that are directly related to the predictors, such as the $\mathbf{r}$ in Equation (8.44).

With SIMPLS, it is not $\mathbf{X}$ and $\mathbf{Y}$ that are deflacted but $\mathbf{S} = \mathbf{X}^{\mathrm{T}}\mathbf{Y}$. The scores $\mathbf{t}_i$ are orthogonal, and both scores and weights are directly referred to the original variables in $\mathbf{X}$-block. Additionally, the solution does fulfil the criterion in Equation (8.44), and it is computationally more efficient.

When there is a single response variable in the **Y**-block, the solutions obtained with NIPALS and SIMPLS coincide. Otherwise, only the first latent variable is the same.

Irrespectively of the algorithm, the PLS model with A latent variables computed with a training set $\{\mathbf{X}_{NV_X}, \mathbf{Y}_{NV_Y}\}$ produces a bilinear decomposition of **X** matrix and another of **Y** matrix in terms of A loadings and scores in the **X**- and **Y**-blocks. The bilinear decomposition of a matrix is detailed in Section 7.1.3 of Chapter 7.

Let **Q** denote the $V_Y \times A$ matrix whose columns are the **q** vectors in the previous expressions, and **T** and **U** the $N \times A$ matrices whose columns are the corresponding $\mathbf{t}_i$ and $\mathbf{u}_i$. In addition, recalling that **P** is the $V_X \times A$ matrix whose columns are the PCs loading vectors, it holds that

$$\mathbf{X}_{NV_X} = \mathbf{T}_{NA}(\mathbf{P}^{\mathrm{T}})_{AV_X} + \mathbf{EX}_{NV_X},$$
$$\mathbf{Y}_{NV_Y} = \mathbf{U}_{NA}(\mathbf{Q}^{\mathrm{T}})_{AV_Y} + \mathbf{EY}_{NV_Y}, \qquad (8.45)$$
$$\mathbf{U}_{NA} = \mathbf{T}_{NA}\mathbf{D}_{AA} + \mathbf{EU}_{NA}.$$

In Equation (8.45), **T** is an orthogonal matrix whose columns are the scores of the **X**-block, **P** is the matrix of the corresponding loadings, and **EX** contains the corresponding residuals. Analogously, **U**, **Q**, and **EY** are the matrices of scores, loadings, and residuals in the **Y**-block, respectively. **D** is a diagonal matrix with the A regression coefficients b_i between the scores of **X**- and **Y**-blocks, **T** and **U**, respectively, plus the residuals in **EU**.

NIPALS and SIMPLS obtain different **T** matrices, but they define the same projection subspace for the scores of the **X**-block. They also have different weight matrices, the matrix in SIMPLS is **R**, made up of the columns $\mathbf{r}_i$ in Equation (8.44), while in NIPALS it is usually denoted as **W**. Their relation is $\mathbf{R} = \mathbf{W}(\mathbf{P}^{\mathrm{T}}\mathbf{W})^{-1}$.

The equalities in Equation (8.45) describe the internal structure of each block and the relations between them. The analysis of these matrices, with the aim of describing the relationships between blocks, is known as structural PLS [21, 22]. However, PLS is presented in this chapter as a regression method, which is its usual use in chemometrics. In order to simplify the expression of the PLS linear estimator that predicts $\mathbf{Y}$ as a function of $\mathbf{X}$, which is a $(V_X \times V_Y)$ matrix $\hat{\mathbf{B}}_{\mathrm{PLS}}$, the vectors of weights $\mathbf{r}_i$ (columns of $\mathbf{R}$) and also the scores $\mathbf{t}_i$ will be normalised, so that $\mathbf{T}^{\mathrm{T}}\mathbf{T} = \mathbf{I}_{AA}$. This normalisation does not introduce essential changes; it only facilitates the expression of the estimator in the form of Equation (8.46) [18]:

$$\hat{\mathbf{B}}_{\mathrm{PLS}} = \mathbf{R}\mathbf{R}^{\mathrm{T}}\mathbf{X}^{\mathrm{T}}\mathbf{Y} = \mathbf{R}\mathbf{Q}^{\mathrm{T}}. \tag{8.46}$$

More interesting is the expression of the PLS model in Equation (8.45) in terms of $\mathbf{T}$. With this aim, $\mathbf{U}$ in the second equality of Equation (8.45) is substituted by the third equality, and new $\mathbf{Q}^*$ and residuals $\mathbf{EY}^*$ are defined to give

$$\mathbf{Y} = \mathbf{U}\mathbf{Q}^{\mathrm{T}} + \mathbf{EY} = (\mathbf{TD} + \mathbf{EU})\mathbf{Q}^{\mathrm{T}} + \mathbf{EY}$$
$$= \mathbf{TDQ}^{\mathrm{T}} + \mathbf{EY}^* = \mathbf{T}(\mathbf{Q}^*)^{\mathrm{T}} + \mathbf{EY}^*. \tag{8.47}$$

Since the rows of $\mathbf{Q}^*$ are those of $\mathbf{Q}$, each multiplied by the corresponding b_i element of the diagonal of $\mathbf{D}$, the vector space they generate is the same. With this new representation, both the $\mathbf{X}$- and $\mathbf{Y}$-blocks are written in terms of the same $\mathbf{T}$ scores, constructed as the projection of the $\mathbf{X}$ columns onto the subspace generated by the $\mathbf{P}$ loadings. These scores are then used to linearly approximate the variables of the $\mathbf{Y}$ block.

In other words, the linear PLS model comprises two linear functions: a projection of $\mathbf{X}$ onto the subspace (of dimension A) generated by the loadings of the latent variables, $\mathbf{P}$, followed by a linear map of this subspace onto $\mathbf{Y}$. This expression of the PLS model indicates that both predictors and responses are functions of $\mathbf{T}$, which can be understood as the representation space of the phenomenon underlying the problem under study. This is the basis for the use of PLS in the field of process control [23] and in the construction of analytical procedures within the context of analytical quality by design (AQbD) [24, 25].

There are not many systematic studies comparing PCR and PLS. In this regard, Refs. [26] and [27] can be consulted. Both conclude using real and simulated data that, for prediction purposes, both methods give similar results, although it is common for PCR to require more components than latent variables in PLS. This aspect is illustrated in the case study in Section 8.7.5.

8.6.1. *Analysis of the PLS model*

In many respects, the analysis of a PLS model is similar to that of a PCR model previously described in Section 8.5. However, since PLS uses the response $\mathbf{Y}$ to construct its latent variables, its solution path is a nonlinear function of $\mathbf{Y}$. Consequently, the variance of the prediction does not have a simple expression, and the usual confidence intervals and hypothesis tests are not applicable. With a single response $\mathbf{y}$, various approaches to the problem of evaluating the uncertainty in the response estimated by a PLS model have been developed. A review of the methods can be found in Ref. [28]. In essence, most of them either mimic OLS, apply local linearisation, use resampling, or simply calculate the uncertainty empirically. Confidence regions have also been developed for a model with multivariate responses, $\mathbf{Y}$ [29].

In particular, if a univariate response $\mathbf{y}$ is considered, the estimate in Equation (8.46) is the vector $\hat{\mathbf{b}}_{\mathrm{PLS}} = \mathbf{R}\mathbf{R}^{\mathrm{T}}\mathbf{X}^{\mathrm{T}}\mathbf{y}$, which is formally similar to the least-squares regression estimate in Equation (8.7), $\hat{\mathbf{b}}_{\mathrm{OLS}} = (\mathbf{X}^{\mathrm{T}}\mathbf{X})^{-1}\mathbf{X}^{\mathrm{T}}\mathbf{y}$. It may be considered that it suffices to use $\mathbf{R}\mathbf{R}^{\mathrm{T}}$ instead of $(\mathbf{X}^{\mathrm{T}}\mathbf{X})^{-1}$ to obtain the variance of the estimator $\hat{\mathbf{b}}_{\mathrm{PLS}}$ and that of a predicted response. But this is not correct. The radical difference is that the $\mathbf{R}$ matrix of weights of the $\mathbf{X}$-block in PLS is linked to the response vector $\mathbf{y}$ by construction, while in OLS there is no such dependency. Despite the lack of a mathematical foundation, this reasoning (based on the analogy) has been frequently applied by using the "counterparts" of Equations (8.9) and (8.14) for PLS regression. That would lead to the following equations:

$$\mathrm{cov}(\hat{\mathbf{b}}_{\mathrm{PLS}}) = (\mathbf{R}\mathbf{R}^{\mathrm{T}})\sigma^2, \tag{8.48}$$

$$\mathrm{var}(\hat{y}_t) = \tilde{\mathbf{x}}_t \mathrm{cov}(\hat{\mathbf{b}}_{\mathrm{PLS}})\tilde{\mathbf{x}}_t^{\mathrm{T}}\sigma^2 + \mathrm{var}(\bar{y}) + \sigma^2$$

$$= \left[\tilde{\mathbf{x}}_t (\mathbf{R}\mathbf{R}^{\mathrm{T}})\tilde{\mathbf{x}}_t^{\mathrm{T}} + \frac{1}{N} + 1 \right]\sigma^2, \tag{8.49}$$

where $\tilde{\mathbf{x}}_t$ is a test vector centred according to the mean values of the training set, $\tilde{\mathbf{x}}_t = (x_{t1} - \overline{x}_1, \ldots, x_{tV} - \overline{x}_V)$.

Regarding the selection of the appropriate number of latent variables, the entire cross-validation process is conducted in the same way as with PCR. For each PLS model built with A latent variables, $PRESS(A)$, $RMSE(A)$, and $RMSECV(A)$ in Equations (8.38), (8.39), and (8.40) are computed with the predictions made with the corresponding PLS model, and the number of latent variables is selected as the value of A that reaches minimum $RMSECV(A)$. The shortcomings of cross-validation, already mentioned, also apply to the case of PLS. An external validation set can also be used; however, there are no inferential methods to select latent variables based on the significance of their corresponding coefficients, that is, selecting only some latent variables instead of the first A. Although it is true that the need for doing this is reduced by PLS itself, as described in Equation (8.44), it is not completely eliminated. This means that it is still possible to have a direction with high variance in one or both blocks ($\mathbf{X}$ and $\mathbf{Y}$) and low correlation with the $\mathbf{Y}$-block that enters into the model before another direction with higher correlation with $\mathbf{Y}$. For this to occur, it suffices that the product of correlation and variances in Equation (8.44) is greater in the former than in the latter case.

Regarding the projection space, the latent variables of the $\mathbf{X}$-block defined using the PLS model have the same structural characteristics as those found in PCR regression. As a consequence, the projection space defined by the critical values of Q and T^2 statistics are equally used to detect anomalous objects of the $\mathbf{X}$-block (in the model construction stage) and/or test samples, avoiding the incorrect application of the model. Also, the analysis of the residuals makes it possible to identify those objects whose values of Q and T^2 are less than the corresponding thresholds but exhibit a response value significantly different from what it should correspond to.

The significance of the model is established using a *permutation test* because there are no inferential methods available. This procedure gives the probability that the PLS model differs significantly from a model that is built with data taken at random from the training set $\{\mathbf{X}, \mathbf{Y}\}$. Specifically, the vector of responses is reordered repeatedly and randomly, and a PLS model is built for each permutation. In this way, the distribution of responses is maintained; however, they do not necessarily correspond to the linked $\mathbf{x}$ vector. Thus, the probability that the original model is significantly equal from one created by randomly permutating the response $\mathbf{Y}$ is computed.

Sometimes, it is necessary to define which are the most relevant predictors for the structure of a PLS model. This selection of variables can be performed by eliminating some of them, reconstructing the model (and its validation), and evaluating its performance. An example is minimising PRESS through a selection based on a genetic algorithm [30]. Another possibility is to rank the variables according to their importance in the model. From this point of view, the variable importance in projection (VIP) [31] has acquired great diffusion. The VIP score of a predictor variable represents the influence of that $\mathbf{X}$-block variable on the PLS model via its importance for the projections to find A latent variables. VIP is a parameter that varies in a fixed range since the sum of squared VIP values for all variables is equal to the number of variables. Thus, the variables with a VIP value greater than 1 (i.e., greater than the average of squared VIP values) have an above-average influence on the model and are, therefore, considered as the most relevant ones for explaining $\mathbf{Y}$. This variable selection procedure can serve as an alternative to inferential selection based on the significance of the coefficients in OLS or PCR.

The variability of the training set $\{\mathbf{X}, \mathbf{Y}\}$ used by the PLS model is also collected by the percentage of variance explained in the $\mathbf{X}$- and $\mathbf{Y}$-blocks, both by each latent variable and with the A latent variables of the PLS model.

Like in PCR, it is very common to depict the $\mathbf{T}$ scores and $\mathbf{P}$ loadings of each latent variable. Moreover, like in PCR, there is rotational ambiguity (see Section 7.1.8 of Chapter 7), so caution must be exercised in their interpretation. Sometimes, it is even necessary to rotate the projection space defined by the latent variables to have a meaningful interpretation of the scores and the resulting structure [32].

Both PCR and PLS are two procedures used to perform biased regression analysis, resulting in more accurate predictions than OLS. Of course, they do not exhaust the possibilities of biasing, as there is a multitude of statistical methods for doing so.

The complexity of constructing PCR and PLS makes it difficult to decide on the number of PCs or latent variables to be used in the model. The nuances in the application of the procedures are so extensive that it has been defined as the "art of biasing". This makes it impossible to define a sequence of predetermined actions to obtain a final PCR or PLS model.

In the field of chemometrics, and more particularly in that of multivariate calibration, knowledge of the chemical problem and the design of the calibration set are tools that must be used together with the purely formal ones, which are the ones that have been described in this chapter.

These notes of caution should not obscure the usefulness and success that the use of these regression methods has had. Indeed, they are part of ISO and ASTM standards (e.g., ASTM-E1655, Standard Practices for Infrared Multivariate Quantitative Analysis).

PCR is a two-step procedure (first PCA on $\mathbf{X}$, then MLR to predict $\mathbf{Y}$). The two separate steps can lead to the need for more PCs to predict the response. Different responses imply different PCR models.

PLS takes into account the variance of both the $\mathbf{X}$- and $\mathbf{Y}$-blocks and the correlation between them, so fewer latent variables are usually needed for predicting $\mathbf{Y}$.

Contrary to PCR, with PLS, the usual inference for MLR models is not available. On the advantages side, PLS can handle several responses at once, and further, it computes a unique common space for the projection of the $\mathbf{X}$- and $\mathbf{Y}$-blocks that can be very useful.

For a single response, the projection spaces built with PCR and PLS are different. However, the predictions with both models could be the same.

8.7. Worked Examples and Case Studies

8.7.1. *PCR: Determination of the percentage of tallow in adulterated lard*

Goal: The aim is the quantitative determination of tallow in adulterated lard. The analyses were performed using a gas chromatography-flame ionisation detector (GC-FID) with three different chromatographic columns: column A (10% DEGS, WAW 80/100 chromosorb, 3 m, 1/8″), column B (20% DEGS, WAW 80/100 chromosorb, 2 m, 1/8″), and column C (20% DEGS, WAW 80/100 chromosorb, 3 m, 1/8″).

Training set: The predictor $\mathbf{X}$ is a 17×5 matrix. The response $\mathbf{Y}$ is a 17×1 vector. There are thus 17 objects corresponding to seven samples of pure lard and 10 samples prepared in the laboratory by adulterating pure samples with increasing percentages of tallow (three samples with 5% tallow, another three with 10%, and four with 15%). Regarding the variables, although the percentage area of chromatographic peaks of 16 fatty acids was available [33], only five (corresponding to fatty acids C14:0, C16:0, C18:0, C18:1, and C18:2) will be used in the present study.

The training set is in sheet "PCReg" of "fraud.xlsx". The last column "%tallow" contains the percentage of tallow added to the samples of lard, i.e., the response variable. The remaining five columns (the predictors) are labelled with the corresponding fatty acid.

Guided solution: As indicated in the theory section, if the variables are correlated and there are collinearities, it is necessary to resort to regression methods based on latent variables. Following the criteria in Section 8.4, Table 8.1 shows the Pearson product moment correlations between each pair of the original variables, together with the p-value of their individual significance test. All of them are well below 0.05; therefore, all two-to-two correlation coefficients are highly significant at 95% confidence.

Table 8.2 shows the principal component analysis (PCA) performed with the 17×5 autoscaled data matrix of predictors.

Table 8.1. Pearson product moment correlations r between each pair of variables ($n = 17$). In brackets are the p-values for the test: H_0: $r = 0$; H_1: $r \neq 0$.

	C16:0	C18:0	C18:1	C18:2
C14:0	0.7825	0.8408	−0.5426	−0.9272
	(0.0002)	(0.0000)	(0.0244)	(0.0000)
C16:0		0.6638	−0.7835	−0.7043
		(0.0037)	(0.0002)	(0.0016)
C18:0			−0.6650	−0.9582
			(0.0036)	0.0000
C18:1				0.6065
				(0.0099)

Table 8.2. Principal component analysis for autoscaled dataset, fraud. Eigenvalues and the explained and cumulative variance when adding principal components (PCs) are shown.

PC number	Eigenvalues	Explained variance (%)	Cumulative variance (%)
1	4.0032	80.06	80.06
2	0.6199	12.4	92.46
3	0.3016	6.03	98.49
4	0.0567	1.13	99.62
5	0.0185	0.38	100.00

The quotient between the greatest and smallest eigenvalues in the second column is 216.4, which is greater than 10, so it can be concluded that there is collinearity in the data.

To perform PCR, initially, the maximum number of PCs is considered and then these new variables are used as predictors in a multiple linear regression (MLR) using least squares to predict the percentage of tallow in lard. The coefficient estimates of the fitted model are shown in Table 8.3.

To choose the necessary PCs in the regression model, the criterion used is that the coefficient b_i multiplying PC_i is significantly different from zero. The p-values in the last column in Table 8.3 correspond

Table 8.3. Parameter estimates (coefficients b_i) and p-values of the significance test for the coefficients of the MLR model to predict the percentage of tallow as a function of the scores of the five PCs.

Parameter	Estimate	p-value
Constant	6.18	0.0000
PC_1	2.83	0.0000
PC_2	3.03	0.0000
PC_3	−1.73	0.0003
PC_4	−0.45	0.5600
PC_5	−0.21	0.8711

Table 8.4. Analysis of variance, determination coefficient R^2, and standard error of estimation ($\hat{\sigma}$) of the PCR model with three PCs.

Source	Sum of squares	Df	Mean squares	F-ratio	p-value
Model	620.6	3	206.8	455.2	0.0000
Residual	5.9	13	0.454		
Total	626.5	16			
R^2	99.1%				
$\hat{\sigma}$	0.674				

to the hypothesis test H_0: $b_i = 0$ and H_1: $b_i \neq 0$ with the statistics in Equation (8.37). Therefore, only the constant and the first three components have a coefficient significantly different from zero at 95% confidence (p-value <0.05).

With these three PCs, after verifying that there are no anomalous data, the next step is to analyse whether the regression is significant, using the analysis of variance (ANOVA) in the first rows of Table 8.4. The test has the null hypothesis H_0: the regression does not explain the variance of the response, and the alternative hypothesis H_1: the regression explains it. With a p-value less than 10^{-4}, the decision is that the model is valid. The coefficient of determination R^2 (explained variance of the response) is 99.1%, and the standard deviation of the residuals (the square root of the one in Equation (8.17)) is 0.674.

The conclusion is that with 98.49% of the variance of the predictors (Table 8.2), 99.1% of the variance of the amount of tallow is explained (R^2). The permutation test confirms that the model is correct. With the coefficients in Table 8.3, the fitted model in Equation (8.50) expresses the relationship between the percentage of tallow and the three PCs chosen:

$$\text{Tallow } (\%) = 6.18 + 2.83 \ PC_1 + 3.03 \ PC_2 - 1.73 \ PC_3. \quad (8.50)$$

Applying the model to the samples in the training set, the predictions in Table 8.5 are obtained, with the respective errors, absolute and relative. The mean of the absolute values of the relative errors is 5.917%.

Table 8.5. Percentage of tallow, predicted tallow, and absolute and relative errors.

Tallow (%)	Predicted tallow (%)	Absolute error (%)	Relative error
0	−0.213	0.213	−
0	0.522	−0.522	−
5	5.693	−0.693	13.84
5	4.131	0.869	−17.38
5	5.293	−0.293	5.87
10	10.456	−0.456	4.56
10	10.394	−0.394	3.94
10	10.257	−0.257	2.57
15	14.221	0.779	−5.192
15	15.241	−0.241	1.60
15	14.636	0.364	−2.43
15	14.732	0.268	−1.79
0	0.069	−0.069	−
0	1.221	−1.221	−
0	−0.401	0.401	−
0	−1.140	1.140	−
0	−0.111	0.111	−

Table 8.6. Predicted values of tallow (%) using PCR, and endpoints of the 95% confidence intervals.

Tallow (%)	Predicted tallow (%)	Lower (%)	Upper (%)	Lower (%)	Upper (%)
0.0	0.058	−1.350	1.467	−0.402	0.519
5.0	5.011	3.639	6.382	4.682	5.340
10.0	9.963	8.579	11.349	9.582	10.345
15.0	14.916	13.469	16.365	14.347	15.486

Although the regression model seems to be completed, from an analytical point of view, accuracy should be studied. The accuracy line with the predictions in Table 8.5 has the following equation: %predicted-tallow $= 0.058 + 0.991 \times$ %tallow. The joint test (see, e.g., Ref. [34]), with H_0: intercept $= 0$ and slope $= 1$ and H_1: this is not the case, has a p-value of 0.9372, so there is no evidence to reject the null hypothesis; therefore, it is concluded that the constructed PCR model has no bias, neither constant nor proportional.

Using the accuracy line, Table 8.6 shows the average predicted values of the percentage of tallow in the samples together with the endpoints of the 95% confidence prediction intervals for new observations (columns 3 and 4) and for the mean (last two columns).

8.7.2. *PLS: Quantification of the concentration of two additives in food colourants*

Goal: To quantitatively evaluate the amount of two additives present in a mixture of tartrazine (E-102) and sunset yellow (E-110) from standards prepared in the laboratory and measured by UV–vis molecular absorption spectrophotometry. These additives are used as food colourants. The relative errors for the calibration standards and the amount contained in two test samples obtained at a local supermarket will be calculated.

Training set: The predictors are in a 16×47 data matrix $\mathbf{X}$. The 16 objects are samples that contain mixtures of both colourants prepared with four levels of concentration (0, 2, 4, and 6 mg kg^{-1})

of E-102 and, in every level, with increasing concentrations of E-110, also in four levels (0, 2, 4, and 6 mg kg^{-1}). The 47 variables are the absorbance recorded by the spectrophotometer between 340 and 570 nm, every 5 nm. The predictor variables are the same as in Chapter 7 (Section 7.1.10.1 and Figure 7.3). The amounts of the two colourants constitute the two responses in matrix $\mathbf{Y}$ (16 × 2).

Test set: There are two samples of different food colourants purchased in a local market, which are prepared in triplicate for measurement in the laboratory, recording their spectra in the same range as the 16 calibration mixtures.

The data are available in sheet "PLS" of the Excel file "colorants". The first column is the label of the samples, the same as in Figure 7.3 in Chapter 7. Then, the columns are labelled with the wavelength number, and the two final columns form matrix $\mathbf{Y}$, first the content of tartrazine (E-102) and then that of yellow sunset (E-110). The first 16 samples are calibration standards, while the test samples are samples 17–22.

Guided solution: A PLS2 model is fitted with autoscaled predictor variables in $\mathbf{X}$ (16 × 47) and mean centring in the responses $\mathbf{Y}$ (16 × 2). The overall model requires three latent variables because the fourth latent variable does not reduce the error in cross-validation, as can be seen by comparing the last two columns of Table 8.7. Leave-one-out cross-validation has been performed due to the specific characteristics of the design of the calibration standards.

Table 8.7. Coefficient of determination, R^2, in training and prediction (estimated by cross-validation) for the two responses Y_1 (E-102) and Y_2 (E-110) fitted with PLS2 with three latent variables (LVs). Also shown are the root-mean-squared error, *RMSE*, and that in cross-validation, *RMSECV*.

Response	R^2 (%)	R^2_{CV} (%)	*RMSE* (3 LV)	*RMSECV* (3 LV)	*RMSECV* (4 LV)
Y_1	99.97	99.93	0.0398	0.0613	0.0681
Y_2	99.86	99.71	0.0838	0.1211	0.1249

Table 8.7 contains the coefficient of determination R^2, the coefficient of determination in prediction R^2_{CV} (estimated with leave-one-out CV), together with the *RMSE* and *RMSECV* of the model made with three latent variables, as well as the values obtained with four latent variables. It is evident that the cross-validation error (*RMSECV*) worsens when adding the fourth latent variable, so three latent variables have been retained. The permutation test carried out also provides a p-value of less than 0.05, concluding that the model is significant. Moreover, the similarity of the explained variance in fitting and prediction points to a highly predictive model.

The variance explained by each latent variable for both predictors and responses is shown in Table 8.8. The three latent variables capture 99.83% of the variance of the spectra ($\mathbf{X}$) to explain 99.91% of the variance of the responses (concentration of both colourants). It is evident how the first two latent variables explain a similar percentage of variance of the response (around 50%) with a rather different percentage of variance in $\mathbf{X}$.

The bar plot in Figure 8.1 shows the loadings on each wavelength for the three latent variables. The loadings of the first latent variable, represented with blue bars, are all positive and practically equal, which is interpreted as a "size" factor. The same happened when a PCA was carried out in Section 7.1.10.1 of Chapter 7. This latent variable explains 49.79% of the variance of the response with 78% of the variance of the predictors.

Table 8.8. Variance captured in predictor $\mathbf{X}$ and response $\mathbf{Y}$ blocks when adding latent variables in the PLS2 model.

Num LV	Variance captured in $\mathbf{X}$ (%)	Cumulative variance captured in $\mathbf{X}$ (%)	Variance captured in $\mathbf{Y}$ (%)	Cumulative variance captured in $\mathbf{Y}$ (%)
1	78.01	78.01	49.79	49.79
2	18.03	96.03	49.92	99.71
3	3.79	99.83	0.20	99.91

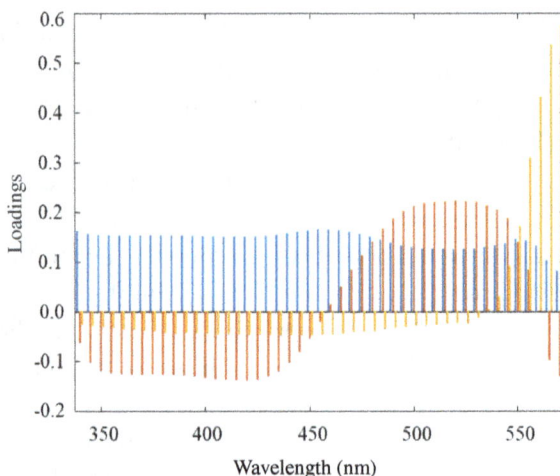

Figure 8.1. Loadings of the PLS2 model with three latent variables. Blue, red, and yellow bars correspond to the first, second, and third latent variables, respectively.

The second latent variable, with its loadings represented by the red bars, explains 18.03% of the variability of the spectra and 49.92% of the response. It somehow distinguishes between tartrazine and yellow sunset through the procedure of multiplying by negative loadings the absorbance up to approximately 460 nm and positive loadings afterwards. The third latent variable, loadings in yellow in Figure 8.1, only explains 0.2% of the response with 3.79% of the variance of the spectra, and taking into account that all but seven of the loadings (on the last wavelengths) are practically null, it seems to be related to some nuances of the upper tail of the spectra.

Before drawing conclusions about the content of the measured samples, the adequacy of applying the model to the samples must be checked for both calibration and test samples. This is done with Q and T^2 statistics, whose values are depicted in Figure 8.2 together with their corresponding threshold values for 95% confidence, which are computed only with the calibration standards during the building of the PLS2 model.

The calibration standards, indicated by black circles, do not show any anomaly in any of the statistics; therefore, the model chosen with

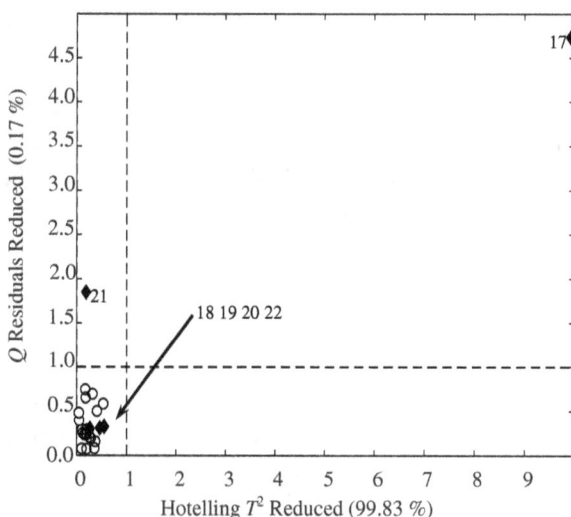

Figure 8.2. Reduced Q-residual and Hotelling T^2 for the PLS2 model with three latent variables. Samples 1–16 in circles are calibration standards, and 17–22 in rhombuses are the commercial test samples. The dashed lines mark the scaled threshold values at 0.95 confidence level.

three latent variables is considered valid to calibrate the mixture of the two colourants. However, it is seen from Figure 8.2 that a test sample, number 17, greatly exceeds the limits for both statistics. That means the model cannot be applied to this sample; it obviously has an orthogonal contribution to the subspace defined with the three latent variables chosen to calibrate due to its large value of Q, and it is also very far from the centroid of the calibration samples due to its very high value of T^2. On the other hand, test sample number 21 only exceeds the threshold for the Q statistic.

Like in the previous case, from an analytical point of view, the validation of the calibration method is performed by computing the least-squares regression line relating the concentrations predicted with the PLS2 model to the true concentrations, that is, the accuracy line $Y_{predicted}$ versus Y_{true}, which is validated in the usual way (i.e., through a hypothesis test on the significance of regression and lack of fit, as well as the normality, independence, and homoscedasticity of residuals).

Both models (one per colourant) are significant (p-values less than 10^{-4}). Since there are four replicates per concentration of each colourant, the lack of fit can be tested. Again, the corresponding p-values were 0.677 and 0.277 for Y_1 and Y_2, respectively, so there is no evidence of lack of fit at the 0.05 significance level. There was no evidence either to reject the hypotheses on the residuals.

Table 8.9 shows the parameters of the accuracy lines. Once the regression model has been validated, the hypothesis test to jointly test whether the slope is 1 and the intercept is zero is used to ensure the absence of bias in the calibration models for both analytes. Table 8.9 contains the corresponding p-values, 0.9603 and 0.9974 for E-102 and E-110, respectively, concluding that both calibration methods have no bias, and the predicted concentration can be considered equal to the true concentration because their accuracy lines are significantly equal to the identity line $y = x$.

The PLS2 calibration model is then used to predict the concentration of the 22 samples, even though it should not be applied to sample 17, as was explained above. Figure 8.3 depicts the predicted values as a function of the sample number. Remember that samples 1–16 are calibration standards, 17–19 are the three replicates of a commercial sample, and 20–22 are the three replicates of the other commercial sample.

For colourant E-102 in Figure 8.3(a), the replicates of the second test sample (20–22) have concentration well inside the calibration

Table 8.9. Parameters of the accuracy line $Y_{\text{predicted}}$ (denoted as y) versus Y_{true} (denoted as x) for E-102 and E-110 obtained with the PLS2 calibration model.

	Y_1 (E-102)	Y_2 (E-110)
Accuracy line	$y = 0.999\,x + 9.6 \cdot 10^{-4}$	$y = 0.998\,x + 4.2 \cdot 10^{-3}$
p-value of the hypothesis test with H_0: slope = 1 & intercept = 0	0.9603	0.9974
Correlation coefficient	0.9998	0.9993
Residual standard deviation	0.0425	0.0895

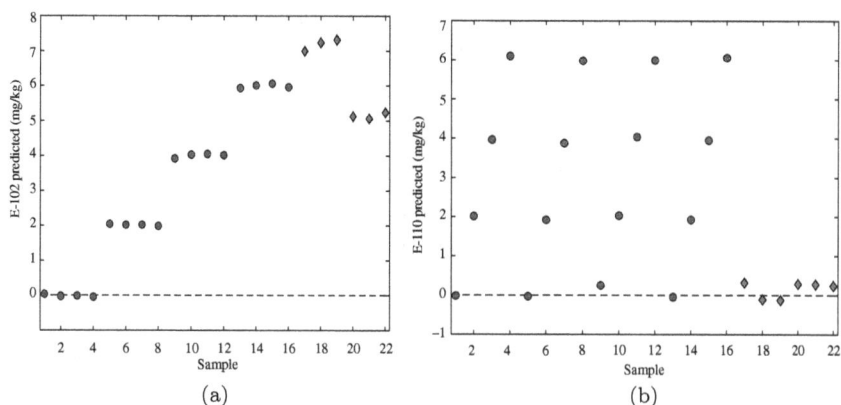

Figure 8.3. Concentrations predicted with the PLS2 calibration model for (a) colourant E-102 and (b) colourant E-110. Samples 1–16 are standards (circles), while 17–22 are commercial test samples (rhombuses).

range (from 0 to 6 mg kg^{-1} in both analytes); however, this is clearly not the case for the other test sample with all replicates (17–19) outside the calibration range.

Note that the prediction of test sample number 17 is also plotted in the graph just to observe that it is near its corresponding replicates without any indication about its outlying position with respect to the projection space detected in Figure 8.2. Analogously, sometimes a "deficiency" in the response cannot be seen by simply observing the value of the response.

Regarding the predicted concentration of sunset yellow E-110 in Figure 8.3(b), all the test samples in rhombuses have concentrations of around zero. The predicted concentrations of the test samples are listed in Table 8.10.

8.7.3. *PLS: Adulteration of whisky*

Goal: The aim is quantitative detection of the percentage of adulteration of a good-quality whisky when mixed with another of poorer quality and/or when adding water.

Training set: The predictors **X** are in a 29 × 91 data matrix. The 29 objects are mixtures prepared with whisky of two different

Table 8.10. Predicted concentration with the PLS2 model for two commercial samples of different food colourants (prepared in triplicate).

	Number	E-102 (mg kg^{-1})	E-110 (mg kg^{-1})
Test sample 1	17	–	–
	18	7.24 (out)	−0.11
	19	7.37 (out)	0.12
Test sample 2	20	5.17	0.28
	21	5.06	0.27
	22	5.24	0.24

brands, one of them an excellent whisky, while the other is a poor-quality whisky, and a small proportion of water, in the percentages detailed in the first columns of Table 8.11. The 91 variables are the absorbances recorded in the UV range between 220 and 400 nm, every 2 nm. Responses are in a 29×3 matrix **Y** with the already mentioned percentages of high-quality whisky, low-quality whisky, and water, respectively.

The data are available in the "whisky" Excel file. The first three columns are the responses in the order just mentioned. The rest of the columns are the predictor variables, labelled with the wavelength number.

Guided solution: A PLS2 model is fitted with autoscaled predictor variables in **X** and mean-centred responses in **Y**. Leave-one-out cross-validation is used to compute *RMSECV* for the three responses, with the values in Figure 8.4.

It is observed how the *RMSECV* values stop decreasing after the third latent variable, or with two for the percentage of water, Y_3 in yellow. The third latent variable explains 0.12% of the variance of **Y** in the overall model, according to the decomposition summarised in Table 8.12, which shows the variance explained by the predictors and the response when including latent variables in the PLS2 model.

With three latent variables, 99.97% of the variance of the predictors (spectra) is used to explain 99.85% of the variance of responses (percentages of both whiskeys and water). The PLS2 model

Table 8.11. Percentage of high-quality whisky (HQ), low-quality whisky (LQ), and water, and their prediction with the PLS2 model with three latent variables.

Sample number	HQ whisky (%)	LQ whisky (%)	Water (%)	Predicted HQ whisky (%)	Predicted LQ whisky (%)	Predicted water (%)
1	100	0	0	100.38	0.77	−1.15
2	100	0	0	102.07	−2.52	0.46
3	93.33	6.67	0	91.93	8.73	−0.65
4	86.67	13.33	0	88.66	11.44	−0.10
5	83.33	16.67	0	83.41	15.85	0.74
6	80.00	20.00	0	79.75	20.02	0.23
7	73.33	26.67	0	72.29	27.74	−0.03
8	66.67	33.33	0	67.46	32.01	0.53
9	60.00	40.00	0	58.75	40.66	0.59
10	53.33	46.67	0	53.22	46.46	0.32
11	40.00	60.00	0	39.76	59.62	0.62
12	33.33	66.67	0	32.92	67.75	−0.67
13	20.00	80.00	0	19.69	80.12	0.19
14	13.33	86.67	0	12.88	86.55	0.57
15	0	100	0	−1.11	101.79	−0.69
16	0	100	0	−0.18	99.80	0.38
17	0	96.67	3.33	0.03	96.54	3.42
18	0	86.67	13.33	0.12	87.31	12.57
19	0	83.33	16.67	0.73	82.11	17.16
20	96.67	0	3.33	96.97	−1.04	4.07
21	86.67	0	13.33	84.82	1.09	14.09
22	83.33	0	16.67	80.57	1.95	17.48
23	66.67	26.67	6.67	67.41	26.69	5.91
24	63.33	20.00	16.67	62.96	23.54	13.49
25	46.67	46.67	6.67	47.26	46.61	6.12
26	43.33	43.33	13.33	44.36	42.25	13.39
27	40.00	40.00	20.00	41.50	37.97	20.53
28	26.67	66.67	6.67	27.95	65.35	6.70
29	20.00	63.33	16.67	20.07	62.87	17.06

with the three selected latent variables has the characteristics shown in Table 8.13. The concordance between the explained variance in the overall model and the individual determination coefficients in

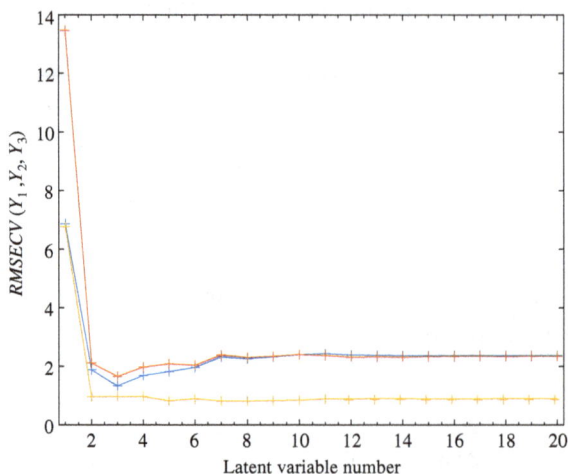

Figure 8.4. *RMSECV* for the three responses in **Y** as a function of the number of latent variables. Blue refers to Y_1 (high-quality whisky), red is for Y_2 (low-quality whisky), and yellow is for Y_3 (water).

Table 8.12. Variance captured in predictor **X** and response **Y** matrices when adding latent variables in the PLS2 model.

Latent variable number	Variance captured in **X** (%)	Cumulative variance captured in **X** (%)	Variance captured in **Y** (%)	Cumulative variance captured in **Y** (%)
1	99.64	99.64	89.08	89.08
2	0.27	99.91	10.65	99.73
3	0.06	99.97	0.12	99.85
4	0.01	99.98	0.04	99.89

training and prediction (estimated via cross-validation) ensures a good predictive model for the three responses.

With the calibration model fitted, the predicted percentages of each compound of the 29 mixtures are in the last three columns of Table 8.11. There is a good agreement between the predicted percentages (both whiskeys and water) and the ones prepared in the laboratory.

Table 8.13. Coefficient of determination in training R^2 and prediction, R^2_{CV} fitted for the three responses (Y_1, Y_2, and Y_3). *RMSE* and *RMSECV*, obtained with three latent variables.

Response	R^2 (%)	R^2_{CV} (%)	*RMSE*	*RMSECV*
Y_1	99.90	99.84	1.069	1.311
Y_2	99.83	99.75	1.329	1.634
Y_3	98.62	98.14	0.812	0.943

Table 8.14. Parameters of the accuracy line $Y_{\text{predicted}}$ versus Y_{true} for Y_1, Y_2, and Y_3 with the predictions obtained using the PLS2 calibration model. p-value when jointly testing slope = 1 and intercept = 0, correlation coefficient (r), and residual standard deviation ($\hat{\sigma}$).

	Y_1 (high-quality whisky)	Y_2 (low-quality whisky)	Y_3 (water)
Acc.line	$y = 0.999\ x + 0.05$	$y = 0.998\ x - 0.07$	$y = 0.986\ x + 0.07$
p-value	0.986	0.965	0.824
r	0.999	0.999	0.993
$\hat{\sigma}$	1.107	1.375	0.836

The accuracy of the method is checked by computing the three accuracy lines, whose characteristics are shown in Table 8.14. All the models are significant, with no significant lack of fit and unbiased at the 0.05 significance level since their slopes and intercepts are significantly equal to 1 and 0, respectively (see the p-value of the joint test in the second row of Table 8.14). All the correlation coefficients are high, and the residual standard deviations (in percentage because this is the unit of measurement) roughly surpasses 1%.

In a PLS model, it is also possible to analyse the regression coefficients (regression vectors) of the model, which indicate the contribution of the predictor variable in the fitted model to explain the responses (three in our case). Figure 8.5 is a bar plot to see the magnitude of the coefficients for the three responses of the PLS2 model as a function of the predictor variables (wavelength number).

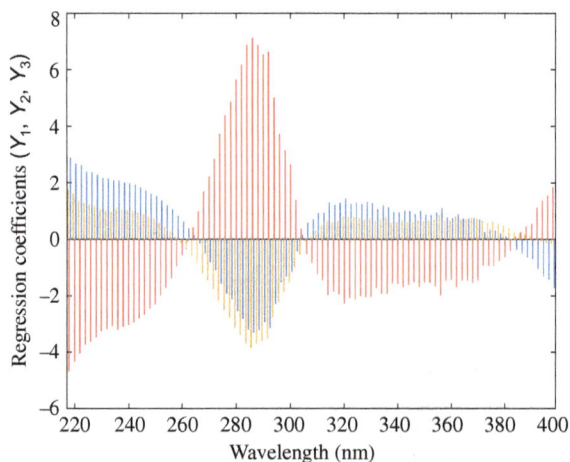

Figure 8.5. Coefficients of the PLS2 model (regression vectors) as a function of the wavelength number for the three responses: blue for Y_1 (high-quality whisky), red for Y_2 (low-quality whisky), and yellow for Y_3 (water).

The blue, red, and yellow bars represent high-quality whisky, low-quality whisky, and water, respectively.

According to the magnitude of the coefficients, the most important region of the spectra is the one that corresponds to wavelengths between 260 and 310 nm, approximately. It is also clear that blue and red coefficients always have opposite signs, distinguishing between high- and low-quality whiskeys, and that the coefficients to predict the percentage of water (yellow bars) have the same sign as those to predict the percentage of high-quality whisky (blue bars), though a little larger (in absolute value) for the region previously mentioned (between 260 and 310 nm) and smaller in the remaining wavelengths.

8.7.4. *PCR: Prediction of height in adult males*

Goal: To predict the height of 44 men based on 14 measurements made on the long bones, namely the humerus and femur, on the right side of the body.

Training set: Predictor matrix $\mathbf{X}$ is a 44×14 data matrix, whereas $\mathbf{y}$ is a 44×1 vector with a single response variable, the height. The data are available in file "forensic.xlxs". The first column is the response "height" (in cm), and the remaining columns (the predictors) are labelled according to the measurement (in mm). These data are part of the file in Ref. [35].

Guided solution: The reader can compute the correlation matrix with the 15 predictor variables to see the high correlation between one another. Also, multicollinearity is present according to the criteria of the quotient between the largest and smallest eigenvalues in Table 8.15, which is 7570, much greater than 10. Therefore, a regression method that handles latent variables is necessary, which, in this case, is PCR.

Table 8.15 contains the characteristics of the PCA decomposition performed with autoscaled **X** up to the total 14 PCs, which are initially considered the new predictor variables for fitting the height of the individuals in **y** with a multilinear least-squares regression.

Table 8.16 shows the coefficient estimates of the regression model with the 14 PCs together with the corresponding standard error, as given by Equation (8.35). Regarding the comments in Section 8.5 about this equation, note the increasing standard error when

Table 8.15. Eigenvalues, and percentages of explained and cumulative variance when adding principal components (PCs).

PC number	Eigenvalues	Explained variance (%)	Cumulative variance (%)
1	7.57	54.08	54.08
2	1.84	13.17	67.25
3	1.15	8.19	75.44
4	0.88	6.26	81.70
5	0.69	4.91	86.61
6	0.54	3.90	90.51
7	0.43	3.11	93.62
8	0.27	1.94	95.56
9	0.25	1.78	97.34
10	0.14	0.97	98.31
11	0.11	0.79	99.10
12	0.07	0.47	99.56
13	0.06	0.43	99.99
14	0.001	0.006	100.00

Table 8.16. Parameter estimates, p-values, and 95% confidence intervals for the MLR model height versus the scores on the 14 PCs ($n = 44$).

Parameter	Estimate	Standard error	p-value	Lower limit	Upper limit
Constant	175.75	0.67	0	174.39	177.11
PC_1	2.34	0.24	0	1.84	2.84
PC_2	−1.72	0.50	0.0017	−2.74	−0.71
PC_3	0.34	0.63	0.5891	−0.94	1.63
PC_4	0.62	0.72	0.3996	−0.86	2.09
PC_5	−0.79	0.81	0.3370	−2.45	0.87
PC_6	−0.71	0.91	0.4431	−2.58	1.16
PC_7	−3.96	1.02	0.0006	−6.05	−1.87
PC_8	1.40	1.29	0.2872	−1.24	4.05
PC_9	2.63	1.35	0.0612	−0.13	5.39
PC_{10}	1.56	1.83	0.4012	−2.18	5.30
PC_{11}	−0.91	2.03	0.6571	−5.05	3.24
PC_{12}	3.14	2.63	0.2431	−2.25	8.53
PC_{13}	0.61	2.74	0.8259	−5.00	6.22
PC_{14}	−13.59	23.72	0.5711	−62.10	34.92

moving from the first to posterior PCs as a consequence of the smaller eigenvalues in Table 8.15.

According to the test on the significance of the individual coefficients, whose p-values are in the fourth column of Table 8.16, only the constant and the coefficients of PC_1, PC_2, and PC_7 are significantly non-null at the 0.05 significance level (p-value < 0.05). This is also seen in the fact that zero belongs to the 95% confidence interval on the remaining coefficients (last two columns of Table 8.16). Finally, no outlier samples have been detected, neither in the PCA decomposition (Q and T^2 statistics) nor in the regression model.

Now, following the procedure to select significant PCs described in Section 8.5.1, the PCR model is re-computed with only PC_1, PC_2, and PC_7 as predictor variables (Table 8.17). It is worth noting that, due to the orthogonality of the PCs, the coefficient estimates are the same as in Table 8.16 regardless of whether all or only some

Table 8.17. Parameter estimates, p-values, and 95% confidence intervals for the MLR model height versus the scores of the three selected PCs ($n = 44$).

Parameter	Estimate	Standard error	p-value	Lower limit	Upper limit
Constant	175.75	0.66	0.00	174.42	177.08
PC_1	2.34	0.24	0.00	1.85	2.83
PC_2	−1.72	0.49	0.00	−2.72	−0.73
PC_7	−3.96	1.01	0.00	−6.01	−1.92

Table 8.18. Parameter estimates, p-values, and 95% confidence intervals for the MLR model height versus the scores of the three selected PCs ($n = 43$, without object 37).

Parameter	Estimate	Standard error	p-value	Lower limit	Upper limit
Constant	176.05	0.60	0.00	174.85	177.26
PC_1	2.36	0.22	0.00	1.92	2.80
PC_2	−1.80	0.44	0.00	−2.69	−0.91
PC_7	−3.95	0.90	0.00	−5.77	−2.12

components are used as long as the number of objects remains the same.

Another detail to note is that, with a lower number of components, the confidence intervals on the coefficients of these components are shorter (compare the corresponding intervals in Tables 8.16 and 8.17), despite the fact that when three PCs are considered, the residual variance is greater than that obtained with all PCs. The effect might be more evident if an external estimate of the residual variance was available, that is, an estimate that did not depend on the number of components chosen.

In any case, the residuals are not the same either, and in this case, the analysis of the new model leads to removing object number 37 with a studentised residual of −3.30. The estimates of the fitted model with 43 objects and the three predictor variables corresponding to the three selected PCs are listed in Table 8.18.

A word of caution: although it is true that in PCR the construction of the components is independent of the response, it is also true that the effect of an anomalous sample can modify the new PCs, even if the "anomaly" is not in the predictors. In this case, object 37 was detected as an outlier because of its response value (as it had an unusually large residual in absolute value), but its removal slightly modified the components – compare Tables 8.17 and 8.18.

The validation of the PCR model with 43 objects is performed as in the previous case studies. The residuals are independent and normally distributed, while the homogeneity of variances cannot be tested because there are no replicates.

Regarding the functional part of the model, Table 8.19 contains the ANOVA table for its validation (without the lack of fit test, again because there were no replicates), as well as the determination coefficient R^2 and the residual standard deviation.

The p-value in Table 8.19 is less than 10^{-3}, implying that the model is significant. It explains near 80% of the variance of height, according to the determination coefficient, by using the 70.36% of the variance of the original predictor variables (PCs 1, 2, and 7 in Table 8.15) and with a standard deviation of 3.91 cm.

Then, the validated model to predict the height of an individual from the scores on the chosen PCs in the following Equation (8.51)

Table 8.19. ANOVA table, coefficient of determination R^2 and residual standard deviation $\hat{\sigma}$ of the MLR model to predict height from the scores of the three selected PCs ($n = 43$ after removing object 37).

Source	Sum of squares	Degrees of freedom	Mean squares	F-ratio	p-value
Model	2357.65	3	785.89	51.40	0.000
Residual	596.35	39	15.29		
Total	2954.00	42			
$R^2(\%)$	79.81				
$\hat{\sigma}$	3.91				

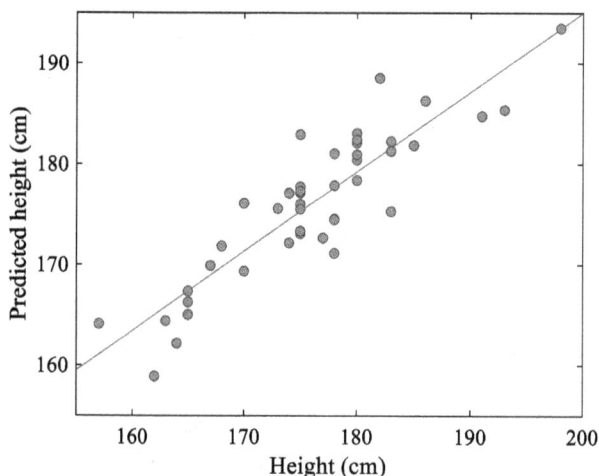

Figure 8.6. Regression line predicting height versus height with the predictions computed using the PCR model with three principal components (first, second, and seventh).

is used to compute the height with the scores in the training set, obtaining $\mathbf{y}_{\text{predicted}}$:

$$\text{Height} = y = 176.05 + 2.36 PC_1 - 1.80 PC_2 - 3.95 PC_7. \qquad (8.51)$$

Figure 8.6 depicts the values of $(y, y_{\text{predicted}})$, that is, (height, predicted height), with circles and the line that fits them, $y_{\text{predicted}} = 35.53 + 0.798\ y$, which is significant ($p$-value $< 10^{-4}$) and with no significant lack of fit (p-value $= 0.175$), and has a correlation coefficient of 0.893 with a standard error of estimate equal to 3.41.

It is left as an exercise for the reader to apply cross-validation for the selected number of PCs to check that, with this criterion, only the first two components are selected for the PCR model, with no studentised residual greater than three in absolute value. Certainly, with this criterion, it is impossible to select PC_7 without considering all the previous ones, despite the fact that a fraction of the information of the predictors (corresponding to PC_7) has a significant correlation with height.

8.7.5. *PLS versus PCR: Prediction of height in adult males*

Goal: To predict the height of 44 men based on 14 measurements made on the long bones, namely the humerus and femur, on the right side of the body.

The previous section tackled the issue of how many and which PCs must be selected for a proper PCR model. Statistical inference was used in a two-step procedure: first, fit the model with all the PCs, then take only those with significant coefficients, and finally rebuild the OLS model with the reduced set of PCs. In the case under study, that leads to selecting the first, second, and seventh PCs.

Technically, a PLS regression method searches for the directions in the space of the predictors with the highest variance while avoiding those that are not correlated with the response, see Equation (8.44). Therefore, a PLS model is built with autoscaled predictors and mean-centred response, as in the case of PCR discussed in Section 8.7.4.

The values of *RMSECV* and *RMSE* as a function of the number of latent variables are shown in Figure 8.7, where it is clear that two latent variables must be considered. In this case, there are enough

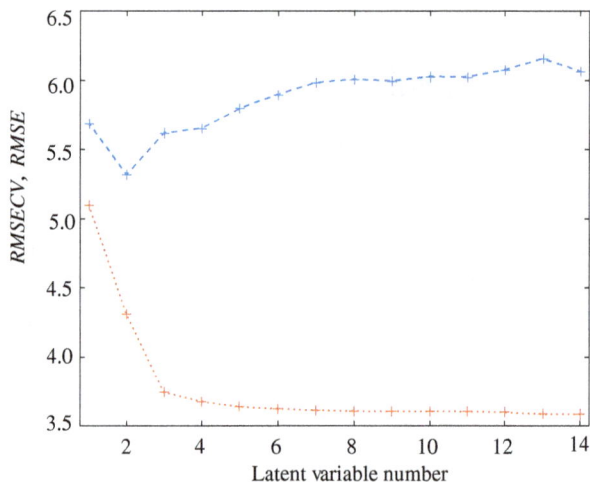

Figure 8.7. Evolution of *RMSECV* (dotted line) and *RMSE* (dashed line) as a function of the number of latent variables of a PLS model to predict height.

samples for the estimates in prediction to be obtained with cross-validation using the venetian blinds method with six splits and a blind thickness equal to one (each test set is determined by selecting every six objects in the dataset).

With two latent variables, the permutation test gives p-values well below 0.05, indicating that the model is significant. As can be seen in Figure 8.7, *RMSE* and *RMSECV* are similar to each other, 4.31 and 5.31, respectively. There is also no studentised residual greater than three in absolute value and no sample with both Q and T^2 greater than the corresponding 95% limits. Therefore, the PLS model with two latent variables is adequate to predict the height from the measurements of the long bones.

Table 8.20 shows some details of the fitted model. The two latent variables explain 73.38% of the variance of height with 65.86% of the variance of the predictors. Comparing with the PCA decomposition in Table 8.15, it is evident that these two latent variables do not coincide with the first two PCs.

It is also clear that the two-dimensional projection space computed using PLS is different from the three-dimensional projection space used by PCR. However, despite this structural difference, the predictions made with both models are significantly similar.

To verify this, a paired samples t-test has been used to compare the differences between the predictions of each ith sample for both models: the PCR model with three PCs and the PLS model with two latent variables, $d(i) = \hat{y}_{\text{PCR}}(i) - \hat{y}_{\text{PLS}}(i)$.

Table 8.20. Variance captured by the PLS regression model with two latent variables (LVs) to predict height.

LV number	Variance captured in **X** (%)	Cumulative variance captured in **X** (%)	Variance captured in **Y** (%)	Cumulative variance captured in **Y** (%)
1	53.72	53.72	62.81	62.81
2	12.14	65.86	10.57	73.38

In the current case, it is not reasonable to compare the predictions made with both methods by fitting them with a regression line and testing if it is significantly equal to the identity function because both series of calculated values have random errors of practically the same magnitude. In other words, the OLS assumption that the response is affected by random errors to a greater extent than the predictor is no longer valid. The assumption is acceptable when the predictor is a quantity measured in the laboratory (e.g., the concentration of a calibration standard) and the response is the value predicted for that sample using PCR or PLS models, as in the previous examples.

In any case, the set of 43 differences $\{d(i)\}$ has a normal distribution with mean $\overline{d} = 0.287$ and standard deviation $s_d = 1.938$. The normality has been tested using the χ^2, Shapiro–Wilk, and Kolmogorov–Smirnov tests (with the null hypothesis that the differences are compatible with a normal distribution) with p-values of 0.213, 0.482, and 0.985, respectively.

Under normal distribution, the Student's t-test is applied to test whether the mean of the differences is zero (H_0) or not (H_1). The corresponding p-value is equal to 0.337, which does not provide evidence for rejecting the null hypothesis. Therefore, statistically, PCR and PLS provide the same predictions for height.

References

[1] Diveev, A. and Shmalko, E. (2021). *Machine Learning Control by Symbolic Regression*. Cham, Swizterland: Springer International Publishing, Chapter 4.

[2] Petersen, B. K., Larma, M. L., Mundhenk, T. N., Santiago, C. P., Kim, S. K., and Kim, J. T. (2021). Deep symbolic regression: Recovering mathematical expressions from data via risk-seeking policy gradients. In *Proceedings of the International Conference on Learning Representations*, Virtual, 3–7 May 2021.

[3] La Cava, W. and Moore, J. H. (2022). Learning feature spaces for regression with genetic programming. Genetic *Programming and Evolvable Machines*, 21, 433–467.

[4] Rivero, D., Fernández-Blanco, E., and Pazos, A. (2022). DoME: A deterministic technique for equation development and symbolic regression. *Expert Systems with Applications*, 198, 116712.

[5] Draper, N. R. and Smith, H. (1998). *Applied Regression Analysis*, 3rd edn. New York: John Wiley & Sons Inc.

[6] Agresti, A. (2015). *Foundations of Linear and Generalized Linear Models*. Hoboken, N.J.: John Wiley & Sons Inc.

[7] Rao, C. R. (1973). *Linear Statistical Inference and Its Applications*, 2nd edn. New York: John Wiley & Sons Inc.

[8] Hastie, T., Tibshirani, R., and Friedman, J. (2009). *The Elements of Statistical Learning. Data Mining, Inference and Prediction*, 2nd edn. New York: Springer.

[9] Friedman, J. (1986). Personal communication. Statistics 201B (Spring 1986) smoothing and curve estimation. Class notes 3, Standford University.

[10] Stone, J. R. N. (1945). The analysis of market demand. *Journal of the Royal Statistical Society* (New Series), 108, 286–382.

[11] Kendall, M. G. (1957). *A Course in Multivariate Analysis*. London, England: Charles Griffin and Company, Ltd.

[12] Massy, W. F. (1965). Principal components regression in exploratory statistical research. *Journal of the American Statistical Association*, 60, 234–256.

[13] Esbensen, K. H. and Geladi, P. (2010). Principles of proper validation: Use and abuse of re-sampling for validation. *Journal of Chemometrics*, 24, 168–187. doi: 10.1002/cem.1310.

[14] Lee, H., Park, Y. M., and Lee, S. (2015). Principal component regression by principal component selection. *Communications for Statistical Applications and Methods*, 22, 173–180. doi: 10.5351/CSAM.2015.22.2.173.

[15] Bates, S., Hastie, T., and Tibshirani, R. (2022). Cross-validation: What does it estimate and how well does it do it? *Preprint* Arxiv:2104.00673v4 [stat.ME]. doi: 10.48550/arXiv.2104.00673.

[16] Valencia, O., Ortiz, M. C., and Sarabia, L. A. (2021). Principal component regression that minimizes the sum of the squares of the relative errors: Application in multivariate calibration models. *Journal of Chemometrics*, 35, e3341. doi: 10.1002/cem.3341.

[17] Wold, S. (2015). Chemometrics and Bruce: Some fond memories. In Lavine *et al.* (eds.), *40 Years of Chemometrics – From Bruce Kowalski to the Future*. ACS Symposium Series. Washington, DC: American Chemical Society.

[18] de Jong, S. (1993). SIMPLS: An alternative approach to partial least squares regression. *Chemometrics and Intelligent Laboratory Systems*, 18, 251–263.

[19] Rosipal, R. and Krämer, N. (2006). Overview and recent advances in partial least squares. In Saunders, C., Grobelnik, M., Gunn, S.,

and Shawe-Taylor, J. (eds.), *Subspace, Latent Structure and Feature Selection, Statistical and Optimization Perspectives Workshop, SLSFS 2005.* Berlin: Springer-Verlag, pp. 34–51.

[20] Sarabia, L. A., Ortiz, M. C., Sánchez, M. S., and Herrero, A. (2003). Partial least squares fine-tuning by a bootstrap estimated signal-noise relation to weight the loadings. *Chemometrics and Intelligent Laboratory Systems*, 68, 83–96. doi: 10.1016/S0169-7439(3)00090-X.

[21] Lohmóller, J. B. (1989). *Latent Variable Path Modelling with Partial Least Squares.* Physica-Verlag. Reissued (2013). Springer Science & Business Media.

[22] Esposito Vinzi, V., Chin, W. W., Henseler, J., and Wang, H. (eds.) (2010). *Handbook of Partial Least Squares. Concepts, Methods and Applications.* Berlin: Springer-Verlag.

[23] Ruiz, S., Ortiz, M. C., Sarabia, L. A., and Sánchez, M. S. (2018). A computational approach to partial least squares model inversion in the framework of the process analytical technology and quality by design initiatives. *Chemometrics and Intelligent Laboratory Systems*, 182, 70–78. doi: 10.1016/j.chemolab.2018.08.014 (Open Access).

[24] Arce, M. M., Ruiz, S., Sanllorente, S., Ortiz, M. C., Sarabia, L. A., and Sánchez, M. S. (2021). A new approach based on inversion of a partial least squares model searching for a preset analytical target profile. Application to the determination of five bisphenols by liquid chromatography with diode array detector. *Analytica Chimica Acta*, 1149, 338217. doi: 10.1016/j.aca.2021.338217 (Open Access).

[25] Arce, M. M., Sanllorente, S., Ruiz, S., Sánchez, M. S., Sarabia, L. A., and Ortiz, M. C. (2021). Method operable design region obtained with a partial least squares model inversion in the determination of ten polycyclic aromatic hydrocarbons by liquid chromatography with fluorescence detection. *Journal of Chromatography A*, 1657, 462577. doi: 10.1016/j.chroma.2021.462577.

[26] Wentzell, P. D. and Vega Montoto, L. (2003). Comparison of principal components regression and partial least squares regression through generic simulations of complex mixtures. *Chemometrics and Intelligent Laboratory Systems*, 65, 257–279.

[27] Lin, Y. W., Deng, B. C., Xu, Q. S., Yun, Y. H., and Liang, Y. Z. (2016). The equivalence of partial least squares and principal component regression in the sufficient dimension reduction framework. *Chemometrics and Intelligent Laboratory Systems*, 150, 58–64. doi: 10.1016 /j.chemolab.2015.11.003.

[28] Zhang, L. and Garcia-Munoz, S. (2009). A comparison of different methods to estimate prediction uncertainty using partial least squares (PLS): A practitioner's perspective. *Chemometrics*

and Intelligent Laboratory Systems, 97, 152–158. doi: 10.1016/ j.chemolab.2009.03.007.

[29]   Lin, W., Zhuanga, Y., Zhanga, S., and Martin, E. (2013). On estimation of multivariate prediction regions in partial least squares regression. *Journal of Chemometrics*, 27, 243–250.

[30]   Arcos, M. J., Ortiz, M. C., Villahoz, B., and Sarabia, L. A. (1997). Genetic-algorithm-based wavelength selection in multicomponent spectrometric determinations by PLS: Application on indomethacin and acemethacin mixture. *Analytica Chimica Acta*, 339, 63–77. doi: 10.1016/S0003-2670(96)00438-2.

[31]   Wold, S., Johansson, E., and Cocchi M. (1993). PLS: Partial least squares projections to latent structures. In Kubinyi, H. (ed.), *3D QSAR in Drug Design: Theory, Methods and Applications*. Leiden: ESCOM Science Publishers, pp. 523–555.

[32]   Meléndez, E., Ortiz, M. C., Sarabia, L. A., Íñiguez, M., and Puras, P. (2013). Modelling phenolic and technological maturities of grapes by means of the multivariate relation between organoleptic and physicochemical properties. *Analytica Chimica Acta*, 761, 53–61. doi: 10.1016/j.aca.2012.11.021.

[33]   Sarabia, L. A., Ortiz, M. C., and Checa, M. A. (1989). Pattern recognition for detection of tallow in lard. In *Agriculture, Food Chemistry and the Consumer*, Vol. 2. Paris: l'Institute Nationale de la Recherche Agronomique, pp. 602–606.

[34]   Ortiz, M. C., Sánchez, M. S., and Sarabia, L. A. (2020). Quality of analytical measurements: Univariate regression. In Brown, S., Tauler, R., and Walczak, B. (eds.), *Comprehensive Chemometrics: Chemical and Biochemical Data Analysis*. Elsevier, pp. 71–105. Available at: https://doi.org/10.1016/B978-0-12-409547-2.14869-3.

[35]   Jantz, R. J. and Moore-Jansen, P. H. (2006). Database for forensic anthropology in the United States, 1962-1991. Ann Arbor, MI: Inter-University Consortium for Political and Social Research [distributor], 30 March 2006. Available at: https://doi.org/10.3886/ ICPSR02581.v1.

Index

A

A-optimality, 143
accuracy line, 284
adjusted R-square, 93
alignment, 171
analytical chemistry, epistemological
  definition of, 32
autopredictive models, 225
autoscaling, 159, 190, 249
  row autoscaling, 161

B

baseline, 161
  correction, 162, 172
  drift, 162
  offsets, 169
Bayesian non-negative factor analysis
  (BNFA), 293
bias, 32, 60, 64
  biased regression methods, 60
  unbiased method, 60
bias–variance dilemma, 314
biased regression analysis, 338
bilinear decomposition, 289
  model, 245, 275, 291, 333
block-scaling, 176
broken stick, 251

C

candidate points, 143
centroid, 201–202, 208–209, 219
chemical rank, 250
class centroid, 215
class distance plot, 214–215
classification, 186
  ability, 223
  methods, 188
coded levels, 88
coefficient of multiple determination,
  61
collinearity, 61
  multicollinearity, 68–69
commutability, 32
composite designs, 112
contour plot, 93
core consistency diagnostic
  (CORCONDIA), 281, 295
correlation, 35
  causality, 36
  cause-effect relation, 35
covariance, 250
  coefficients, 90
  matrix, 237, 309
cross-validation, 177, 224, 252, 324,
  326
customer, 32–33

www.ingramcontent.com/pod-product-compliance
Lightning Source LLC
Chambersburg PA
CBHW050536190326
41458CB00007B/1802